Genome Management in Eukaryotes

BOOKS IN THE BIOTOL SERIES

The Molecular Fabric of Cells
Infrastructure and Activities of Cells

Techniques used in Bioproduct Analysis
Analysis of Amino Acids, Proteins and Nucleic Acids
Analysis of Carbohydrates and Lipids

Principles of Cell Energetics
Energy Sources for Cells
Biosynthesis and the Integration of Cell Metabolism

Genome Management in Prokaryotes
Genome Management in Eukaryotes

Crop Physiology
Crop Productivity

Functional Physiology
Cellular Interactions and Immunobiology
Defence Mechanisms

Bioprocess Technology: Modelling and Transport Phenomena
Operational Modes of Bioreactors

In vitro Cultivation of Micro-organisms
In vitro Cultivation of Plant Cells
In vitro Cultivation of Animal Cells

Bioreactor Design and Product Yield
Product Recovery in Bioprocess Technology

Techniques for Engineering Genes
Strategies for Engineering Organisms

Principles of Enzymology for Technological Applications
Technological Applications of Biocatalysts
Technological Applications of Immunochemicals

Biotechnological Innovations in Health Care

Biotechnological Innovations in Crop Improvement
Biotechnological Innovations in Animal Productivity

Biotechnological Innovations in Energy and Environmental Management

Biotechnological Innovations in Chemical Synthesis

Biotechnological Innovations in Food Processing

Biotechnology Source Book: Safety, Good Practice and Regulatory Affairs

BIOTECHNOLOGY BY OPEN LEARNING

Genome Management in Eukaryotes

PUBLISHED ON BEHALF OF :

Open universiteit and **University of Greenwich (formerly Thames Polytechnic)**

Valkenburgerweg 167
6401 DL Heerlen
Nederland

Avery Hill Road
Eltham, London SE9 2HB
United Kingdom

Butterworth-Heinemann Ltd
Linacre House, Jordan Hill, Oxford OX2 8DP

A member of the Reed Elsevier group

OXFORD LONDON BOSTON
MUNICH NEW DELHI SINGAPORE SYDNEY
TOKYO TORONTO WELLINGTON

First published 1993

British Library Cataloguing in Publication Data
A catalogue record for this book is
available from the British Library

Library of Congress Cataloguing in Publication Data
A catalogue record for this book is
available from the Library of Congress

ISBN 0 7506 0558 8

Composition by University of Greenwich
(formerly Thames Polytechnic)
Printed and Bound in Great Britain

The Biotol Project

The BIOTOL team

OPEN UNIVERSITEIT, THE NETHERLANDS
Prof M. C. E. van Dam-Mieras
Prof W. H. de Jeu
Prof J. de Vries

UNIVERSITY OF GREENWICH (FORMERLY THAMES POLYTECHNIC), UK
Prof B. R. Currell
Dr J. W. James
Dr C. K. Leach
Mr R. A. Patmore

This series of books has been developed through a collaboration between the Open universiteit of the Netherlands and University of Greenwich (formerly Thames Polytechnic) to provide a whole library of advanced level flexible learning materials including books, computer and video programmes. The series will be of particular value to those working in the chemical, pharmaceutical, health care, food and drinks, agriculture, and environmental, manufacturing and service industries. These industries will be increasingly faced with training problems as the use of biologically based techniques replaces or enhances chemical ones or indeed allows the development of products previously impossible.

The BIOTOL books may be studied privately, but specifically they provide a cost-effective major resource for in-house company training and are the basis for a wider range of courses (open, distance or traditional) from universities which, with practical and tutorial support, lead to recognised qualifications. There is a developing network of institutions throughout Europe to offer tutorial and practical support and courses based on BIOTOL both for those newly entering the field of biotechnology and for graduates looking for more advanced training. BIOTOL is for any one wishing to know about and use the principles and techniques of modern biotechnology whether they are technicians needing further education, new graduates wishing to extend their knowledge, mature staff faced with changing work or a new career, managers unfamiliar with the new technology or those returning to work after a career break.

Our learning texts, written in an informal and friendly style, embody the best characteristics of both open and distance learning to provide a flexible resource for individuals, training organisations, polytechnics and universities, and professional bodies. The content of each book has been carefully worked out between teachers and industry to lead students through a programme of work so that they may achieve clearly stated learning objectives. There are activities and exercises throughout the books, and self assessment questions that allow students to check their own progress and receive any necessary remedial help.

The books, within the series, are modular allowing students to select their own entry point depending on their knowledge and previous experience. These texts therefore remove the necessity for students to attend institution based lectures at specific times and places, bringing a new freedom to study their chosen subject at the time they need and a pace and place to suit them. This same freedom is highly beneficial to industry since staff can receive training without spending significant periods away from the workplace attending lectures and courses, and without altering work patterns.

Contributors

AUTHORS

Dr A.T.H. Burns, University of Wolverhampton, Wolverhampton, UK

Dr J.J. Gaffney, Manchester Metropolitan University, Manchester, UK

Dr J.S. Gartland, Dundee Institute of Technology, Dundee, UK

Dr K.M.A. Gartland, Dundee Institute of Technology, Dundee, UK

Dr N.W. Scott, De Montfort University, Leicester, UK

Dr A. Slater, De Montfort University, Leicester, UK

Dr C.A. Smith, Manchester Metropolitan University, Manchester, UK

EDITOR

Dr A. Slater, De Montfort University, Leicester, UK

SCIENTIFIC AND COURSE ADVISORS

Prof M.C.E. van Dam-Mieras, Open universiteit, Heerlen, The Netherlands

Dr C.K. Leach, De Montfort University, Leicester, UK

ACKNOWLEDGEMENTS

Grateful thanks are extended, not only to the authors, editors and course advisors, but to all those who have contributed to the development and production of this book. They include Mrs A. Allwright, Miss J. Skelton, and Professor R. Spier.

The development of this BIOTOL text has been funded by **COMETT, The European Community Action Programme for Education and Training for Technology**. Additional support was received from the Open universiteit of The Netherlands and by University of Greenwich (formerly Thames Polytechnic).

Contents

How to use an open learning text

An open learning text presents to you a very carefully thought out programme of study to achieve stated learning objectives, just as a lecturer does. Rather than just listening to a lecture once, and trying to make notes at the same time, you can with a BIOTOL text study it at your own pace, go back over bits you are unsure about and study wherever you choose. Of great importance are the self assessment questions (SAQs) which challenge your understanding and progress and the responses which provide some help if you have had difficulty. These SAQs are carefully thought out to check that you are indeed achieving the set objectives and therefore are a very important part of your study. Every so often in the text you will find the symbol Π, our open door to learning, which indicates an activity for you to do. You will probably find that this participation is a great help to learning so it is important not to skip it.

Whilst you can, as an open learner, study where and when you want, do try to find a place where you can work without disturbance. Most students aim to study a certain number of hours each day or each weekend. If you decide to study for several hours at once, take short breaks of five to ten minutes regularly as it helps to maintain a higher level of overall concentration.

Before you begin a detailed reading of the text, familiarise yourself with the general layout of the material. Have a look at the contents of the various chapters and flip through the pages to get a general impression of the way the subject is dealt with. Forget the old taboo of not writing in books. There is room for your comments, notes and answers; use it and make the book your own personal study record for future revision and reference.

At intervals you will find a summary and list of objectives. The summary will emphasise the important points covered by the material that you have read and the objectives will give you a check list of the things you should then be able to achieve. There are notes in the left hand margin, to help orientate you and emphasise new and important messages.

BIOTOL will be used by universities, polytechnics and colleges as well as industrial training organisations and professional bodies. The texts will form a basis for flexible courses of all types leading to certificates, diplomas and degrees often through credit accumulation and transfer arrangements. In future there will be additional resources available including videos and computer based training programmes.

Preface

A major challenge to biologists is to understand how the expression of the information contained within the genome of eukaryotes is controlled to enable the production and organisation of distinctive cell types that make up these complex and diverse organisms. This challenge is not simply limited to the acquisition of knowledge. There are enormous practical implications that arise from this knowledge. The ability to exploit the plasticity of plant development, to make new bioproducts using manipulated eukaryotic cells and organisms, to diagnose and treat genetic disorders are all dependent upon understanding how genes are organised and controlled.

In recent years, there have been many advances in our knowledge of eukaryotic gene organisation and expression. Many, hitherto seemingly unrelated facts and impenetrable problems have been replaced by rational explanations and a picture of efficient and elegant molecular mechanisms is emerging. There is, however, still much to be done before the story is complete.

This text is designed to set out the facts as we know them. Inevitably, there has had to be a careful selection of material from the vaste amounts available but the authors have distilled the essential ingredients to provide readers with an up to date understanding of the core concepts in the management of genomes in eukaryotes. They have written the text on the assumption that readers are familiar with the fundamentals of biological chemistry and cell biology and are aware that eukaryotic cells are organised, in multicellular organisms, into tissues and organs.

The text begun by discussing multicellularity, cell differentiation and inheritance patterns in eukaryotes to provide a context in which gene organisation and expression is described. The emphasis of the text is, however, molecular and mainly considers the issues of gene organisation and expression from the point of view of individual cells. In Chapter 3, the concepts of mutation and mutagenesis are introduced and the consequences of mutation examined. Subsequent chapters, explain how the genomes of eukaryotes are physically organised and the processes of transcription and translation. Subsequent chapters deal with the regulation of gene expression, transcription factors and post-translational modification of gene products. The final chapter examines the organisation and expression of genetic material in the major cellular organelles, the mitochondria and chloroplasts.

Throughout the text, examples of the experimental approaches which enabled the elucidation of these molecular and cellular mechanisms are described. The authors have also included many helpful in text activities that will help learning.

An understanding of how the genomes of eukaryotes are organised and controlled underpins a very wide range of scientific and biotechnological activities. This text, therefore, provides a valuable resource to those who seek to contribute to these important endeavours.

We thank the author:editor team for their excellent contributions to the preparation of this text. These thanks will undoubtedly be echoed by its readers.

Scientific and Course Advisors: Professor M. C. E. van Dam-Mieras
Dr C. K. Leach

Multicellularity, cell differentiation and gene regulation

Multicellularity, cell differentiation and gene regulation

1.1 Introduction

This brief introductory chapter is designed to provide a context in which you may study the management of genetic information in eukaryotes. It has been written on the assumption that you are familiar with the major structural differences between eukaryotic and prokaryotic cells and with the management of genetic information in prokaryotic cells. This latter aspect is the central topic of the partner BIOTOL text 'Genome Management in Prokaryotes' while the general properties of prokaryotic and eukaryotic cells are described in the BIOTOL text 'Infrastructure and Activities of Cells'.

1.1.1 Molecular and cellular themes in biology

There are common molecular themes that run through all living cells. For example, the key macromolecular constituents of all cells are variants of just a few major chemical classes:

- genetic information is always stored within the sequence of bases of the nucleic acids, DNA and RNA;

- proteins shape out essential cell structures and the specialised microenvironments of enzymes that are needed to catalyse the particular chemical reactions that sustain and propagate life.

Another example of a strong common theme is the use of proteins and lipids in the construction of the membrane boundaries which conserve cells and their organelles.

The 'central dogma' of molecular biology holds that genetic information flows from DNA to proteins via RNA in all organisms. Proteins represent the prime organisers of structure and function in all cell types, but they are gene products, the nature of which is determined by genes coded into DNA.

cells are diverse

Surprising uniformity is also seen in many fundamental cellular processes. The essential details of DNA replication and transcription, protein synthesis and metabolism are remarkably similar and the molecules involved usually highly conserved. But diversity has been superimposed upon this fundamental molecular framework that is shared by all cell types. This diversity relates both to species - eg rat cells differ from rabbit cells - and also to the specialised role of each cell of the organism. A nerve cell needs characteristics which are far removed from those needed by liver cells, and farther yet from those required by bacteria existing in a specialised ecological niche. Diversity between species is explained by the inherited DNA of each species containing different genes. Diversity of cell types within a particular species cannot be explained in the same way. By-and-large the genetic information in DNA of all the different cell types in an organism is the same - it is the differential control of the genes that explains the distinction between cell types. Deriving a large number of different, specialised cell types from the parental cell is defined as differentiation. The means whereby cells can maintain or change their differentiated state remain largely unknown.

differential gene activity is the basis of cellular differentiation

The attributes that determine the evolutionary success of single-celled organisms differ markedly from those governing multicellular forms. Put simply, single celled organisms just need to grow and divide when food is available to favour the successful propagation of the species. The main control task is to survive by using nutrient sources efficiently - thereby competing effectively against other species - and to channel the derived energy into growth activity and replication. Only the supply of foodstuff need govern the rate of growth. But the most important task in more complex organisms is to control the identity and function of each cell over time and space. Growth and replication must be controlled very tightly, otherwise tissue abnormalities, cancer or some other dysfunction will occur. Not infrequently, the progressive differentiation of cells by successive replication leads to a final, full-differentiated form of cell which is unable to replicate further. Such irreversible progression contrasts to the situation in viable single-celled organisms which inevitably retain the ability to self-replicate.

unicellular strategy: economy adaptation and flexibility in changing environment

Sophisticated gene control is the essence of this process whereby cell-types achieve their specialised identities in a highly programmed way. Cells from multicellular organisms must each play their individual roles in the highly complex and intricately orchestrated processes of differentiation and development that lead to the co-ordinated growth and maturation of the whole organism. Thus cells in multicellular organisms all share fundamental constraints not imposed upon unicellular organisms: they need to adopt their own individual identity and role, and they need to interact socially and co-operatively with other cells. Co-operation is needed between cells in physical contact with each other to ensure co-ordination of growth and function. Cancer is an example of a disease which comes about when the co-operation between cells breaks down (see Chapters 3 and 7). Further, multi-celled organisms need to control how whole communities of cells function and communicate in order to provide for internal homeostasis and for higher functions like movement. This control can be exerted over long distances by chemical factors like hormones and by nervous impulses. Chapter 7 investigates the interplay between hormone signals and a set of transcription-regulating proteins known as transcription factors.

multicellular strategy: specialisation, long term commitment, internal homeostasis minimising fluctuation bacteria are highly evolved organisms

It would be naive and incorrect to view present day, unicellular micro-organisms such as bacteria and yeast as simple, more primitive progenitors of the higher organisms. The common evolutionary ancestors of all todays life forms have disappeared. Bacteria and other single-celled organisms have evolved over the aeons to adapt to their specialised niches just as have more complex organisms like ourselves. Indeed since they have passed through vastly greater numbers of generations than more slowly growing eukaryotes, they can be considered as more highly evolved forms, streamlining their genomes from any unnecessary, surplus DNA, under extreme selection for efficient growth and rapid multiplication rates over the last 3500 million years or so. Nor is multicellularity confined to eukaryotes. The cyanobacteria in particular are rich in multicellular forms. Filamentous forms, maintaining connections between cells, probably appeared not long after unicellular forms, presumably when cells failed to separate after division and are certainly represented early in the prokaryote fossil record. The fossil cyanobacteria also show differentiation of their end cells which can be either round or conical. Such communities of microbial cells, arising so early in evolution, allow the possibility of division of labour amongst cells; multiple cells can perform multiple tasks. Modern bacterial examples of morphological and physiological specialisation include spore production in *Clostridium* and *Bacillus*, heterocyst cells devoted to nitrogen fixation in cyanobacteria filaments, complex fruiting bodies in the mxyobacteria, complex hyphae in *Streptomyces*.

differentiation and division of labour

The earliest organisms, their DNA incessantly threatened by ultraviolet radiation may ultimately have bequeathed the mechanisms of genetic recombination used in mitosis, meiosis and cellular differentiation by developing DNA repair mechanisms. The critical role of such mechanisms in eukaryotes and the consequences of mutation are the subject of Chapter 3. Gene splicing and rejoining mechanisms, originally derived from ancient ultraviolet protective mechanisms have evolved to make possible new DNA combinations, in the view of Margulis and Sagan (1986).

We will examine, in the remainder of this chapter, the fundamental similarities and differences which characterise single-celled and multi-celled organisms and speculate on the origins of multicellularity.

SAQ 1.1

Are the following statements true or false?

1) The diversity of cell types within an organism can be explained by each cell type containing a different set of genes.

2) The process of deriving a number of different specialised cell types from the parent cell is defined as homeostasis.

3) Bacteria are highly evolved organisms.

4) All prokaryotes are unicellular.

5) Differentiation occurs only in eukaryotes.

1.2 Prokaryotes and eukaryotes

nucleus

Virtually all multi-celled organisms are eukaryotes - ie the DNA which holds the genetic information of the cell is packed into linear chromosomes, which in turn are located in a separate body, the cell nucleus, which is delimited by a distinct nuclear membrane (see Chapter 4). Some unicellular species such as yeasts, algae and protozoa are also eukaryotes, but most single-celled organisms are prokaryotes, in which the DNA exists free in the cytoplasm, unsegregated from the rest of the cell by any such membrane.

eukaryotes have membrane bound organelles

Apart from the existence of nuclei, eukaryotic cells have other features which distinguish them clearly from prokaryotic cells. Perhaps the most immediately obvious of these is size: a typical eukaryotic cell has a diameter more than 10-times that of a typical prokaryote. This means, of course, that eukaryotic cells are usually hundreds of times larger in terms of cell volume and mass. Eukaryote cytoplasm can contain other discrete compartments or organelles such as mitochondria, lysosomes, peroxisomes, vesicles, Golgi bodies and the distinctive photosynthetic organelle of higher plants: chloroplasts. The envelope surrounding chloroplasts and the nucleus is made up of two concentric membrane bilayers. Folds and stocks of this lipid bilayer extending from the outer nuclear membrane make up the endoplasmic reticulum and Golgi body. In contrast to this picture of discrete compartments, prokaryotic cells consist of a single cell compartment. Extensive cellular architecture is lacking as are membrane bound organelles, nor is the genetic material enclosed by any boundary material although it is organised into a compact nucleoid zone.

1.3 DNA and chromosomes

bacterial DNA is circular

All of the genes within bacterial cells are present on a single, giant molecule made up of a closed loop of circular DNA. A typical bacterial DNA molecule contained in a typical bacterial cell of 1 μm, would extend to around 1 mm in length if fully stretched. This is sufficient DNA to code for around 3000-4000 different protein gene products. In fact, the DNA is not fully extended, but compacted by supercoiling into irregular RNA rich bodies called nucleoids which exist free in the cytoplasm, occupying around 10% of the cytoplasmic volume. Histones are not used in the compaction process in prokaryotes, unlike the situation in eukaryotes.

eukaryotic DNA is linear and organised into nucleosomes

The simplest eukaryotes like yeasts have around 3 times as much DNA as this; higher eukaryotes have around 1000-times more. This greater amount of DNA is distributed between different chromosomes: eg cells of the yeast, *Saccharomyces cerevisiae*, have 16 chromosomes, each comprised of single, linear DNA molecule, complexed with histones. Histones complex DNA into globular subunits, called nucleosomes, containing 146 base pairs of nucleic acids. Some protozoa such as dinoflagellates lack mitotic spindles, centromeres and nucleosomes suggesting that the nuclear membrane precedes these features in evolution, although clearly these organisms have taken an evolutionary route independent of all other eukaryotes.

Genes in eukaryotes can be rendered inactive by changes in the local chromosome structure brought about by methylation of DNA, by histone modification and by the tight grouping of nucleosomes into solenoid structures. The packaging of eukaryotic genomes is examined in Chapter 4, whilst the regulation of transcription at the DNA level is looked at in detail in Chapter 6.

1.4 Gene structure: Introns and exons

expressed regions = exons

non-expressed regions = introns

Much of the additional DNA in eukaryotes is not expressed into actual gene products such as proteins. The structure of eukaryotic genes is more complex than that of prokaryotes. A single eukaryotic gene usually comprises alternating regions of DNA called exons - ie expressed regions - which contain coding sequences for parts of the polypeptide gene product, and intervening sequences, introns, which do not contribute to the coding of the product. Typically, exons may comprise only around 1% of the total DNA of higher eukaryotes, and introns may account for a further 8-9%. The discovery of such interrupting sequences almost exclusively in the genes of higher eukaryotes is one of the great fundamental surprises of molecular biology.

structural role of DNA

Not all of the great excess of DNA in eukaryotes compared to prokaryotes is explained by the existence of introns. There are other factors, too, such as the existence of large amounts of highly repetitive DNA, which has a structural role. This unexpected complexity and the techniques that first revealed it are examined in Chapter 4.

Introns may be evolutionary remnants. Some simple eukaryotes such as yeasts have introns only in a minority of their genes, as do the prokaryotic *Archaebacteria*. A consensus is emerging that introns may have been present in the very first organisms. They have been retained in the eukaryote and archaebacterial lines of descent but eliminated from the eubacterial lineage, all extra DNA being lost as the genome became streamlined.

1.5 Gene expression: Transcription and translation

concurrent transcription and translation in prokaryotes

At any given moment in bacterial cells around 1000-2000 RNA polymerase molecules can be active in transcription of the genetic code of the DNA into complementary strands of RNA. Around half of these polymerase molecules will be producing mRNA, for ultimate translation into polypeptide gene products. The mRNA produced has a very short life span, which on average is only around 2 minutes. In prokaryotes like bacteria, translation of the mRNA by ribosomes is concurrent with transcription. Very shortly after the synthesis of the mRNA molecules has started, ribosomes attach to the 5′-end of the mRNA and systematically work their way towards the 3′-end. New ribosomes will continue to attach until the mRNA has been degraded (degradation starts from the 5′-end) or until blocking factors bind to the mRNA: an average of 10-15 ribosomes will attach in this way to each mRNA molecule.

separation of transcription and translation in eukaryotes

However, with a set of genetic instructions sequestered in a nuclear compartment, additional processing steps to those of prokaryotes are inevitable in eukaryotes. In eukaryotes, the transcription of DNA to mRNA and translation of mRNA to polypeptide do not take place at the same time. Rather, pre-mRNA molecules are produced directly by transcription in the nucleus. But pre-mRNA includes unwanted RNA sequences complementary to introns and gene flanking sequences. Such sequences are not to be translated and must first be removed. The resulting mRNA is then passed into the cytoplasm for translation by ribosomes.

processing of RNA in eukaryotes

The processing and modification of the protein coding sequences are examined in Chapter 5. Introns are removed by a cutting and splicing process involving multi-ribonucleoprotein particles containing small nuclear RNAs (snRNAs). The trailing 3-'ends of mRNA molecules are cleaved at a site near to the termination sequence by an endonuclease. A polyriboadenylate tail, up to 200 adenosine bases long, is then added to the 3'-end of the mRNA. The function of this poly(A) tail is not altogether clear (not all mRNA molecules have it - eg histone mRNA forms a well-known exception to this rule), but it may confer mRNA stability. The 5'-end of the pre-mRNA is capped by a GTP residue linked in reverse direction. The guanine base is then methylated. This cap structure serves to ensure the correct positioning of the mRNA on the ribosome, as well as acting as an anti-nuclease device. Eukaryote cells also go in for a 'division of labour' with particular RNA polymerases involved in transcribing particular classes of genes. Details of this, splicing and processing are examined in Chapter 5.

1.6 Gene control in prokaryotes

We have seen that for single celled species to thrive they must gear themselves, when appropriate, to rapid growth and replication by the efficient use of nutrients. As particular local nutrient sources may become exhausted, the ability to adapt to new energy sources confers a selective advantage to any cells that can rapidly re-equip with the necessary enzyme machinery to exploit the new source. In prokaryotes, it is the rate at which new mRNA can be produced by transcription that governs the speed with which new enzymes can be manufactured by translation. The rate of transcription is controlled in turn by signals, such as the presence or absence of certain foodstuffs. These signals are interpreted by protein molecules that control which genes are to be transcribed. These regulatory proteins are called repressors if they reduce the rate at which synthesis of new mRNA molecules begins; they are termed activators if they act to increase this rate. Both activators and repressors act at regions of DNA called operators. One operator can often control more than one gene: clusters of adjacent genes controlled by one operator are called operons. Long polycistronic (ie incorporating more than one gene) mRNA molecules are transcribed from such operons.

repressors and activators

operons and polycistronic mRNA

Note that eukaryotes do not have polycistronic operons. For this reason, the co-ordinated regulation of multiple genes is more complex. Indeed, co-ordinated genes need not even be present on the same chromosomes in eukaryotes.

Operators often, but not always, reside on the DNA next to the promoter region for the operon. It is the productive binding of RNA polymerase molecules to promoters that initiates transcription in prokaryotes: the more frequently such initiation takes place the more rapid is the production of mRNA. Note that some gene expression in prokaryotes is regulated by control of the termination step of transcription at specific positions on the mRNA: this is called attenuation. These regulatory concepts have been examined in detail in Chapter 7 of a partner book in this series, 'Genome Management in Prokaryotes'.

1.7 Gene control in eukaryotes

transcriptional control is important

The rate at which the DNA of individual genes is transcribed is also of primary importance to overall gene control in eukaryotes. Although this is the predominant means of control in eukaryotes, at least in higher eukaryotes, additional mechanisms which act after transcription can sometimes be significant in the regulation of gene expression. The fact that the synthesis of mRNA and its translation into proteins do not occur concurrently affords further opportunities for control. The rate at which pre-mRNA is processed in nuclei is important to regulation, as is the differential stabilisation of mRNA in the cytoplasm. The rate at which the mRNA is translated by ribosomes is also a point of further control.

post-transcriptional control also happens

gene amplification

In some specialised cases particular genes are copied many times. This amplification can affect the rate at which particular genes are expressed.

The mechanisms whereby gene control is exerted are much less understood at the molecular level than are those in prokaryotes. Eukaryotes do not respond to nutritional signals by turning particular genes on or off; rather they respond to co-ordination singles such as hormones, growth factors or cell-to-cell contact.

transcription factors

Such signals elicit changes in gene regulation often via regulatory proteins in the nucleus called transcription factors. The signals may influence the balance between active and inactive forms of such factors, or may act to change their rate of synthesis - usually by acting at the post-transcriptional level of gene regulation. The overall pattern of gene expression in each cell reflects the activity of these regulatory proteins. The development of organisms can be determined by sequential cascades of transcription factors.

Transcription factors act to influence transcription rates by affecting initiation by RNA polymerase. In many cases control sites interacting with transcription factors may be close to where transcription starts, but in other cases can also be far removed. It is likely that DNA looping can bring even distant sites into close physical proximity to transcription sites. Most control sites are upstream of where transcription starts, but downstream elements, some even in introns, have been described.

Transcription factors are often classified into families by common molecular features within their peptide chain structure. Such structural features include: helix-turn-helix or homeobox structures; zinc fingers; helix-loop-helix structures and leucine zippers (Chapter 7).

Generally, most transcription factors act positively to encourage gene transcription. Often a concerted series of such positive influences is needed to switch on a gene (which must also be present in a section of the DNA rendered active by under-methylation - we will discuss this in Chapter 6). Compare this to the simplicity of gene activation in prokaryotes which can be achieved by the action of a single repressor or activator.

The initiation of transcription of genes which code for mRNA is carried out by RNA polymerase II, (see also Chapter 5). A pre-initiation complex is assembled from the polymerase and general initiation factors such as $TF_{II}D$, $TF_{II}B$ and $TF_{II}E$. Transcription starts from a site around 30 nucleotides downstream from the highly-conserved TATA box sequence in DNA.

TATA box sequence

1.8 Strategies for progress: Sex and polyploidy

We are accustomed to accept the Darwinian ideas about the evolution of species and natural selection without necessarily thinking about the molecular and cellular mechanisms that drive these processes forward. It is perhaps not hard to understand that the low mutation rate that occurs spontaneously for most genes leads occasionally, by chance, to more successful gene forms (Chapter 3). The spontaneous duplication of genes, or of portions of genes, can also be seen to fuel the possibilities for progression in the gene complement of species. Clearly, these events operating in unicellular organisms which can rapidly acquire millions of progeny by fission will occasionally throw up successful changes and additions to their gene complement. But how do the genetic complements of the less prolific and more intricate higher organisms evolve?

sex and genetic recombination

Sex is one partial answer. Even lower organisms often use the recombination events that accompany fusion, in one cell, of both maternal and paternal genes as a way of creating new combinations of gene sequences. The prevalence of sexual reproduction amongst virtually all eukaryotic species underlines the significance of this advantage to their success.

polyploidy and its consequences

Higher organisms have an additional device - polyploidy - which permits them a much greater degree of tolerance for the negative consequences of spontaneous gene mutations than is available to haploid species. Polyploidy is having multiple copies of genes and chromosomes present in one cell. The cells of most higher organisms are diploid for most of the time. Only after the process of meiosis are haploid germ cells created. The fact that cells in the diploid state normally contain two copies - one maternal, one paternal - of each gene means that if one copy is mutated, the other copy can normally be expected to provide a functional gene product which will protect the viability of the organism. The mutated gene may prove to encode a new gene product which is beneficial and confers some selective advantage. If not immediately of value, further spontaneous mutations may render it so. The diploid state permits a greater degree of latitude for this kind of spontaneous experiment than would be possible in species which are predominantly haploid. Such experiments in haploids are more likely to end in failure since the gene that has mutated might well be important for viability in its original form.

1.9 The genes of organelles

circular DNA in organelles

Surprisingly, circular extranuclear DNA can be found in the eukaryotic organelles; the mitochondria and chloroplasts, suggesting extraordinary cross fertilisation of eukaryotic genomes from prokaryotic sources. Primitive unicellular organisms such as microsporidians, and amoeba such as *Pelomyxa palastris* and *Entamoeba histolytica* lack mitochondria and presumably are relics of an ancestral state prior to the acquisition of mitochondria. This discovery strengthens the concept that chloroplasts and mitochondria have originated endosymbiotically, derived from engulfed prokaryotic organisms which entered into a perpetual symbiotic relationship with eukaryote cells. Probably *Prochloron*-type species and those of cyanobacteria such as *Gynecoccus* became chloroplasts, whilst (aerobic) purple eubacteria supplied oxidative phosphorylation properties. All eukaryote mitochondria are uniform in their oxygen-accepting, cell respiration pathways, unlike the great diversity shown in prokaryote respiratory enzymes, suggesting ancestry of mitochondria from a very restricted number of sources.

present day bacteria present plausible examples for mitochondrial symbiont theory

The overwhelming evidence is such that, although many of the proteins comprising each organelle are now encoded by genes resident in the cell nucleus, both organelles possess a vestigial independent genome which codes for organelle specific proteins and nucleic acids that often closely resemble bacterial molecules. For example the RNA and proteins of mitochondria are very similar to those of the bacterium *Paracoccus denitrificans*, suggesting a more recent ancestry between these two than between mitochondria and nuclear sequences.

In the eukaryotic lineage that led to land plants, the two organelles may have cohabited for something like a billion years. Their genetic regulation and biogenesis are now heavily under the control of the nucleus, reflecting a long, successful and intimate association (Chapter 9). However, since the inheritance patterns of nuclear genes are complex enough in themselves (Chapter 2) these are probably best mastered before considering the extranuclear phenomena such as cytoplasmic and maternal inheritance described in Chapter 9.

1.10 The origins of multicellularity in eukaryotes

Eukaryotes can of course be unicellular or multicellular and their success is such that they constitute all the macroscopic forms of life known, ie animals, plants and fungi. In considering eukaryote organisation we have emphasised the problems that follow from the organisation of cells into compartments, interactions between cytoplasmic organelle genomes and that of the nucleus and we must emphasise, in the case of multicellular organisms, co-operation between many cells, which may have quite different contributions and specialities to offer to the individual organism's survival.

However, how did multicellularity arise in eukaryotes? No mitotic organism is known that lacks a nucleus. One fascinating proposal advanced by Margulis and Sagan (1986) focuses on the role of eukaryotic microtubules which they argue can be used to separate chromosomes or alternatively, to the exclusion of mitotic events, used to provide cellular mobility through making up cilia and flagella, the whip-like appendages that propel protozoa and spermatozoa.

microtubule organising centre

The initiating sites associated with the appearance and assembly of such molecules are known as microtubule organising centres (MTOCs). In animal cells for example, they consist of a pair of centrioles and satellite proteins. All eukaryote mitotic spindles are made up of microtubules. Undifferentiated MTOCs provide genetic continuity in this scheme, but specialised MTOCs if used to produce cellular motility, result in the final cell being unable to divide further. In other words, we have a situation of mutual exclusion, animal cells produce either the motility appendages or the mitotic segregation apparatus but not both at once. The scheme visualises two-celled protozoans, one cell motile, the other capable of mitosis, giving rise to both plants and animals. Multicellular organisms could arise by repeated mitotic divisions, the daughter cells held together in a single mass. Specialised MTOCs would form ultimately additional motility structures, excluding mitotic options in these other cells, but ultimately leading to the development of multiciliated cell layers and nonmitotic internal cell motility. In this mode, it is crucial to retain some functional undifferentiated MTOCs capable of mitosis. Rather strikingly fertilisation reflects the fusion of an egg cell with an uncommitted MTOC with a motile cell (sperm) containing a specialised MTOC.

Meiosis, also a distinctive feature of eukaryotes, is then seen as derived from a cyclical check-up or inventory process, to ensure that the complex, eukaryote cell, seen as a set of interacting microbial genomes, contains the appropriate genetic components. These include nucleus, mitochondria, chloroplast (at least one of each for plant cells) and undifferentiated MTOCs for mitosis. Such a check-up state means producing a haploid cell and makes the cycle of fertilisation - meiosis obligatory for multicellular organisms. Thus the development of sexuality, multicellular organisms and cellular differentiation may all be intrinsically linked. These connections may be reflected in the unique ability of eukaryote cells to engulf bacteria and protozoa by phagocytosis, literally 'cell eating'. The plastic, deformable, vacuole-forming cell membrane of eukaryotes made endosymbiosis possible, giving rise to mitochondria and chloroplasts whilst also making eukaryote cell fusion possible, reflected in eukaryote cell fertilisation. The subsequent evolution of consciousness in at least one species allows us to reflect on the dictum 'you are what you eat'!

SAQ 1.2

Answer true or false to the following questions:

1) Prokaryote cells contain a number of discrete compartments.

2) Yeast cells contain a nucleus, mitochondria and other organelles.

3) Bacterial DNA is packaged with histones to form a compact nucleoid zone.

4) Introns appeared in genes relatively recently during the evolution of eukaryotes.

5) Intron sequences are removed from pre-mRNA in the nucleus.

6) Operons are transcribed by transcription factors.

7) In eukaryotes, transcription is generally activated by several transcription factors.

8) Chloroplasts and mitochondria probably originated from bacterial cells engulfed by primordial eukaryote cells.

Summary and objectives

This introductory chapter has been designed to provide a context in which the remainder of the text can be studied. In this chapter, we have explained that although there are many common molecular and cellular themes in biology, there are fundamental differences between the organisation and expression of genetic information in prokaryotic and eukaryotic organisms. We emphasised that the compartmentalisation of eukaryotic cells and the packaging of DNA in such cells generated opportunities and imposed certain constraints on the regulation of gene expression in these systems. We also reflected on the importance of these features on the generation of polyploidy and the need to develop systems of genetic recombination (sex). We concluded the chapter by speculating that the development of sexuality, multicellularity and cell differentiation may be intrinsically linked.

Now that you have completed this chapter you should be able to:

- use a wide variety of terms applicable to a description of the cellular states of eukaryotes (eg multicellular differentiation);
- show an understanding of why the packaging and segregation of DNA in nuclei in eukaryotes enable post transcriptional processing;
- explain why polyploidy and sexual recombination are favoured in eukaryotes;
- understand the biological significance of the detailed molecular events described in subsequent chapters;
- appreciate the diverse nature of the different genomes found in modern eukaryotic cells.

Inheritance patterns in eukaryotes

Inheritance patterns in eukaryotes

2.1 Introduction

2.1.1 DNA and eukaryotic inheritance

In eukaryotes (as in prokaryotes), deoxribonucleic acid (DNA) is the genetic material. The sequence of bases that make up the DNA of animals and plants must:

- carry the genetic information;
- be capable of expression to produce biologically useful molecules;
- be replicated faithfully;
- be transmitted from generation to generation;
- be capable of variation.

In this chapter, we shall see how the genes which make up DNA affect physical appearance, or phenotype, and how these genes are inherited.

2.2 Particulate inheritance

2.2.1 Genes are pairs of alleles

homologous chromosomes

genetic loci

The genes which make up each of the eukaryotic chromosomes may be considered as beads on a string. Each of these beads represents a genetic locus, which, via its sequence of bases (A, T, G or C) encodes a single polypeptide chain. Typically, each eukaryote cell contains two copies of each chromosome, known as homologues. Both homologous chromosomes have the same order of beads, or genetic loci, along its length. On each copy of a chromosome, alternative forms of the gene, known as alleles, may be found. So we can see that a given cell may actually encode two different types of, for example, an enzyme at a particular genetic locus. These allelic forms may be reflected in different phenotypes, depending on the combination of alleles that make up the genotype of the organism.

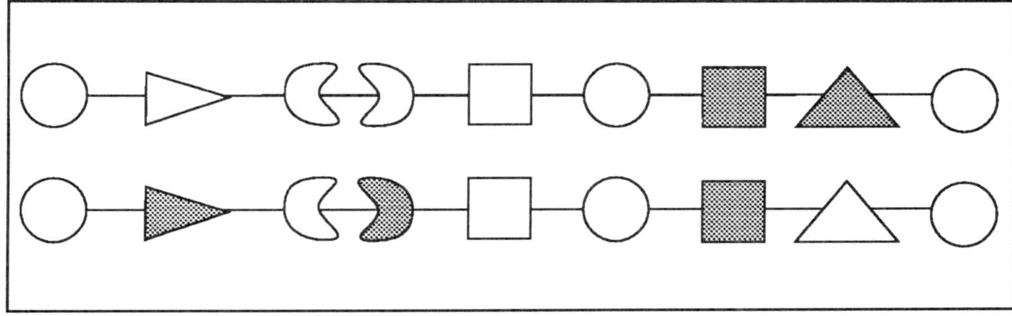

Figure 2.1 A stylised representation of genes on a chromosome as beads along a string. Open and shaded forms of the same beads represent different alleles.

2.2.2 Dominance and recessivity

The genotype, or genetic make-up, of an individual at a particular locus, may be:

- homozygous, if only one type of allele is present;
- heterozygous, if two types of allele are present.

alleles may be dominant or recessive

Alleles may be either dominant, if their phenotypic characteristics are expressed both in homozygous and heterozygous genotypes, or recessive, if expressed only in the homozygous state. Each gamete formed as a product of meiosis will contain only one allele from a particular genetic locus. An example of this relationship between dominant and recessive alleles at a single genetic locus is seed coat texture in peas. The dominance and recessivity of seed coat texture alleles in pea are described in Figure 2.2. The convention in eukaryotic genetics is to use upper case symbols for the dominant allele and lower case for the recessive, ie *S* is the smooth coated pea allele and *s* the wrinkled coat allele.

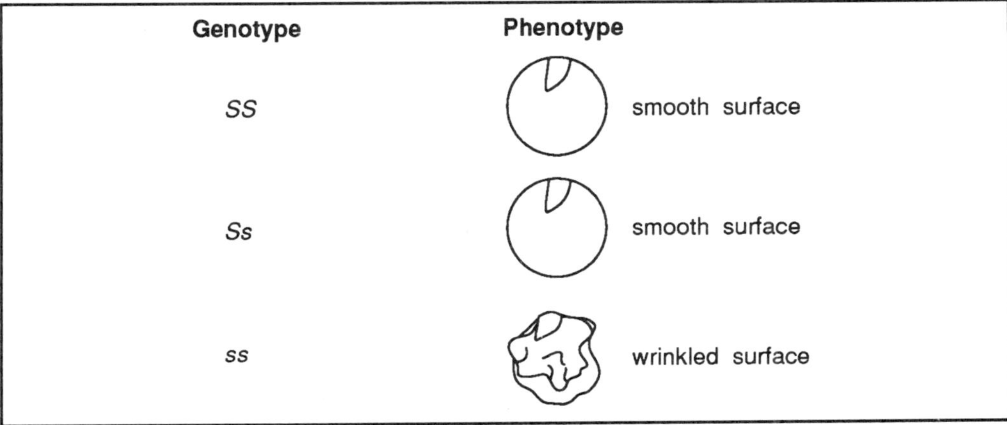

Figure 2.2 Smooth and wrinkled coat texture alleles in peas. The *S* smooth allele is dominant to the *s* wrinkled allele at the seed surface locus in pea plants. Therefore the heterozygous pea with the genotype Ss (ie one smooth allele S and one wrinkled allele s present) will have a smooth surface, not an intermediate wrinkled surface.

2.2.3 Monohybrid inheritance

In peas, plant height is controlled by alternative alleles at a single locus. The tall allele, which we can call *T*, is dominant to the dwarf allele *t*. When homozygous tall peas are crossed with homozygous dwarf pea plants, the first progeny generation (F1) are all tall. If the F1 plants are crossed amongst themselves, the second progeny generation (F2) plants occur in the ratio of 3 tall: 1 dwarf. No intermediate forms are found amongst the progeny generations. This is because definite structures are transmitted from the parents to the progeny pea plants. This is known as particulate inheritance.

We can follow this cross using the allelic symbols *T* and *t* to represent the tall and dwarf alleles, respectively.

Parental Phenotypes	**Homozygous Tall**		**Homozygous Dwarf**
Genotypes	*TT*	x	*tt*
Gametes	*T*	x	*t*

In forming these gametes, we must assume that mating is at random, and that gamete formation and fertilisation are normal.

Π Write down the possible genotypes of the F1 generation. From this, work out the phenotypes of the F1 generation.

The only possible genotype for the F1 generation is *Tt*. Since *T* is dominant to *t*, all of the F1 generation will be tall pea plants.

Now look at a F1 x F1 cross:

Phenotypes	**Tall**		**Tall**
Genotypes	*Tt*	x	*Tt*
Gametes	*T*, *t*	x	*T*, *t*

Π What types of gametes are formed by each of the F1 parents?

Each F1 parent can produce two types of gametes, *T* or *t*. We should expect these to occur with approximately equal frequency.

In forming the F2 generation, each gamete can combine with each gamete from the other F1 member.

Π Complete the table shown below with the genotypes and phenotypes of the F2 generation:

Genotype	Phenotype
TT	Tall
?	?
?	?
?	?

From your table, what is the ratio of Tall: Dwarf phenotypes in the F2 generation?

You should find that the phenotypic ratio is 3 Tall : 1 Dwarf, as shown in the table below.

F2 Generation:	Genotype	Phenotype
	TT	Tall
	Tt	Tall
	tT	Tall
	tt	Dwarf

Each of the 4 F2 groups occur equally often, so the F2 phenotypic ratio is:

3 Tall : 1 Dwarf.

We can see from these results that dwarfness is hidden in the F1 generation, but reappears in the F2 generation.

Mendel's First Law

These finding gave rise to Mendel's First Law, which describes the behaviour of chromosomes during meiosis:

Alleles segregate randomly

2.2.4 Dihybrid inheritance

Sometimes crosses involving more than one genetic locus are performed. Each of the loci involved will have its own dominant and recessive alleles. For example, in pea plants, the purple flower colour allele (P) is dominant to the white flower colour allele (p). The tall and dwarf plant height alleles are as described in Section 2.2.3.

When homozygous tall, homozygous purple flowered pea plants are crossed with homozygous dwarf, homozygous white flowered peas, the F1 progeny are all tall, with purple coloured flowers. When the F1 are crossed amongst themselves, four phenotypic groups are found amongst the F2 generation.

These events can be explained in terms of allelic symbols:

Parental phenotypes	**Tall purple flowers**	x	**Dwarf white flowers**
Genotypes	*TT PP*	x	*tt pp*

In forming the gametes, any one allele from the T/t locus can combine with any one allele from the P/p locus.

Gametes	*TP*	x	*tp*
F1 Genotype		*Tt Pp*	
Phenotype		All Tall, Purple flowered	

If the F1 progeny are crossed amongst themselves:

F1 Genotype	*Tt Pp*	x	Tt Pp
Gametes	*TP, Tp, tP, tp*	x	*TP, Tp, tP, tp*

In forming gametes from the F1 members, remember that any one allele from the T/t locus may form a gamete with any allele from the P/p locus, and that these combinations occur equally often.

Punnett square

The combinations of F1 gametes to produce the F2 may best be carried out using a Punnett square, with the gametes from one F1 member listed horizontally, and those from the other vertically.

Π Before reading on see if you can work out all of the combinations of these gametes which might occur, by covering up the table within the bold line.

Gametes	*TP*	*Tp*	*tP*	tp
TP	*TTPP*	*TTPp*	*TtPP*	*TtPp*
Tp	*TTPp*	*TTpp*	*TtPp*	*Ttpp*
tP	TtPP	*TtPp*	*ttPP*	*ttPp*
tp	TtPp	*Ttpp*	*ttPp*	*ttpp*

Remembering the rules of dominance and recessivity, we can assign phenotypes to each of the 16 F2 groups shown in the Punnett square above. The *T* (tall) allele is dominant to the *t* (dwarf) allele, and the *P* (purple flowers) allele is dominant to the *p* (white flowers) allele.

Π Any progeny with one or more *T* and one or more *P* alleles will be tall, with purple flowers. Look through the Punnett square above, and count the number of F2 groups with this phenotype.

You should find 9/16ths of the F2 have this phenotype. Progeny with at least one *T* allele and two *p* alleles will be tall with white flowers.

Π Look through the Punnett square and count how many of the F2 groups will be tall, with white flowers.

You should find 3/16ths of the F2 generation have this appearance.

Π Progeny with two *t* alleles and with at least one *P* allele will be dwarf, purple flowered plants. Look through the Punnett square and count the number of groups with this phenotype.

You should find 3/16ths of the F2 are dwarf, purple flowered peas.

Finally, those plants with two *t* alleles and two *p* alleles will be dwarf, white flowered plants. Look carefully through the Punnett square, and you will find that just 1/16th of the F2 generation are dwarf, white flowered plants.

Mendel's Second Law

The 9:3:3:1 F2 phenotypic ratio obtained is characteristic of dihybrid inheritance, and forms the basis of Mendel's Second Law:

Different pairs of alleles are inherited independently

If the phenotypic ratios associated with Mendel's First and Second Laws are borne in mind, understanding the genetic basis of inheritance becomes straightforward.

2.3 Genetic crosses

Often much information on the relationship between alleles and between different loci can be obtained by carrying out controlled genetic crosses.

These controlled crosses include:

testcrossing

- testcrossing with the homozygous recessive genotype;

backcrossing

- backcrossing to one or other parent.

These types of controlled matings are very useful in forming conclusions about the physical positions of genes on chromosomes, and in producing maps of individual chromosomes.

SAQ 2.1

A heterozygous tall pea plant is testcrossed.

1) What will be the genotype of the testcross parent?

2) What phenotype will the testcross parent exhibit?

3) What types of gamete will be formed by the heterozygous tall pea plant?

4) What type of gamete will be formed by the testcross parent?

5) Identify the F1 phenotypic groups obtained.

6) What phenotypic ratio will be found amongst the F1 generation of peas?

SAQ 2.2

A dihybrid tall, purple flowered pea plant is crossed with a dwarf, white flowered plant.

1) What is the genotype of the dihybrid tall, purple flowered parent?

2) What is the genotype of the dwarf, white flowered parent?

3) List the different gametes that can be formed by the dihybrid tall, purple flowered parent.

4) What gametes can be produced by the dwarf, white flowered parent?

5) List the F1 genotypes produced.

6) What phenotypic ratio is observed amongst the F1 pea plants?

2.4 Incomplete dominance

heterozygote gives unique phenotype in incomplete dominance

In the case of some genetic loci, the normal rules of dominance and recessivity between alleles do not apply. This means that the heterozygous genotype has its own phenotype, distinct from either the homozygous dominant, or the homozygous recessive phenotypes. This pattern of inheritance is known as incomplete dominance, due to the heterozygous genotype having a unique phenotype.

An example of incomplete dominance is flower colour in snapdragons. When homozygous red flowered snapdragons are crossed with homozygous white flowered plants, the F1 progeny resemble neither parent, but instead are all pink flowered. When the F1 snapdragons are crossed amongst themselves, all three flower colours are found in the F2 generation, but the pink flowered heterozygous phenotype occurs twice as frequently as the parental forms.

We can understand how this happens more fully by following what happens at the genotypic level:

Parents	Homozygous Red	x	Homozygous White
Genotypes	*RR*		*rr*
Gametes	*R*		*r*
Genotype		*Rr*	
F1 Phenotype		All pink flowered	

This pink flowered phenotype is different to either of the parental phenotypes, and demonstrates incomplete dominance at this locus. We can go on to see what happens when the F1 generation are crossed amongst themselves, to produce the F2 generation.

F1 Mating	*Rr*	x	*Rr*
F1 Gametes	*R, r*		*R, r*

Each gamete can combine with each gamete from the other parent.

Π Fill in the blank spaces in the table below which describes the F2 generation:

Genotype	**Phenotype**
RR	?
Rr	?
rR	?
rr	?
F2 phenotypic ratio:	? Red : ? Pink : ? White

What ratios do you find?

You should have found 1 Red : 2 Pink : 1 White, as shown in the completed table below.

F2 Genotypes	*RR*	*Rr*	*rR*	*rr*
F2 Phenotypes	Red	Pink	Pink	White
F2 Phenotypic ratio	1 Red : 2 Pink : 1 White			

The 1:2:1 ratio amongst the F2 from homozygous, or true-breeding parents, with the heterozygous genotype providing the largest group, is characteristic of incomplete dominance.

2.5 Codominance

MN blood group as example of codominance

The alleles at some genetic loci do not show dominance or recessivity. Each allele at such loci can be expressed to some degree when in the heterozygous genotype. This is known as codominance, and means that the heterozygote is often intermediate in character between the two homozygous phenotypes. This can often take the form of both alleles being expressed at once. When considering codominant alleles, we should use two letter symbols, with a common first letter for each locus, to avoid confusion. The MN blood group system in Man is a good example of codominance. Red blood cells will be agglutinated by a particular antisera (anti-M, or anti-N), if the cells express the alleles LM for the M, or LN for the N antigen on their surface. As we can see from Table 2.1, three distinct phenotypes can be distinguished by their agglutination reactions with anti-M and anti-N antibodies.

Genotype	anti-M reaction	anti-N reaction	blood group
LM LM	+	-	M
LM LN	+	+	
LN LN	-	+	N

Table 2.1 MN blood group analysis. + = positive agglutination test; - = negative agglutination test.

Π What is the blood group of the LM LN individual?

Since the LM LN individual gives a positive agglutination test with both the anti-LM and anti-LN reactions, this individual's blood group is MN. This shows codominance.

distinctiveness of third phenotype allows distinction between codominance and incomplete dominance

We have seen that when codominance occurs, a third phenotype, due to the heterozygous genotype is found. This third phenotype is distinct from either of the homozygous phenotypes, but both types of allele involved are contributing to their maximal extent. In our example, alleles for both the M and N type cell surface antigens are being fully expressed. The distinctiveness of the third phenotypic group permits codominance to be distinguished from incomplete dominance. Codominance, like incomplete dominance will also produce a 1:2:1 phenotypic ratio amongst the F2 from true-breeding parents for a single genetic trait. Comparing the heterozygotes with the two parental groups permits us to decide whether incomplete dominance, or codominance, is occurring. If the heterozygotes appear halfway between the two parental phenotypes, then incomplete dominance is likely to be occurring. If however, we find that the heterozygotes have some of the different properties shown by each set of parents, then codominance is more likely.

SAQ 2.3

In the Venezuelan beaver, black fur colour is produced by the homozygous genotype *FBFB*, grey fur by the heterozygous genotype *FBFW* and white fur by the homozygous genotype *FWFW*. When grey furred beavers are mated:

1) What are the genotypes of the F1 generation?

2) What is the phenotypic ratio amongst the F1 progeny?

2.6 Multiple alleles and allelic series

At some genetic loci, we can find that many more than the usual two alleles are possible. At these loci, we find great variety in the amount of gene product produced. This means that these multiple alleles can produce a whole series of different phenotypes. This is known as an allelic series.

We can see how an allelic series of phenotypes can arise by looking at coat colour in rabbits. At a single genetic locus, four different alleles can influence fur colour. A hierarchy of dominance exists, with the wild type, or fully coloured allele (C) being dominant to all other alleles, the 'Chinchilla' allele (c^{ch}) encoding silver grey fur, is dominant to the 'Himalayan' allele (c^{h}) encoding white fur, with black tips to the ears, feet and nose, and to the albino (*c*) allele, which lacks colour. The 'Himalayan' allele is, itself, dominant to the albino allele.

We can represent this in the following way:

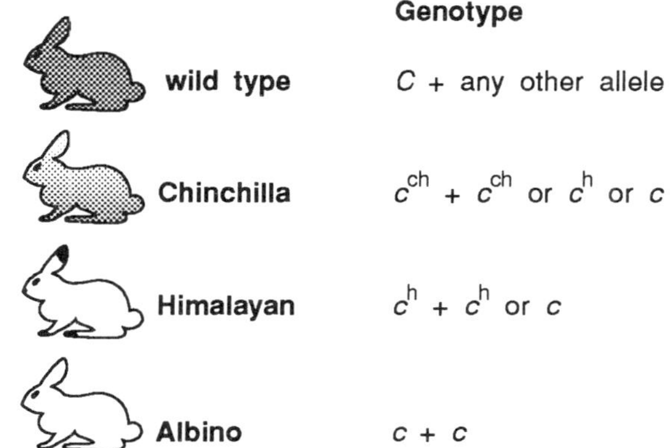

Genotype
wild type
C + any other allele
Chinchilla
c^{ch} + c^{ch} or c^{h} or c
Himalayan
c^{h} + c^{h} or c
Albino
c + c

We can demonstrate the full range of fur colour phenotypes in rabbits in the following cross:

Parental Phenotypes	Himalayan	x	Wild type
Genotypes	$c^{h}\,c$	x	$C\,c^{ch}$
Gametes	c^{h} and c		C and c^{ch}
F1 Genotypes	$c^{h}\,C$, $c^{h}\,c$, $c\,C$, $c\,c^{ch}$		
F1 Phenotypes	Wild type, Himalayan, Wild type, Chinchilla		
Phenotypic ratio	1:1:1:1		

So we can see that the Chinchilla phenotype appears in the F1 generation.

Π Perform the same analysis for the mating of the Chinchilla and Himalayan F1 progeny rabbits:

Genotypes	$c\,c^{ch}$	x	$c\,c^{h}$
Gametes			
Progeny			
Phenotypes			

What phenotypic ratio do you find?

You should have found that there were three phenotypic groups amongst the F2 rabbits, in the ratio shown in the table below:

Gametes	c, c^{ch}	c, c^{h}
Progeny	$c\,c,\quad c\,c^{h},\quad c^{ch}c,\quad c^{ch}\,c^{h}$	
Phenotypes	Albino, Himalayan, Chinchilla, Chinchilla	

The $c^{ch}\,c^{h}$ genotypic group has the Chinchilla phenotype because c^{ch} is dominant to the Himalayan allele c^{h}.

Since each of the progeny groups occurs with equal frequency, the progeny phenotypic ratio is: 1 Himalayan : 2 Chinchilla : 1 Albino

We can see how the fourth phenotype, albino furred rabbits, can be obtained, when gametes carrying *c* alleles are united.

Allelic series of this type can be extremely complicated, and probably represent the true situation at a great many genetic loci.

SAQ 2.4

1) Identify the dominance order for alleles at the fur colour locus in rabbits.

2) Without looking through the text, list the genotypes that can encode each of the fur colour phenotypes, wild type, Chinchilla, Himalayan and albino.

2.7 Pleiotropy

phenylketonuria

Sometimes the alleles at a particular genetic locus can have more than one effect. This situation, which is known as pleiotropy, occurs for the recessive allele for phenylketonuria in humans. The primary effect, in homozygous recessive people, is for metabolism of the amino acid phenylalanine to be defective, and for phenylalanine levels to be abnormally high, particularly in the brain. Secondary effects of this allele include a reduction in intelligence quotient, smaller head size and lighter hair colour than might otherwise be expected.

cystic fibrosis

Another example of alleles at a single locus having more than one effect is the allele for cystic fibrosis in humans, The primary effect is for the secretion of an abnormally thick mucus in the lungs. Expression of this allele, however, can also lead to an inability to digest food properly, and to fatal secondary lung infections. The secondary effects of pleiotropic alleles are often consequences of the primary effect, occurring further along a developmental, or metabolic pathway.

2.8 Lethal alleles

At certain genetic loci, alleles can manifest themselves phenotypically by causing death between fertilisation and sexual maturity. These are known as lethal alleles, and can be divided into three types; dominant, recessive and conditional lethals.

2.8.1 Dominant lethal alleles

If a dominant allele is lethal, any individuals possessing that allele will die before reaching reproductive maturity. Such alleles, being dominant, will always be expressed, and will tend to disappear from a population once they have arisen by mutation, because they cannot be passed on to the next generation.

2.8.2 Recessive lethal alleles

If an allele kills an individual before reproductive maturity when in the homozygous state, but not in the heterozygous condition, then it is a recessive lethal allele. The presence of such a recessive lethal allele cannot be seen in an heterozygous individual without performing additional crosses.

curly winged allele is lethal recessive

In the fruitfly *Drosophila melanogaster*, the curly winged allele *Cw*, is dominant to the straight winged allele *cw*, but acts as a recessive lethal, in that any homozygous *Cw* flies will die before contributing to the next generation. We can see the recessive lethal effects of the *Cw* allele on viability when two flies heterozygous for the curly winged allele are crossed:

Parental Phenotype	curly	x	curly
Genotype	*Cw cw*	x	*Cw cw*
Gametes	*Cw, cw*		*Cw, cw*

Remember that each gamete from one parent can combine with each of the available gamete types from the other parent.

Π What will the genotypes and phenotypes of the F1 progeny be? What is the phenotypic ratio amongst the F1 generation?

If you think that there is a 3:1 ratio of curly winged : straight winged flies, you have evaluated the potential F1 genotypes correctly, but have forgotten that the Cw Cw individuals will be killed because of the recessive lethal allele. You should have decided upon a 2:1 ratio of curly winged : straight winged flies, as shown below:

F1 Progeny: *Cw Cw, Cw cw, cw Cw, cw cw*

We must remember that the curly winged allele *Cw* is lethal when homozygous, so that any *Cw Cw* individuals will be killed:

F1 Phenotypic ratio	2 Curly winged *Cw cw, cw Cw*	:	1 straight winged *cw cw*

So we can see that a 2:1 phenotypic ratio amongst the progeny from heterozygous parents is characteristic of recessive lethal allele expression. This is in contrast to the 3 curly winged : 1 straight winged phenotypic ratio we might have expected if the *Cw* allele had no effect on fly viability.

2.8.3 Conditional lethal alleles

dropded allele is conditional lethal

Some lethal alleles are only expressed under particular environmental conditions, but not expressed at other times. An example of this type of allele, known as a conditional lethal is the '*dropded*' allele of the fruitfly, *Drosophila melanogaster*.

These alleles are not expressed at temperatures of 28°C, or below, but when the fruitflies are incubated at 30°C or hotter temperatures, the *dropded* alleles gene products cause lethality. Conditional lethal alleles, in general, show their lethal effects at environmental extremes.

2.9 Gene interactions

epistatic gene interaction

When considering pairs of genetic traits, sometimes we can come across situations where the expression of alleles at one locus are masked or prevented, by alleles at a completely different locus. This is known as epistatic gene interaction, and is characterised by less than the expected four phenotypic groups amongst the progeny of heterozygote matings. Instead of the 9:3:3:1 phenotypic ratio which is normal for two independently assorting genetic loci, only two, or at most three phenotypic groups are obtained.

There are six types of epistatic gene interaction, each of which is named after the form that the masking of one locus by expression of alleles at a second locus takes. Table 2.2 describes the various forms of epistatic gene interaction.

Epistatic form	Number of Phenotypes	Phenotypic Ratio
dominant epistasis	3	12:3:1
recessive epistasis	3	9:3:4
duplicate genes with cumulative effect	3	9:6:1
duplicate dominant epistasis	2	15:1
duplicate recessive epistasis	2	9:7
dominant and recessive interaction	2	13:3

These phenotypic ratios should be contrasted with the 9:3:3:1 ratio expected for two independently segregating genetic loci.

Table 2.2 Epistatic gene interactions, amongst progeny from heterozygote matings.

We can see that whenever epistatic gene interaction occurs, fewer than the expected four phenotypic groups are found amongst the progeny of heterozygote matings, and that the type of epistasis involved can readily be identified from the phenotypic ratio obtained.

SAQ 2.5

Consider each of the following sets of progeny data from dihybrid crosses. Identify the phenotypic ratio you consider most likely to explain the data obtained, and name the pattern of inheritance appropriate to this ratio.

	Progeny data	Phenotypic ratio	Inheritance pattern
Example	72:24:24:8	9:3:3:1	Independent assortment
1)	135:45:60		
2)	180:12		
3)	144:48:16		
4)	221:51		
5)	63:49		
6)	126:42:42:14		
7)	171:114:19		

2.10 Penetrance, expressivity and environment

2.10.1 Penetrance

retinoblastoma shows incomplete penetrance

Normally, the alleles at a genetic locus are always expressed, producing a particular phenotype. This is known as complete penetrance, since the phenotype is always exhibited. Occasionally, however, traits may be incompletely penetrant, and skip a generation. This is the case for a type of eye tumour found in children, known as retinoblastoma. Only 9/10 children who might be expected to suffer from retinoblastoma develop these eye tumours. The retinoblastoma trait can therefore, be described as having 90% penetrance. Unfortunately, this trait will go on to reappear in the next generation. Such incompletely penetrant traits, which occasionally seem to miss out a generation, often show wide variations in the extent to which they are expressed within a generation.

2.10.2 Expressivity

The level of expression produced by an allele at a particular locus is known as its expressivity. This can be expressed as a percentage of the maximal expression possible, and can vary tremendously between individuals. The expressivity of the retinoblastoma trait will be either 0% or 100%, since children will either develop, or fail to develop, these eye tumours.

2.10.3 Factors influencing penetrance and expressivity

The main factors influencing whether or not a particular trait is expressed in an individual (its penetrance), or the extent to which a trait is expressed, (its expressivity), are genetic background, and environment. Since individuals inherit a complex collection of pairs of alleles, their overall genotype is very likely to exert some influence on how any one particular genetic trait is expressed. Similarly, the influence of environment, for example nutritional supply, can dramatically alter expression of some traits, eg height. It is often very difficult to unravel the contributions which genetic background and environment make to the variations in penetrance and expressivity often observed.

2.11 Sex linkage

The chromosomes of any animal can be divided into two groups:

- sex chromosomes;
- autosomes.

The sex chromosomes are involved in determining the sex of the individual. The autosomes are all of the other chromosomes possessed by that individual.

Genetic loci found on those chromosomes determining sex in animals can be said to be sex-linked. In this section, we shall see how the sex chromosomes of mammals and many other animals differ. We will also discuss how sex linked traits are inherited, and consider some variations on the basic pattern of sex linkage.

2.11.1 Sex determination in mammals

X and Y chromosomes

In a large number of animals, including the mammals, sex is determined by the X and Y chromosomes, which differ in size, and shape. Each of the cells of the female normally carry two copies of the X chromosome, and is known as the homogametic sex, because all the gametes produced by a female will have an X chromosome. Each of the cells of the male sex, however, carry one X chromosome and one copy of the Y chromosome, and is known as the heterogametic sex, because two different types of gametes, with respect to the sex chromosomes, can be formed in meiosis. The majority of sex-linked genetic loci are found on the X chromosome. The X and Y (or sex) chromosomes are shown in Figure 2.3.

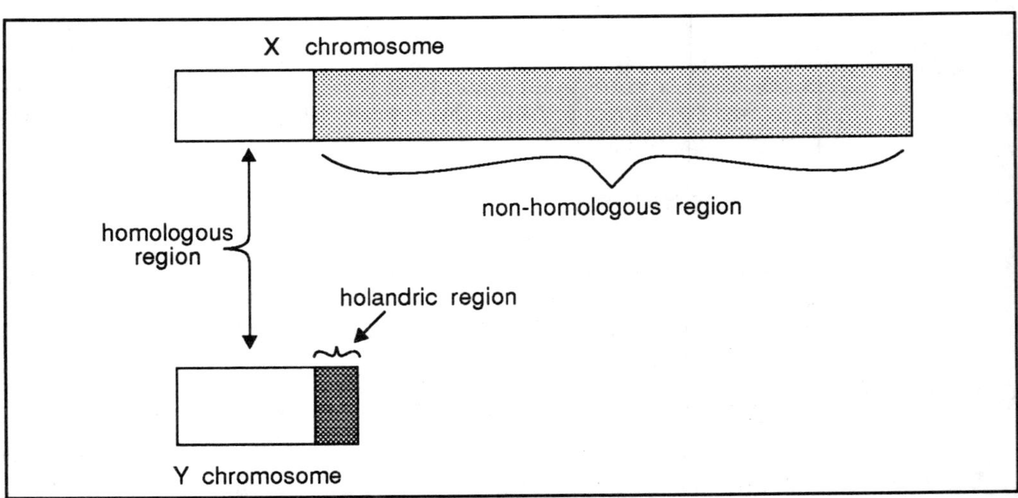

Figure 2.3 Stylised representation of the X and Y chromosomes.

The X and Y chromosomes share a common segment, known as the homologous region. Genetic loci located in this region are incompletely sex-linked, since both X and Y chromosomes carry these loci. Loci on the portion of the X chromosome which has no counterpart of the Y chromosome, known as the non-homologous region, are completely sex-linked. This is because only the X chromosome can influence expression of these genetic traits, since these loci are not found on the Y chromosome. Finally, the differential segment of the Y chromosome, which has no equivalent on the X chromosome, is known as the holandric region. Loci located in this region can only be expressed in males, since only males possess the Y chromosome. The importance of the sex chromosomes in influencing development may be seen from Table 2.3 which describes the consequences of sex chromosome abnormalities in humans.

sex chromosome abnormalities

constitution	syndrome	features
XXX	Triple X	decreased fertility mental retardation female
XO	Turner's	infertile abnormal ovaries female
XXY	Klinefelter's	infertile small testes partial breast development male
XYY		normal fertility low I.Q., aggressive male

Table 2.3 Sex chromosome abnormalities in humans.

2.11.2 Sex linked traits

In the mammals, and those other animals where the male is the heterogametic sex, eg the fruitfly *Drosophila melanogaster*, particular patterns of inheritance are associated with sex linked alleles.

red eyed allele is sex linked

Reciprocal crosses, performed using genotypically identical parents, but with the sex of each genotype reversed, produce dissimilar results. We can see how this occurs using alleles at the X-linked eye colour locus in *Drosophila melanogaster*. The red eyed allele (which we can call *W*) is dominant to the white eye allele (called *w*). When considering crosses involving sex linkage, we should assign alleles to each of the sex chromosomes individually, and separate the sex chromosomes by a '/' sign.

Parental Phenotypes	White eyed female	x	Red eyed male
Genotypes	X^w / X^w	x	X^W / Y
Gametes	X^w		X^W, Y
F1 Genotypes	X^w / X^W	and	X^w / Y
F1 Phenotypes	Red eyed females		White eyed males
Phenotypic ratio	1	:	1

Notice how all of the male F1 generation have white eyes, like their mother, but all of the female F1 generation have red eyes, like their father. Sex linked traits often exhibit this sex swopping pattern of inheritance. Now compare this with what happens when the sex of the parents is reversed, but the overall parental eye colour genotypes remain unchanged.

Parental Phenotypes	White eyed male	x	Red eyed female
Genotypes	X^w/Y	x	X^W/X^W
Gametes	X^w, Y		X^W
F1 Genotypes	X^w/X^W	and	Y/X^W
F1 Phenotypes	Red eyed females		Red eyed males

Π What will be the phenotypes of the F1 generation with respect to eye colour and sex?

In this case, the sex linked white eyed phenotype does not appear in the F1 generation. This is very different to what we saw when the white eyed parent was female.

haemophilia is sex linked

Perhaps the best known example of a recessive sex linked allele in humans is haemophilia A inheritance in the Royal Families of Europe in the nineteenth century. A recessive mutant allele for an abnormal antihaemophilic globulin was passed on by Queen Victoria of England to some of her children. This globulin is normally required for blood clotting, so that each of Victoria's sons had a 50% chance of being haemophiliac, since the X chromosome in these sons must have come from their mother. Of Victoria's four sons, one was a haemophiliac. Amongst her five daughters, two were carriers, and passed on the haemophilia allele to their sons.

In general, the patterns of inheritance and expression followed by sex linked alleles is governed by whether the allele is dominant, or recessive. We can sum up these patterns in Table 2.4.

Recessive sex-linked alleles are: • most often expressed in males; • expressed only in females if carried by the male parent; • rarely expressed in both father and son, and then only if the mother is genotypically heterozygous. Dominant sex-linked alleles are: • expressed most often in females; • expressed in all the daughters of a male expressing the allele; • never transmitted to the sons of a female failing to express the allele.

Table 2.4 Inheritance and expression patterns for sex-linked alleles.

2.11.3 Holandric traits

external ear hair porcupine skin trait

Genetic loci associated with the portion of the Y chromosome for which there is no homologous region on the X chromosome are described as holandric. Of necessity, these traits can only be expressed in males, since only males possess the Y chromosome, and examples of this inheritance pattern in humans include external ear hair, and the very rare porcupine skin trait.

SAQ 2.6

Identify the terms which correctly describe genetic loci found in mammals:

1) on the homologous region of the X chromosome;
2) on the differential segment of the Y chromosome;
3) on the non-homologous region of the X chromosome;
4) on a chromosome other than the X or Y chromosomes.

2.11.4 Sex limited traits

When an allele is expressed in one sex, but not in the other, that trait is sex limited. This is because the penetrance of such an allele differs between the sexes, and is often affected by hormone levels within the body. An example of this is milk production in goats. Male goats posses many of the alleles associated with milk production, since they can pass them on to their daughters, but do not themselves ever produce milk.

2.11.5 Sex influenced traits

pattern baldness

Sometimes an allele may be dominant in one sex, but recessive in another. This can occur for loci on the homologous portion of the sex chromosomes, or any of the autosomes, and is the case for pattern baldness in humans. The b^* allele, when homozygous, produces baldness in men or women. The b^* allele when in the heterozygous state, is dominant in men, but recessive in women. This, like sex limited gene expression, may be due to internal hormonal differences between the sexes.

SAQ 2.7

A pattern bald man and his wife had 10 children, 8 of which were pattern bald boys. One of the two girls was also pattern bald, but the other had normal hair.

1) What was the genotype of the wife?

2) What are the possible genotypes of the pattern bald man?

3) What is the most likely genotype of the pattern bald man?

Justify your answers.

2.12 Linkage and linkage groups

genetic loci on the same chromosome make up a linkage group

In previous sections, we have seen alleles which segregate randomly, and described how different pairs of alleles segregate independently of one another. However, because the chromosome is the physical unit of inheritance, all of the genetic loci on a particular chromosome will not segregate independently but will be inherited together during gamete formation. They may be described as being linked together on the same chromosome. All of the genetic loci on a particular chromosome make up a linkage group.

We can see how linkage between two genetic loci affects genotypic and phenotypic ratios by testcrossing dihybrid (*AaBb*) individuals for loci that are unlinked, or on the same linkage group.

For independently assorting loci:

Parents	*Aa Bb*	x	aa bb
Gametes	*AB, Ab, aB, ab*		ab
F1 Genotypes	*Aa Bb, Aa bb, aa Bb, aa bb*		
Genotypic ratio	1:1:1:1		

Notice how, for unlinked loci, four different types of gametes are produced from the dihybrid genotype. This is because the loci involved are on different chromosomes, and therefore not linked to each other in anyway.

Compare this genotypic ratio with the results for two linked genetic loci, which are on the same chromosome, and therefore in the same linkage group. The alleles at these linked loci on each homologue tend to stay together during meiosis, and consequently, their genotypes are listed with a slash (/) separating the alleles on the two homologous chromosomes.

For completely linked loci:

Parental Genotypes	*AB / ab*	x	ab / ab
Gametes	*AB, ab* only		ab
F1 Genotypes	*AB / ab*		ab / ab
Genotypic ratio		1:1	

When gamete formation occurs for these completely linked loci, the allelic combinations from each of the homologous chromosomes are maintained in the gametes, without any exchange of genetic information between homologous chromosomes. This is in contrast to the unlinked loci situation, where four different gametes can be formed from the dihybrid genotype. Deviations from the 1:1:1:1 phenotypic ratio which we would expect to find in the testcross progeny from a dihybrid are characteristic of linkage.

The exact ratios obtained for two loci within the same linkage group will vary between pairs of loci. This is due to crossing over during the first meiotic division, which, for partially linked genetic loci, will permit some reciprocal exchange of genetic information between homologous chromosomes. Examples of partially linked genetic loci include the purple/red flower colour locus and the pollen grain shape locus in sweet peas. We will discuss these in the next section.

SAQ 2.8

The fungus *Neurospora crassa* has 7 linkage groups. Many of the genetic loci in these linkage groups relate to nutritional requirements for growth, or to the form of fungal growth. Amongst the nutritional loci, are the recessive mutations encoding requirements for sulfonamide (*su*), and for tryptophan (*trp*). The recessive 'button' mutation (*bu*) produces a compact, button like growth habit.

1) When a *Neurospora* line dihybrid for the *su* and *bu* alleles, with both mutant alleles on the same homologue, was testcrossed to a line homozygous recessive at both loci:

   50% of the progeny did not require sulfonamide and grew normally,

   50% required sulfonamide and showed 'button' growth.

   Are the *su* and *bu* loci in the same linkage group?

2) Justify your answer to 1).

3) When a *Neurospora* line dihybrid for the *su* and *trp* alleles, with both mutant alleles on the same homologue was testcrossed to a line homozygous receissve at both loci:

   25% of the progeny did not require sulfonamide or tryptophan;

   25% required both sulfonamide and tryptophan;

   25% required sulfonamide but not tryptophan;

   25% required tryptophan but not sulfonamide.

   Are the *su* and *trp* loci in the same linkage group?

4) Justify your answer to 3).

2.13 Crossing over

meiosis

crossing over between two homologues

During prophase I of meiosis, each homologous chromosome lines up gene for gene with its homologue. The opportunity then exists for a physical breaking and rejoining between the replicated arms, or chromatids, of the two homologues. This is known as crossing over, and provides the potential for recombination, or the exchange of genetic information, between the two homologues of a particular chromosome pair. The variation produced provides raw material, in the form of new allelic combinations, for evolution. The further apart two genetic loci are on the same chromosome, the more possibility there is for crossing over to occur between them. If crossing over does occur, then some of the allelic products in the gametes following the completion of meiosis will be recombinant types. These will always be in the minority, as compared to the parental type gametes, which have maintained the allelic combinations from the individual homologues. Figure 2.4 illustrates the events involved in crossing over.

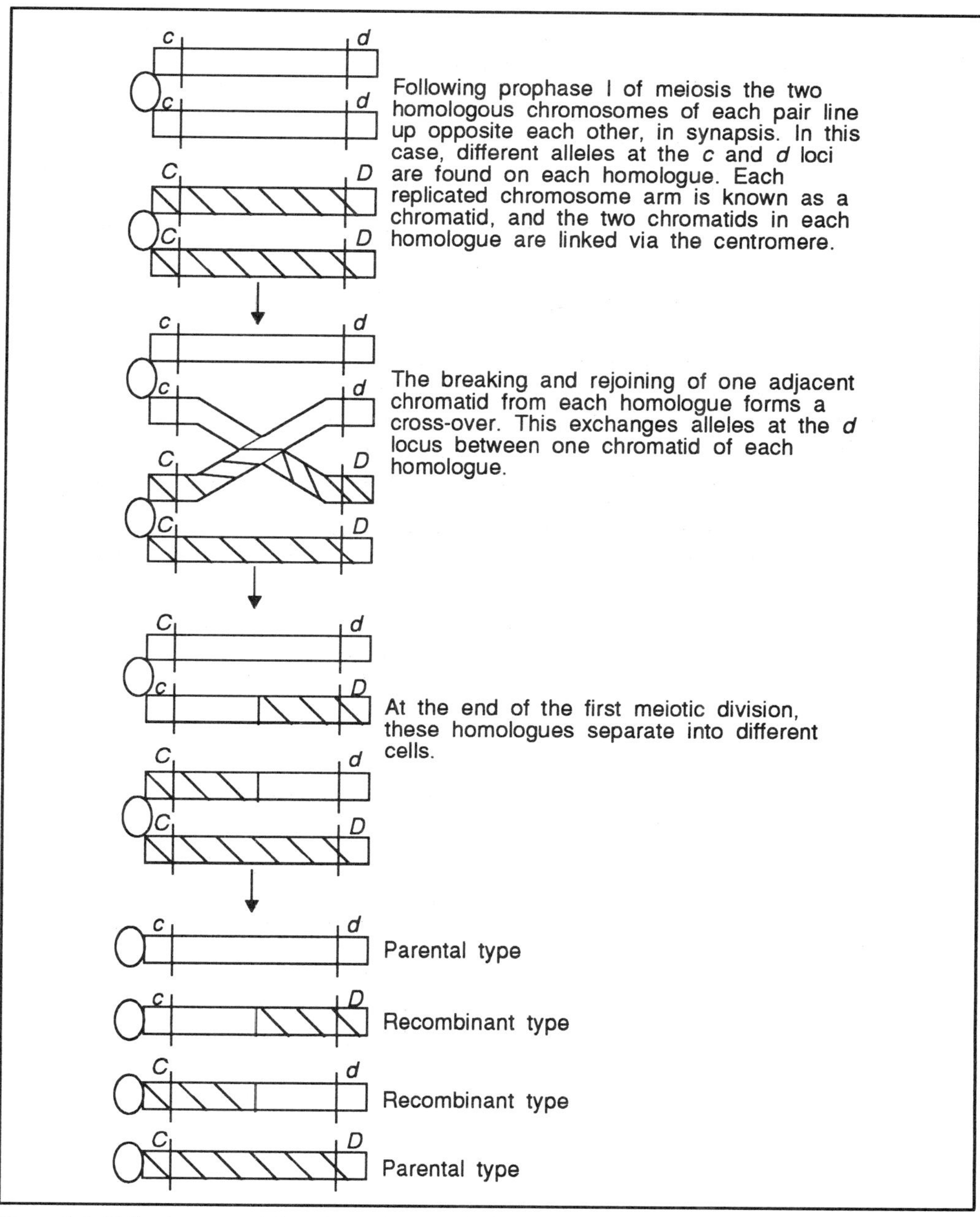

Figure 2.4 Stylised representation of crossing over during meiosis.

We can see that, following a single cross-over, a number of recombinant type gametes will be produced, which can never exceed the number of parental type gametes. A cross-over need not occur, however, amongst all of the pairs of homologous chromosomes. The frequency of crossing over dictates the percentage of the gametes which will be recombinant, but this can never exceed 50%, since two of the four meiotic products from a single cross-over will be of the parental type. The probability of crossing over is related to the distance between two genetic loci on the same

cross-over recombinants can never exceed 50%

chromosome, and therefore within the same linkage group. This means that two loci at opposite ends of a chromosome are more likely to have a cross-over event occur between them, and therefore more likely to have recombinant type gametes formed, than two genetic loci which are adjacent to each other. Thus each pair of loci in a linkage group will have a particular probability, or frequency, of a cross-over occurring between them, and the proportions of recombinant and parental gamete types formed can be predicted from this frequency. This finding provides the basis of genetic mapping studies, and in assigning gene order to the loci in a linkage group.

2.14 Genetic mapping studies

2.14.1 Mapping linkage groups

centiMorgan - unit of cross-over frequency

Each eukaryote chromosome with the genetic loci arranged along its length can be likened to the beads on a piece of string (Figure 2.1). But what is the order of these beads, and how far apart are they along the length of string? These questions can be addressed by the use of genetic mapping. In constructing genetic maps, the basic unit is a 1% frequency of crossing over, which is known as the centiMorgan (cM, not to be confused with the centimetre cm). This genetic mapping unit permits us to decide upon the order of genetic loci, on the basis of the probability of a cross-over event occurring between them.

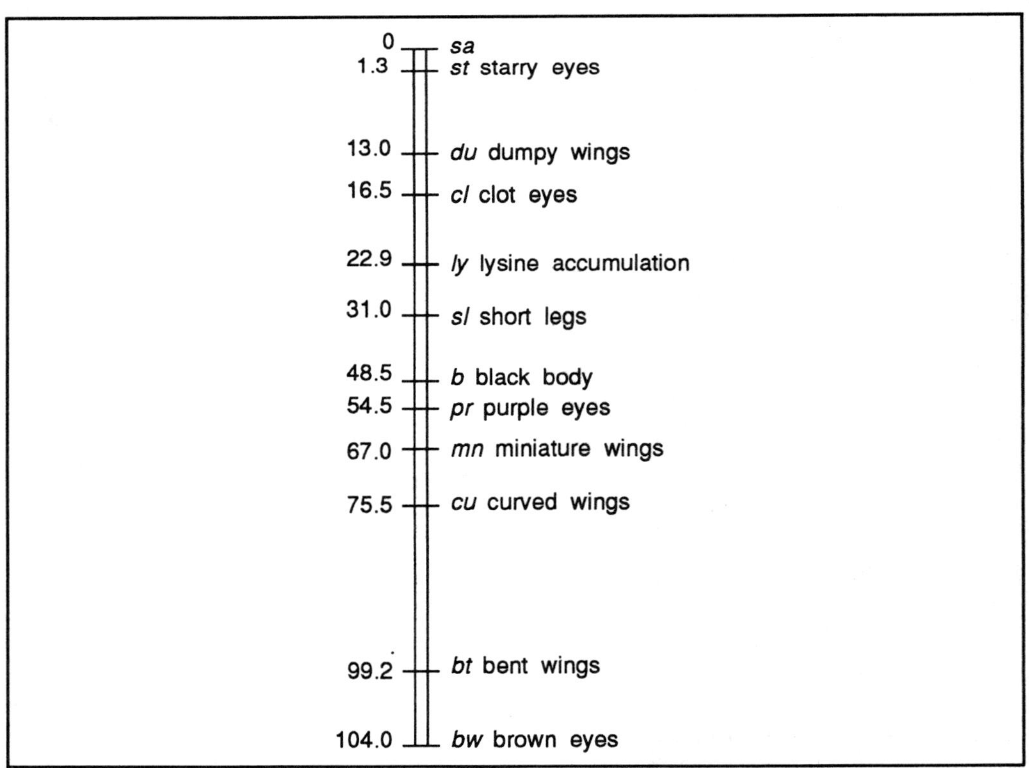

Figure 2.5 *Drosophila* chromosome 2 linkage map.

Figure 2.5 describes a partial map of the *Drosophila* chromosome 2 linkage group. From this, we can see that the short legged locus (*sl*) is 17.5 cM from the black body locus (*b*).

Both of these mutations are recessive, and are given lower case symbols. This means that if, in a testcross, a parent with the genotype *sl b/Sl B* will produce 17.5% recombinant type gametes, and 82.5% parental type gametes, due to a single cross-over occurring between these loci.

We can see how this is brought about more clearly if each of the homologues involved in drawn separately.

original homologues: —— *sl* —————— *b* ——
and
—— *Sl* —————— *B* ——

If a single cross-over occurs between these loci, then two new, or recombinant type chromosomes can be produced:

----- *sl* ------------ *B* ------
and
----- *Sl* ------------ *b* ------

these chromosomes will be included amongst the gametes produced once meiosis is completed.

We should remember that the cross-over occurs only in 17.5% of pairings between these loci, and the remainder of homologous pairings proceed without any crossing over. This means that four different types of gametes will be formed, but that the recombination types will be present less frequently than the parental types, since crossing over does not always occur between these loci.

Meiotic products: ----- *sl* ------------ *b* ------ 41.75%

----- *Sl* ------------ *B* ------ 41.75%
and
----- *sl* ------------ *B* ------ 8.75%

----- *Sl* ------------ *b* ------ 8.75%

Notice that the two largest gametic product groups are the parental types, which have not undergone any crossing over, and that the smallest gametic product groups are the recombinant types, which have undergone a single crossing over. Since the frequency of crossing over is equal to the sum of the recombinant type gametes percentages, we can see that the *sl* and *b* loci on chromosome 2 of the *Drosophila* genome will indeed be 17.5 cM apart.

combining crossing over frequency data to cover greater map distance

Any loci which assort completely independently of one another will be 50 cM apart, as each of the four gametic products possible will occur equally frequently. For large chromosomes, it is very likely that two loci at opposite ends of the chromosome will undergo crossing over, and for this reason, the maximum map distance measurable between any two loci in the same linkage group is 50 cM. However, by combining information on the crossing over frequency between pairs of genetic loci, more complex genetic maps, extending over a total of more than 50 cM can be built up. This is the case for the *Drosophila* chromosome 2 genetic map, where the total map distance covered is 104 cM.

2.14.2 Gene order and complex genetic maps

As we have seen in the previous section, the frequency of recombinant type gamete formation is a reflection of the amount of crossing over between two genetic loci on the same linkage group. The crossing over frequencies obtained can be represented as centiMorgan distances separating the loci on a genetic map of the linkage group. These map distances are additive, so that all of the loci in a linkage group can be placed in a linear order, which is consistent with the crossing over frequency data obtained by studying pairs of loci. This is often done by using a 2 point testcross, in which any recombinant type gametes, produced following crossing over, can easily be detected following union with the gametes from a doubly homozygous recessive individual.

2 point testcross

We can see how these more complex genetic maps can be built up by considering the data in Table 2.5, from the *Drosophila* chromosome 2 linkage group.

Genetic loci	% Recombinant progeny	% Parental progeny	Genetic map distance (cM)
b and *pr*	6.0	94.0	6.0
pr and *mn*	12.5	87.5	12.5
mn and *cu*	8.5	91.5	8.5
b and *cu*	27.0	73.0	27.0
b and *mn*	18.5	71.5	18.5
pr and *cu*	21.0	79.0	21.0

Table 2.5 Linkage data from *Drosophila* chromosome 2.

For each set of 2 point testcross progeny described in Table 2.5, the percentage of recombinant progeny is a reflection of the amount of crossing over that occurs between the 2 loci being considered. From these data, you can build up quite a detailed genetic map for part of the *Drosophila* chromosome 2.

First of all you must decide upon the two loci which are furthest apart, in this region of chromosome 2, from the available data. The highest percentage recombinant progeny was the 27% obtained for the black body (*b*) and curly wings (*cu*) loci. This means that these loci are likely to be the furthest apart in this region. So we can place these loci at the extremities of our genetic map, 27 centiMorgans apart.

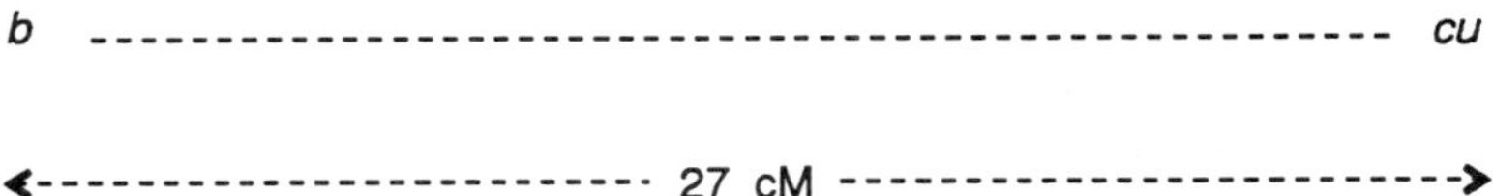

Π You must now decide upon the order of the other two loci, for purple eyes (*pr*) and for miniature wings (*mn*) which may lie in between these outer two loci and draw them on the map above.

If we consider the percentage recombinant progeny in 2 point testcrosses involving these loci and the black body (*b*) locus individually, we see that 6% recombinant progeny were obtained when the purple eyes (*pr*) and black body loci were studied, compared to 18.5% recombinant progeny when the miniature wings (*mn*) and black body loci were examined. Since crossing over is a function of the distance between two loci, it is likely that the miniature wing locus (*mn*) is further from the black body locus, at the left end of our map, than the purple eyes (*pr*) locus. This means that we can now map out the complete order of the four loci from this region of chromosome 2. We can also place the map distances that we have reconciled so far onto this map.

We can gain further confidence in our positioning of the *pr* and *mn* loci by considering their relative map distances from the curly wing (*cu*) locus, at the right end of our map. If our gene order is correct, then there should be more recombinant progeny produced when the *pr* and *cu* loci are considered, in comparison with the numbers of recombinants formed when the *mn* and *cu* loci are testcrossed. We can see from Table 2.5 that this is indeed the case, with 21.5% and 8.5% recombinant progeny being produced respectively, for these testcrosses. Another piece of evidence to be considered in positioning the intermediate loci is the distance between them on our map. Table 2.5 suggests that they are 12.5 cM apart. This too can be added to our genetic map for this part of chromosome 2.

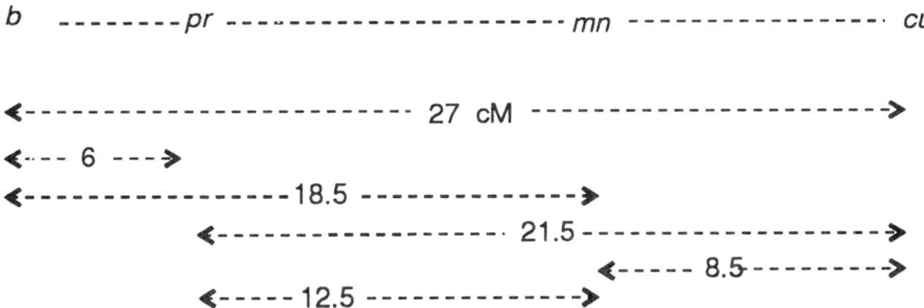

map distances are additive

We can also see from this map that the sum of the genetic map distances for *b-pr* and for *pr-mn* is equal to the recombinant percentage, and therefore map the distance expected for *b-mn*. Since we have found that the sum of *b-pr* and *pr-mn* map distances is equal to the *b-mn* map distance, we can rule out the possibility that *pr* and *mn* are on opposite sides of the *b* locus. This shows the additivity of these map distances. The findings for *pr-mn* and *mn-cu* when compared to the *pr-cu* map distance also support the idea of genetic map distances being additive. We have now reconciled each of the map distances obtained from the 2 point testcrosses, shown in Table 2.5 and our partial map of the *Drosophila* chromosome 2 is complete. This genetic map is, in fact, a subset of the map shown in Figure 2.5.

SAQ 2.9

1) When dihybrid *Drosophila* of genotype *y w/Y W* were crossed with flies homozygous recessive at both loci, the following progeny flies were obtained.

Progeny genotypes	% Progeny
y w/y w	49.25
Y W/ y w	49.25
y W/y w	0.75
Y w/y w	0.75

a) List the parental type progeny genotypes obtained.

b) What is the total percentage of parental type flies produced?

c) List the recombinant type progeny genotypes obtained.

d) What is the total percentage of recombinant type flies produced?

e) What is the genetic map distance between the *y* and *w* loci?

2) From a further series of dihybrid testcrosses, the following percentage recombinant progeny data were obtained:

Loci	% Recombinant progeny
y and *ct*	20.0
w and *ct*	18.5
y and *ec*	5.5
ec and *ct*	14.5
w and *ec*	4.0

Construct a genetic map of this linkage group, including the answer to question 5 above).

2.14.3 Multiple cross-overs

Sometimes two crossing over events can occur between loci on the same chromosome. This will have the effect of producing only the parental type products. Figure 2.6 demonstrates how this can come about.

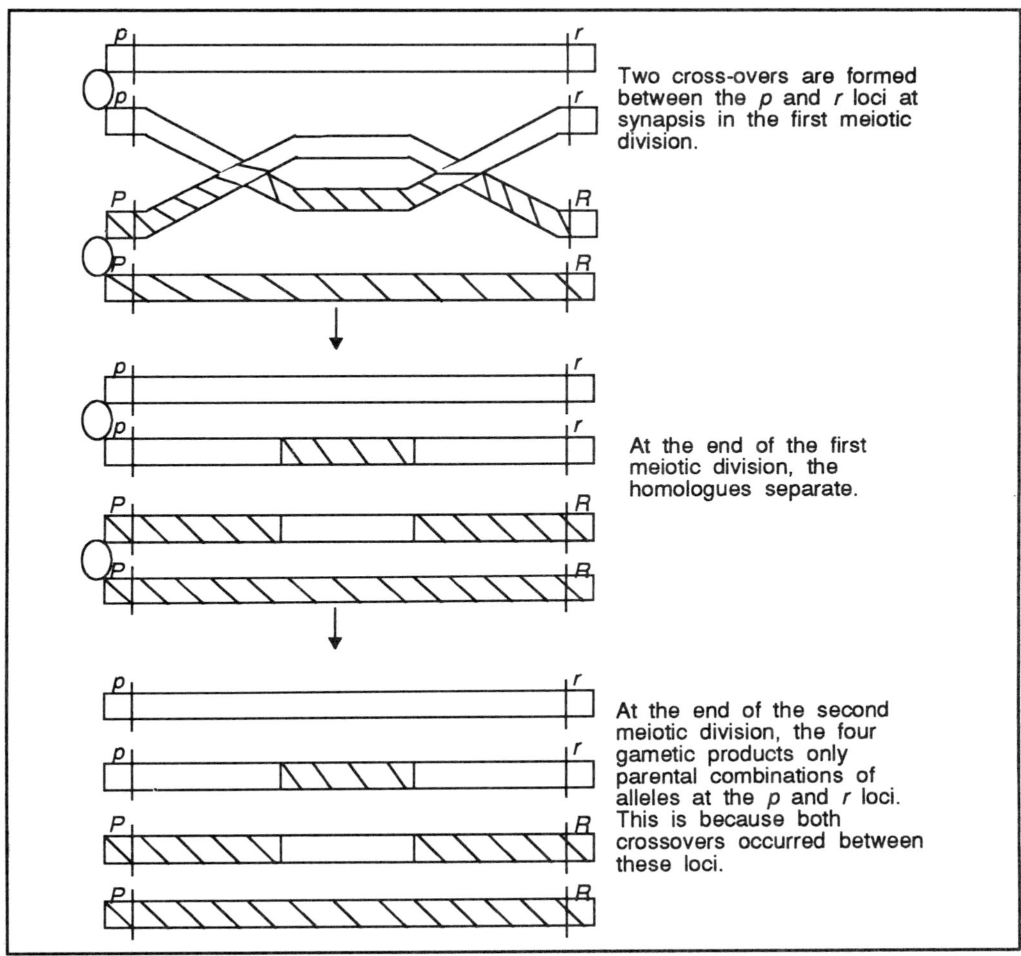

Figure 2.6 Multiple cross-overs.

double cross-overs

These double cross-over products will only be detected if a third locus, located between the cross-over sites, is taken into consideration. The probability of a double cross-over event occurring is the product of the two single cross-over probabilities. This is because they are mutually dependent on one another. These double cross-overs do not normally occur for loci less than 5 cM apart.

2.14.4 Three point testcrosses

Information on double cross-over products can be obtained by testcrossing an individual heterozygous at three loci, in other words a trihybrid testcross. An example of such a three point testcross is given in Table 2.6.

36% *PQR/pqr*	9% *Pqr/pgr*	4% *PQr/pqr*	1% *PqR/pqr*
36% *pqr/pqr*	9% *pQR/pqr*	4% *pqR/pqr*	1% *pQr/pqr*
72% parental: types	18% single P-Q cross-over	8% single Q-R: cross-over	2% double cross-overs

Table 2.6 Three point testcross data (see text for discussion).

Consider the three loci *p*, *q* and *r*, within a linkage group. At each of these loci, either the mutant allele, which is recessive and shown in lower case, or the wild type allele, which is dominant and shown in upper case, can be present. When a trihybrid individual (*PQR/pqr*), with all of the mutant alleles on one homologue is testcrossed to a *pqr/pqr* individual, the progeny were found to be as shown in Table 2.6.

map distance from the sum of single and multiple cross-overs

From the data, the *p-q* map distance is equal to the sum of the percentage of single cross-overs occurring between these loci, and the percentage of double cross-overs, ie 18+2 = 20 cM. The *Q-R* map distance similarly is equal to the sum of the percentage of single cross-overs between the *Q* and *R* loci, and the percentage of double cross-overs, ie 8+2 = 10 cM. We can conclude that the *p-r* map distance must therefore be 30 cM, on the basis of these 3 point testcross data.

double cross-overs not detected if we only use outside markers

If we had considered only the two outside loci, *p* and *r*, we would not have detected the double cross-over products (2% of progeny). This would mean that the *p-r* map distance would be 18+8 = 26% as estimated from a two point testcross. In other words, the double cross-over products are not detected if only the outside markers only are considered, and the true map distance would be underestimated, by the amount of double cross-overs occurring, but not showing up as recombinant type products. For this reason, it is sometimes the case that all of the component map distances for a particular linkage group do not quite add up to the same value as when the two outside loci are investigated. The amount of detectable cross-overs (and therefore of recombinant type products) between the two outer loci (*p* and *r*) if the central locus *Q* is not studied will be equal to the sum of the individual single cross-over percentages, minus twice the double cross-over percentage. The possibility of not detecting multiple cross-overs of this type should be considered when constructing complex genetic maps for linkage groups.

2.14.5 Relationship between recombination percentage and genetic map distance

We have seen how the further apart two loci within the same linkage group are, the more change there is of multiple cross-overs occurring between them. So, for very closely linked loci, the recombination percentage will be an accurate measure of how far apart these loci are on the genetic map. However, beyond about 15 cM, the recombination percentage underestimates the real genetic map distance between two loci.

2.14.6 Physical and genetic maps

When we construct genetic maps for particular linkage groups, we are basing those maps on the probability of recombination occurring between the loci involved. This probability varies enormously, even along the length of a single eukaryote chromosome. This means that although we can rely on the order of genetic loci obtained from a genetic mapping study, the actual physical distances between the loci within that linkage group is not correlated to genetic map distances. We can see this most clearly from comparing the genetic and physical maps of the *Drosophila* chromosomes, as shown in Figure 2.7, in which the number of bands staining intensely is used as an estimate of physical size.

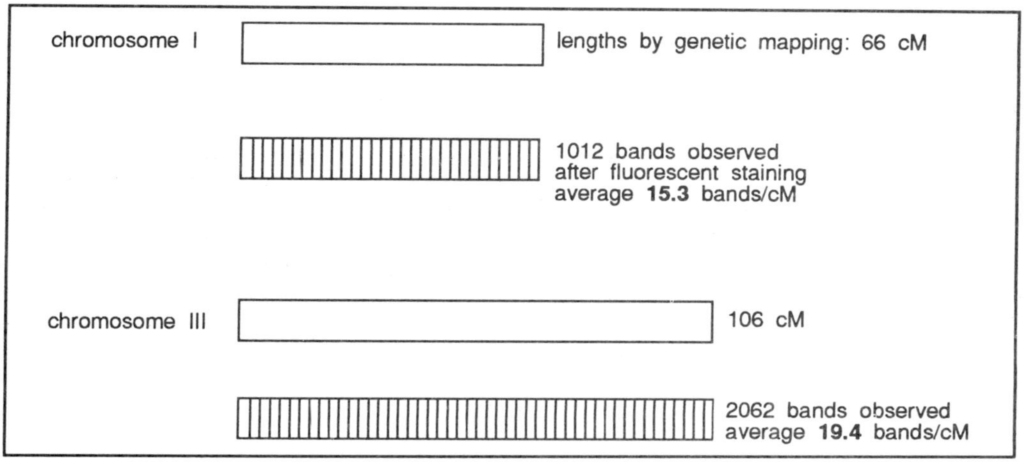

Figure 2.7 Genetic and physical maps of the *Drosophila* chromosome.

2.14.7 Different linkage groups assort independently

We have seen how genetic maps can be built up for particular linkage groups. The number of chromosomes an organism has determines the number of linkage groups. Each of these linkage groups represents a physical unit passed from generation to generation via gamete formation. Just as each chromosome pair assorts independently from the other chromosome pairs in meiosis, so each genetic locus in a particular linkage group assorts independently of the other linkage groups present.

Summary and objectives

In this chapter we have seen how genes are inherited in eukaryotes. We have learnt of dominance, Mendel's Laws, and how to analyse data from genetic crosses. We have seen how exceptions to the rules of dominance and recessivity can come about, how sex can influence gene expression, and how complex genetic maps of linkage groups can be constructed. Armed with the principles of eukaryotic inheritance described in this chapter, understanding the basis of genetic transmission in eukaryotes becomes straightforward.

Now that you have completed this chapter, you should be able to:

- discuss particulate inheritance;
- explain Mendel's Laws of inheritance;
- analyse data from genetic crosses;
- summarise gene interactions;
- discuss genetic linkage;
- summarise genetic mapping.

Mutations

Mutations

3.1 Introduction

In this chapter, we will first remind you of the key features of the genetic code before considering mutation and inborn errors of metabolism. This will lead on to a consideration of origins of polymorphism. We will then consider in some detail how mutations may arise and how damage to DNA may be repaired. The consequences of mutation will be examined together with how mutation may be exploited.

3.1.1 Introduction to mutations in eukaryotes

A mutation is an heritable change in the information stored in the genome. As you know, the sequence of four bases in DNA constitutes the genetic information. In general terms, this sequence of bases corresponds to the sequence of amino acid residues (ie primary structure) in the polypeptides of the organism. Thus changes in the base sequence of DNA could lead to changes in the polypeptides (proteins) in the organism.

point mutation

A mutation is an alteration in the base sequence of a gene. The mutation will form a different allele of the gene. If only a single base is involved it is called a point mutation. The relationship between the sequence of bases in a gene and the sequence of amino acid residues in a protein is mediated through the genetic code. We will remind you of the salient features of the genetic code.

degeneracy

A protein can contain up to 20 different types of amino acid residues. Each amino acid is coded for by a codon of three bases. This genetic code is shown in Table 3.1. It should be noted that the code is described in terms of the RNA which transfers the information between DNA and polypeptides, that is the messenger RNA. In RNA, uracil is used in place of the thymine found in DNA. It can be seen that, for example, tryptophan has only one codon, whereas arginine, leucine and serine have six codons. The existence of multiple codons is called degeneracy. Codons are also required to initiate the synthesis and termination of polypeptide synthesis.

Π As a revision, which amino acids are coded for by the codons: UUC, ACU, GGG and UGA? Also, work out which codons specify the amino acids: proline, asparagine and cysteine.

The four codons encode phenylalanine, threonine, glycine and STOP, respectively. Proline is coded for by CCN (where N = A, U, G or C), asparagine by AAU and AAC, and cysteine by UGU and UGC.

chromosomal mutations

Mutations which occur through changes in the organisation of the genetic material in the chromosome are called chromosomal mutations.

multifactorial inheritance

Multifactorial inheritance is associated with defects in several genes. They particularly contribute to developmental disorders which result in congenital malformation. A combination of individually small deleterious variations may result in a serious clinical condition. Multifactorial disorders do tend to recur in families although they do not show the distinctive inheritance patterns associated with single-gene defects.

	U		C		A		G		
U	UUU	Phe	UCU	Ser	UAU	Tyr	UGU	Cys	U
	UUC	Phe	UCC	Ser	UAC	Tyr	UGC	Cys	C
	UUA	Leu	UCA	Ser	UAA	End	UGA	End	A
	UUG	Leu	UCG	Ser	UAG	End	UGG	Trp	G
C	CUU	Leu	CCU	Pro	CAU	His	CGU	Arg	U
	CUC	Leu	CCC	Pro	CAC	His	CGC	Arg	C
	CUA	Leu	CGA	Pro	CAA	Gln	CGA	Arg	A
	CUG	Leu	CCG	Pro	CAG	Gln	CGG	Arg	G
A	AUU	Ile	ACU	Thr	AAU	Asn	AGU	Ser	U
	AUC	Ile	ACC	Thr	AAG	Asn	AGC	Ser	C
	AUA	Ile	ACA	Thr	AAA	Lys	AGA	Arg	A
	AUG	Met	ACG	Thr	AAG	Lys	AGG	Arg	G
G	GUU	Val	GCU	Ala	GAU	Asp	GGU	Gly	U
	GUC	Val	GCC	Ala	GAC	Asp	GGC	Gly	C
	GUA	Val	GCA	Ala	GAA	Glu	GGA	Gly	A
	GUG	Val	GCG	Ala	GAG	Glu	GGG	Gly	G

Table 3.1 The genetic code. The first nucleotide for each triplet is given in the left hand column, the second nucleotide is given across the top while the third base is on the right. Note three codons are termination codons that specify the end part of translation. AUG is the usual start codon.

Mutations may produce only a slightly altered variant of the protein and biological function is retained. It may be difficult to detect such alterations other than by biochemical analysis. Other mutations may, however, lead to noticeable changes in appearance or physiological functions of the organism or even death. Indeed, many of the inherited disorders of metabolism have arisen through mutation.

3.2 Inborn errors of metabolism

alkaptonuria, cystinuria, pentosuria, albinism

The idea that inborn errors of metabolism, ie mutations, may be the cause of some clinical conditions was first suggested by Garrod, a London physician early this century. Garrod's ideas were formed following his studies of four diseases: alkaptonuria, cystinuria, pentosuria and albinism.

homogentisate

alkapton

Alkaptonuria is usually diagnosed because the urine from patients goes black on contact with air! This is due to the presence of abnormally large amounts of homogentisate in the urine which is oxidised in alkaline conditions to the black coloured polymer, alkapton (Figure 3.1). Alkapton also accumulates in the tissues, notably connective tissues, and may cause their early degeneration. Individuals suffering from alkaptonuria failed to produce an enzyme (homogentisate dioxygenase) necessary for the catabolism of tyrosine.

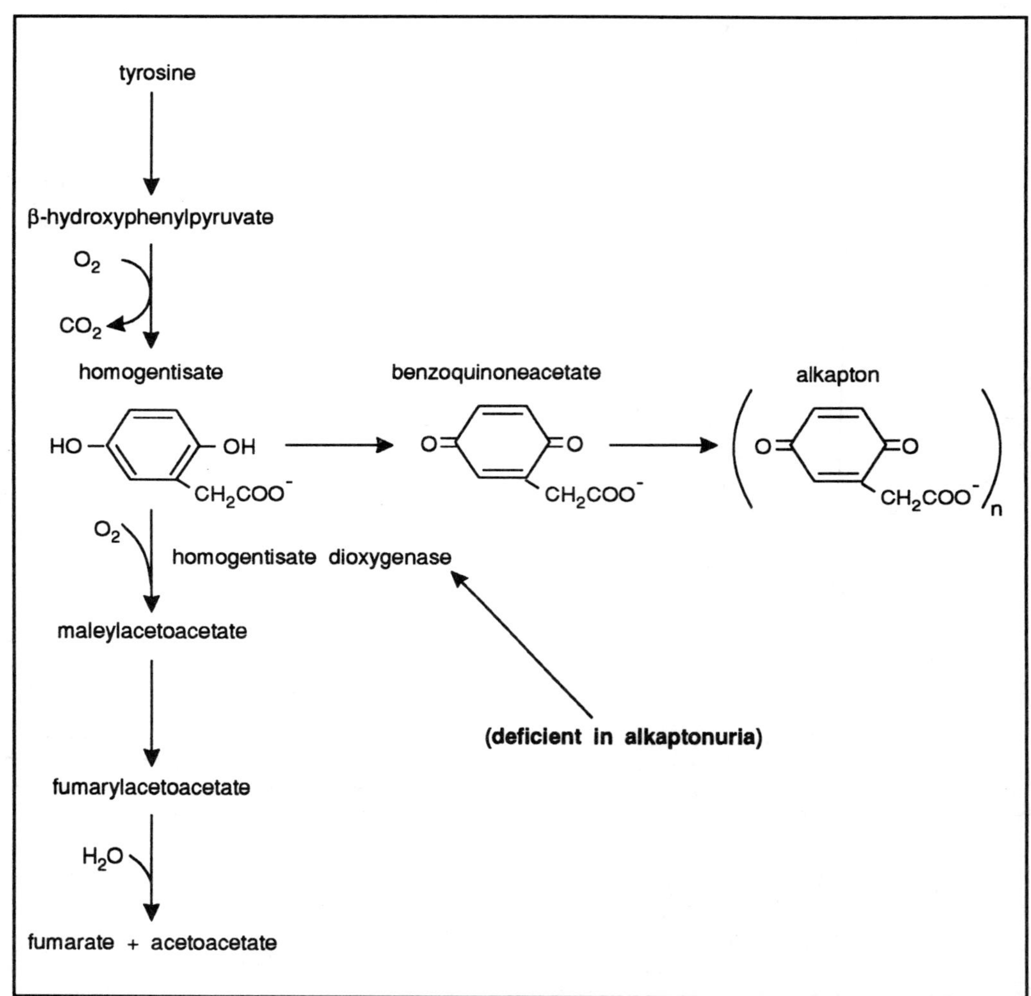

Figure 3.1 The 'normal' catabolism of tyrosine and the site of the defect which gives rise to alkaptonuria.

Π Given that alkaline conditions are necessary to oxidase homogentisate to alkapton, explain in general terms, (ie chemical transformations are not required), how the homogentisate in urine is converted to alkapton.

The presence of excreted ammonia and urea in urine provides the necessary alkaline conditions. Absorption of oxygen from the air then causes the oxidation.

xylose

Pentosuria is characterised by the excretion of excessive quantities of xylose, a pentose sugar intermediate of the pentose phosphate pathway for glucose oxidation. Disease symptoms are not associated with this condition. Cystinuria is the increased excretion of the amino acid, cystine. This is relatively insoluble and crytallises in the urinary tract causing 'stones'. Garrod assumed the condition was due to a disorder of cystine metabolism, but is now recognised to be a defect in cystine transport. Albinism is the well known condition of lack of pigmentation to hair, skin and eyes. The normal pigment, melanin, is synthesised from the amino acid tyrosine.

melanin

Garrod investigations established several common features to these diverse conditions:

- they are all inherited;
- they tend to occur in several children of the same family;
- they do not occur in successive generations and, therefore, are consequences of single recessive Mendelian traits;
- all result from a biochemical abnormality; in three an excess of an abnormal compound occurs, in the other a normal product of metabolism is not formed.

Garrod concluded that each condition could be explained by the presence of a defective enzyme in a particular metabolic pathway, although of course at that little was known of the specific metabolism of the compounds concerned (see, for example, Figure 3.1). He also perceived the direct relationship between a gene and an enzyme.

Garrod's work on inborn errors of metabolism was imaginative and far reaching. Many diseases are now known to be caused by such errors (Table 3.2). Over 600 genetic diseases with a recessive pattern of inheritance corresponding to a single defective gene are known, of these the specific abnormal enzyme or protein is recognised in about 200. Approximately 1-2% of births show inborn errors of metabolism due to single defective genes, although the severity of the conditions vary enormously.

Read through Table 3.2 carefully and then attempt SAQ 3.1.

SAQ 3.1

Sickle cell anaemia is one of a number of inheritable haemoglobinopathies resulting from mutations in the globin genes.

1) Use Tables 3.1 and 3.2 to determine the possible point mutations which can cause sickle cell anaemia.

2) Suggest how the haemoglobins from sickle cells and normal erythrocytes may be conveniently separated as a result of the difference in amino acid sequence between normal and sickle cell β haemoglobins.

Inborn error	Frequency in live births
Sickle cell anaemia (change from Glu to Val residue in β-haemoglobin)	0.0025
β-Thalassemia (defective β-haemoglobin)	0.0025 (in some Mediterranean populations)
Duchenne muscular dystrophy (defective dystrophin causing muscular degeneration)	3.3×10^{-4}
Phenylketonuria (defective phenylalanine hydroxylase)	1×10^{-4}
Haemophilia A (defective clotting factor VIII)	1×10^{-4}
Tay-Sachs disease (defective hexosaminidase A)	5×10^{-4} (in Askenazi Jews)
Von Gierrke's disease (defective glucose-6-phosphatase)	1×10^{-5}
Galactosemia (defective galactose-1-phosphate uridyl transferase)	2.5×10^{-5}
Lesch-Nyhan syndrome (defective hypoxanthine guanine phosphoribosyl transferase)	1×10^{-4} (males)
Metachromatic leukodystrophy (defective arylsulphatase A)	2.5×10^{-5}
Familial hypercholesterolaemia (defective low-density lipoprotein receptors)	0.001
Protoporphyria (defective ferrochelatase)	rare

Table 3.2 Some examples of inborn errors of metabolism.

3.3 Polymorphism

Mutations that occur in somatic cells (cells not involved in reproduction) are only passed to cells that arise by mitotic division of the mutant cell. A 'normal' tissue may thus be a mixture of normal and mutated cells with different biochemical properties. However, the mutation is not passed on to offspring and is eliminated with the death of the organism. If a mutation occurs in a germ cell it may be passed on to offspring and will provide biological variation, (the basis of evolution).

polymorphisms

Mutations that prevent the production of a protein or result in the production of an altered protein produce changes in the appearance of an organism. These changes are called polymorphisms. In diploid organisms, the mutant gene may not be expressed unless it is present in the homozygous form.

Genes which control many of the fundamental processes of biology such as the coordination of development, energy release or gene regulation have their base sequences conserved. The sequence of bases in these essential genes has changed little with time and is similar in organisms which have become widely separated during evolution. Mutations in these genes lead to a nonviable organism and are examples of lethal mutations. Non-lethal mutations may occur without affecting the viability of the organism.

3.4 Spontaneous and induced mutations

spontaneous mutations

tautomeric

Spontaneous mutations are not caused by an external source such as radiation or chemicals, but occur because of the formation of tautomeric forms of the bases. Tautomers arise because a keto group can interconvert to an enol group and an amino group to an imino group (Figure 3.2).

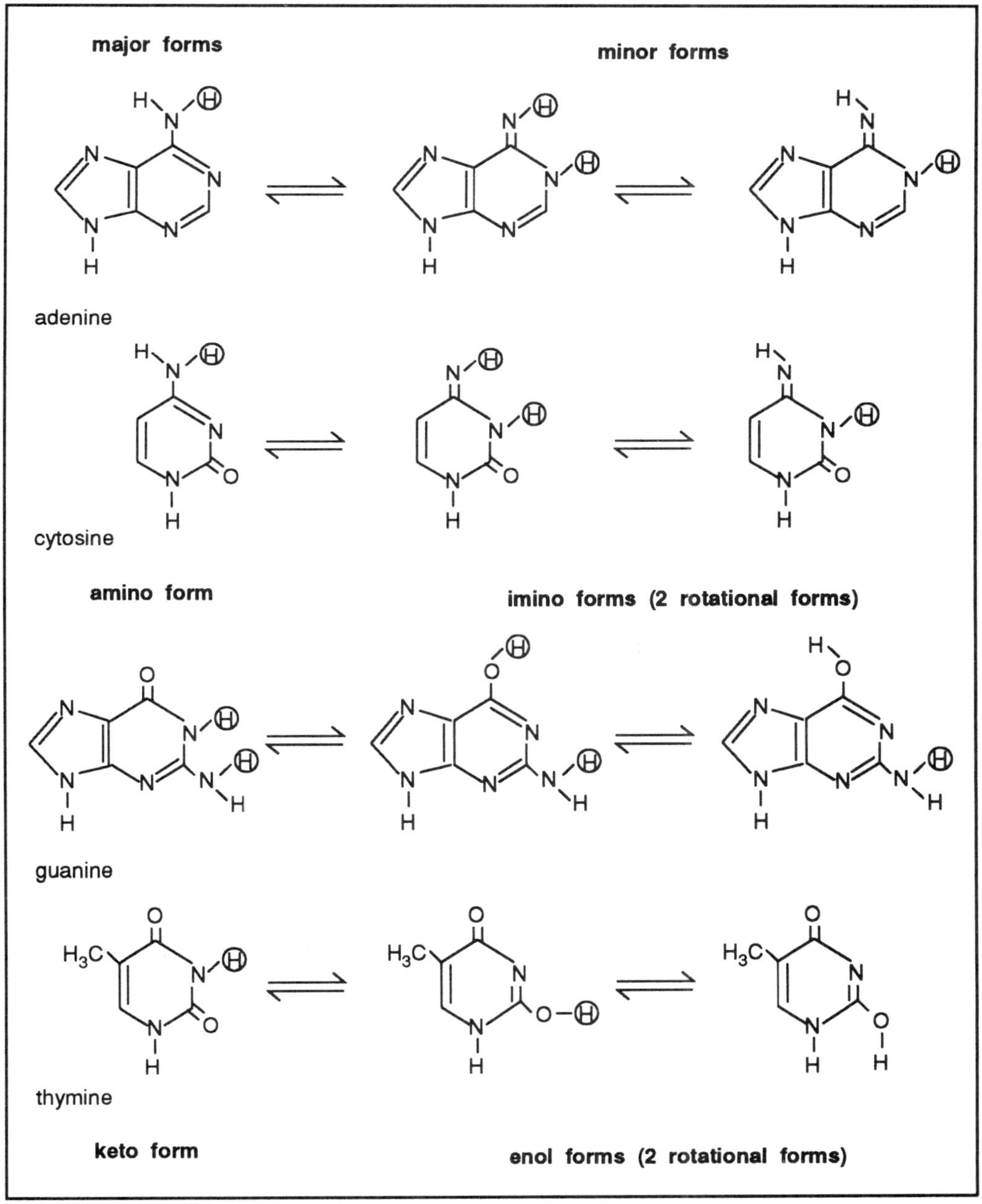

major forms

minor forms

adenine

cytosine

amino form

imino forms (2 rotational forms)

guanine

thymine

keto form

enol forms (2 rotational forms)

Figure 3.2 Tautomeric forms of the bases A, G, C and T. The amino and keto are the major forms, while the imino and enol are minor forms. Hydrogens which take part in hydrogen bonding are indicated by a circle. Note that the different tautomers form different numbers of hydrogen bonds.

keto

enol

The pyrimidine bases normally adopt a keto structure at physiological pH, but can adopt an enol form which allows adenine-cytosine and guanine-thymine basepairs to occur (Figure 3.3).

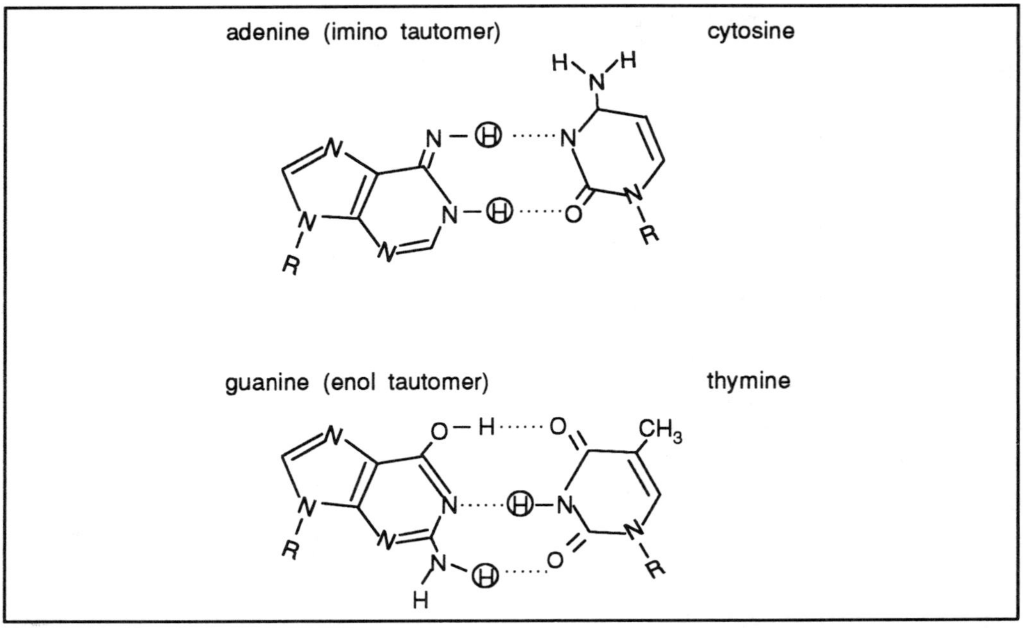

adenine (imino tautomer) cytosine

guanine (enol tautomer) thymine

Figure 3.3 The formation of tautomers allows the basepairings A-C and G-T to form.

Figure 3.4 shows how the occurrence of the 'wrong' tautomer can lead to the replacement of an A-T base-pair by a G-C pair during replication of DNA.

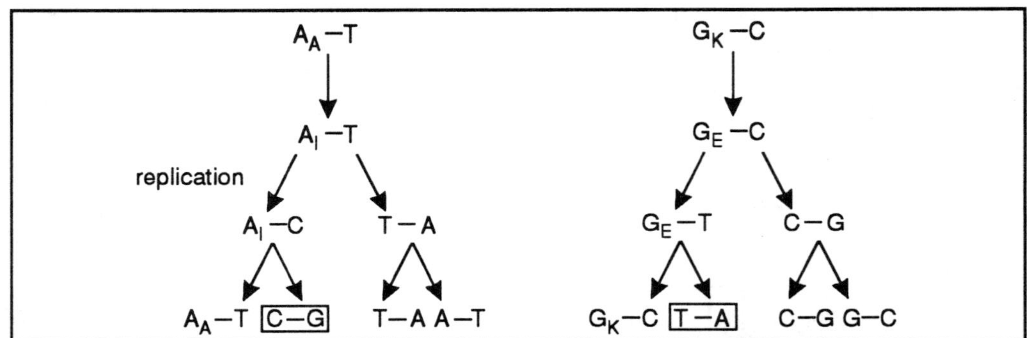

Figure 3.4 The imino tautomer of adenine can pair with cytosine. On replication this will lead to the basepair G-C replacing A-T of the original DNA. Each double arrow indicates a replication of the DNA. Subscripts A = amino, I = imino E = enol and K = keto forms respectively.

Π Figure 3.2 shows the tautomers of the four bases. In order to see the possible basepair combinations more clearly, it is useful to locate the hydrogen bonding acceptor and donor sites on each tautomer. For example, the normal keto form of guanine has three sites, which are shown using an arrow for a H donor site and a wedge for an acceptor site.

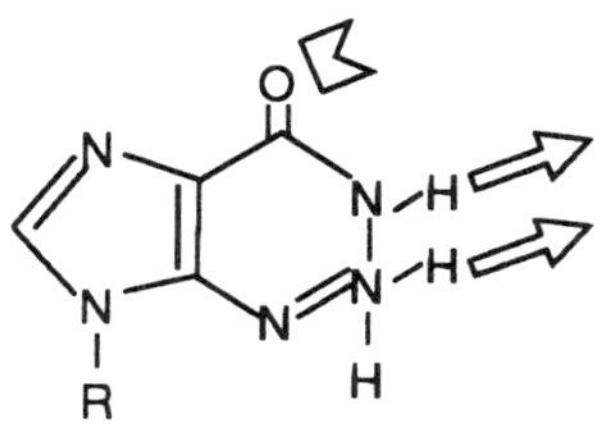

This will hydrogen bond to cytosine because the three hydrogen bonding sites on C are complementary to those on G.

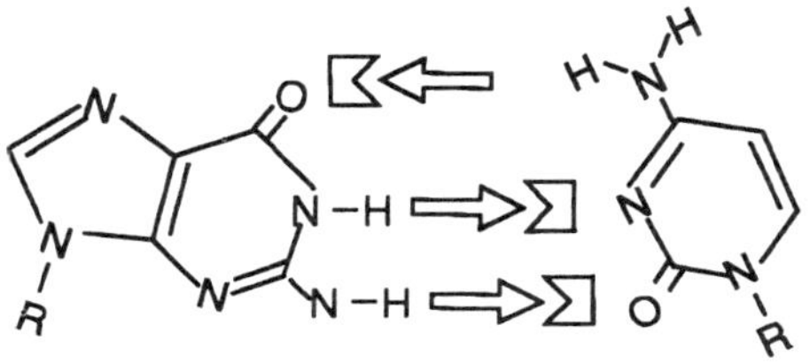

Π Using the same type of analysis, show the potential hydrogen bonding atoms for the two imino forms of guanine and work out the potential basepair combinations which could be formed.

The imino forms can exist in two rotational forms, with different hydrogen bonding possibilities:

a) b)

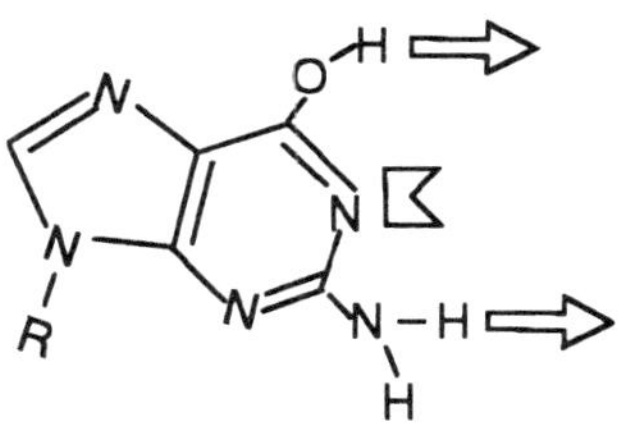

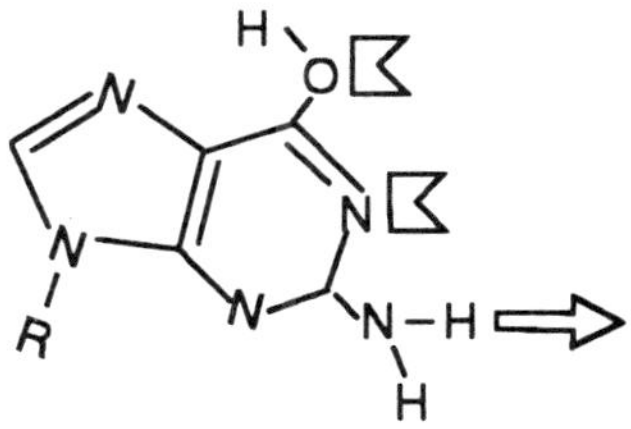

These could basepair in the following way:

Form a) could basepair with the keto form of T.

Form b) could basepair with the imino form of C.

About one base in 10 000 will mispair because of tautomerisation. This would be expected to give a high mutation rate. However, as we shall see later, such mutations can be repaired and so a much lower rates of mutations occurs than might be expected.

3.4.1 Chemical-induced mutations

mutagens

carcinogens

Chemicals which cause mutations are called mutagens. Many mutagens are also carcinogens, that is they can cause cancer. For this reason much study has been devoted to assessing the mutagenecity of chemicals. We shall consider this further in Section 3.13.

Mutagens may be classified into one of three groups:

- analogues of bases of DNA;
- those which chemically modify bases;
- intercalating agents.

Analogues of bases found in DNA

Base analogues are structurally similar to naturally occurring bases of DNA but are chemically modified to increase their likelihood of mispairing. The bases are incorporated into DNA during its replication. 5-Bromouracil, an analogue of uracil, can tautomerise to a structure that can pair with guanine (Figure 3.5a) and, on replication, the basepair A-T will give rise to a G-C pair.

2-Aminopurine (2aPu) is an analogue of adenine. It can pair with thymine or cytosine but in the latter case forms only a single hydrogen bond. This can result in an A-T to G-C basepair change (Figure 3.5b) if it pairs with T, or a G-C to A-T basepair change if it pairs with C when it is first incorporated into DNA. Other mutagenic base analogues are 8-azaguanine, 6-thioguanine and 5-iodouracil.

Π Complete the figure shown below to indicate how incorporation of 2-aminopurine basepaired to cytosine could result in a G-C to A-T basepair change.

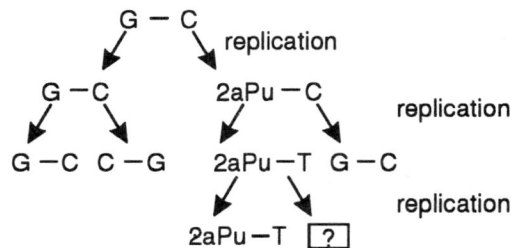

G − C
replication
G − C 2aPu − C
replication
G − C C − G 2aPu − T G − C
replication
2aPu − T [?]

The missing basepair is A-T. In other words, we started with a piece of DNA which had a basepair G-C but one of the progeny molecules has had this pair replaced by A-T.

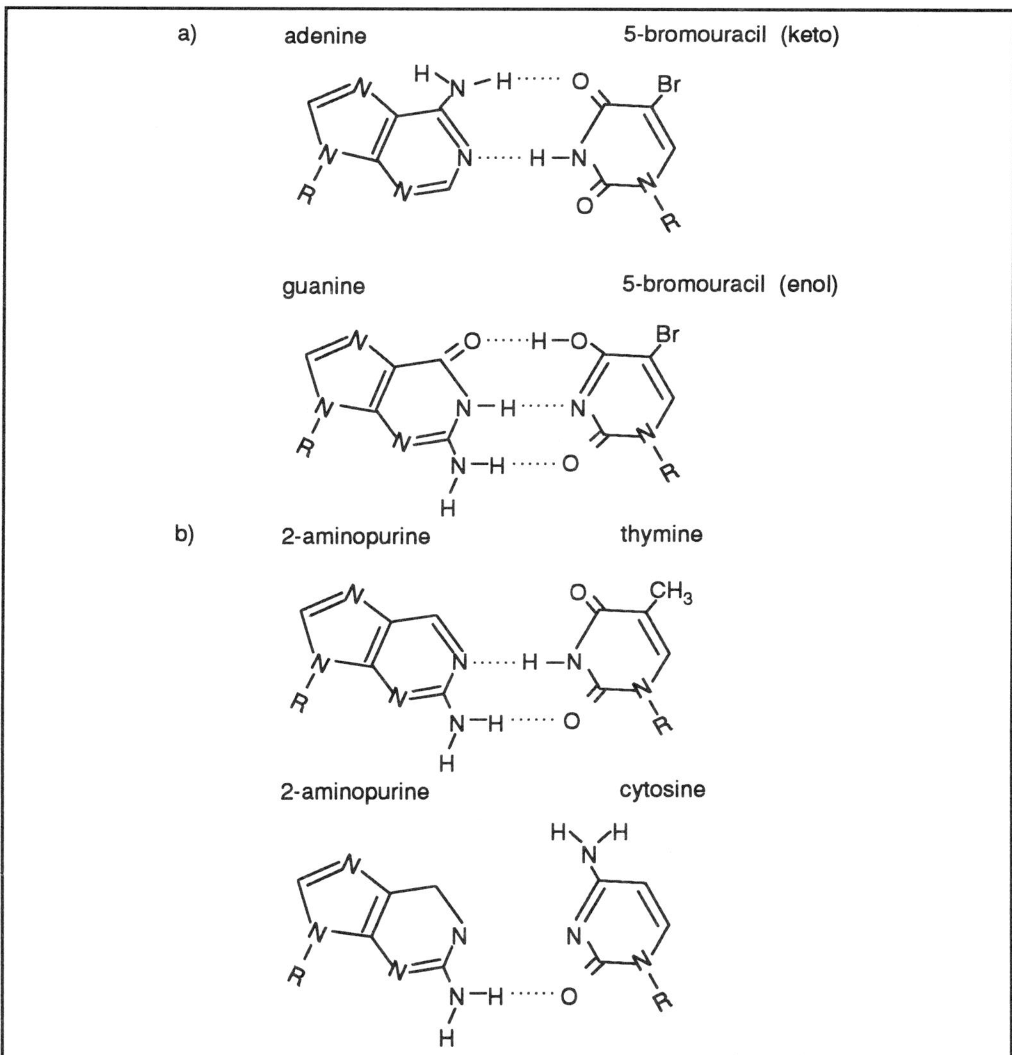

Figure 3.5 a) The structure of 5-bromouracil (5-Bu) and its basepairing with adenine and guanine. When 5-Bu is incorporated into DNA in the enol form it causes a transition leading to a G being incorporated into DNA. If it reverts to the keto form it then pairs with A. b) 2-aminopurine and its basepairing with thymine and cytosine. Only one hydrogen bond forms in the latter case.

Chemical modification of bases

nitrous acid

hydroxylamine

Nitrous acid deaminates bases. This may result in the change of C to U; A to hypoxanthine (which is read as G), and G to xanthine (which is read as G) (Figure 3.6). The net effect is an A-T to G-C transition or *vice versa*. Hydroxylamine deaminates C to U and will produce similar mispairings to nitrous acid. Thus a G-C to A-T basepair change will result.

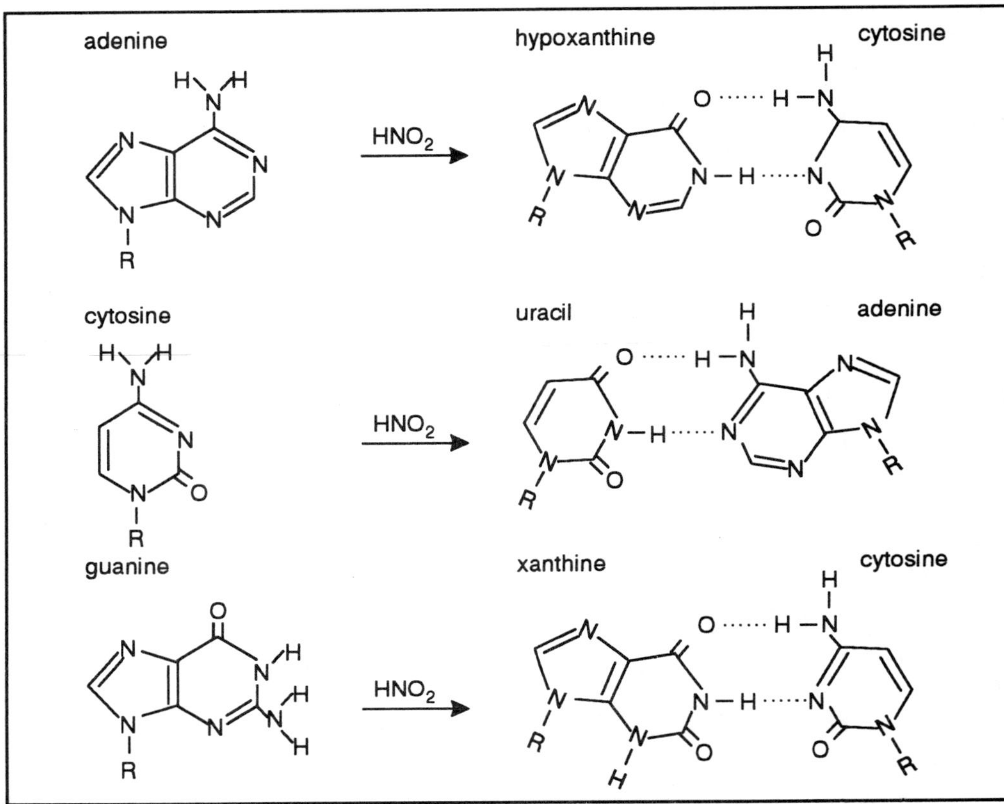

Figure 3.6 The deamination of the bases A, C and G to form hypoxanthine, uracil and xanthine respectively. The deaminated bases may subsequently mispair.

alkylating agents

Alkylating agents are compounds that produce a positively charged alkyl group, commonly CH_3+ or CH_3CH_2+ which react with and chemically modify bases in DNA. Guanine is especially susceptible to alkylation as potential reactive groups are exposed in the major groove of DNA. Alkylation occurs at positions N-7 and O-6. Alkylated guanine pairs with thymine rather than with cytosine and encourages a G-C to A-T transition. Position O-4 of thymine is also affected. Table 3.3 lists some alkylating agents; dimethyl sulphate and N-methyl-N'-nitro-nitrosoguanidine (MNNG) are particularly effective alkylators. Their structures are shown in Figure 3.7.

D1-(2-chloroethyl)-sulphide	(sulphur mustard)
Methylnitrosourea	
N-methyl-N'-nitro-N-nitrosoguanidine	(nitrosoguanidine or MNNG)
Dimethyl sulphate	(DMS)
Ethylmethane sulphononate	(EMS)

Table 3.3 Examples of alkylating agents. Names in brackets are commonly used names.

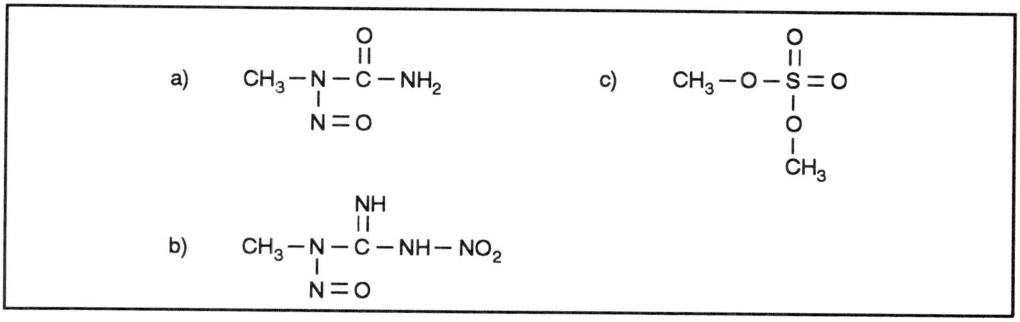

Figure 3.7 The structure of some commonly used alkylating agents a) methylnitrosourea b) N-methyl-N'-nitro-N-nitrosoguanidine and c) dimethyl sulphate.

Intercalating agents

proflavine, acroflavine, fluorescent DNA stain, ethidium bromide

Intercalating agents are planar molecules with dimensions similar to those of a basepair. Thus they are able to insert between basepairs in DNA leading to the addition or deletion of a nucleotide during replication. Common intercalating agents are proflavine, acroflavine and the fluorescent DNA stain, ethidium bromide (Figure 3.8). It is thought that intercalating agents lead to localised swelling and distortion of the double helix resulting in a mutation.

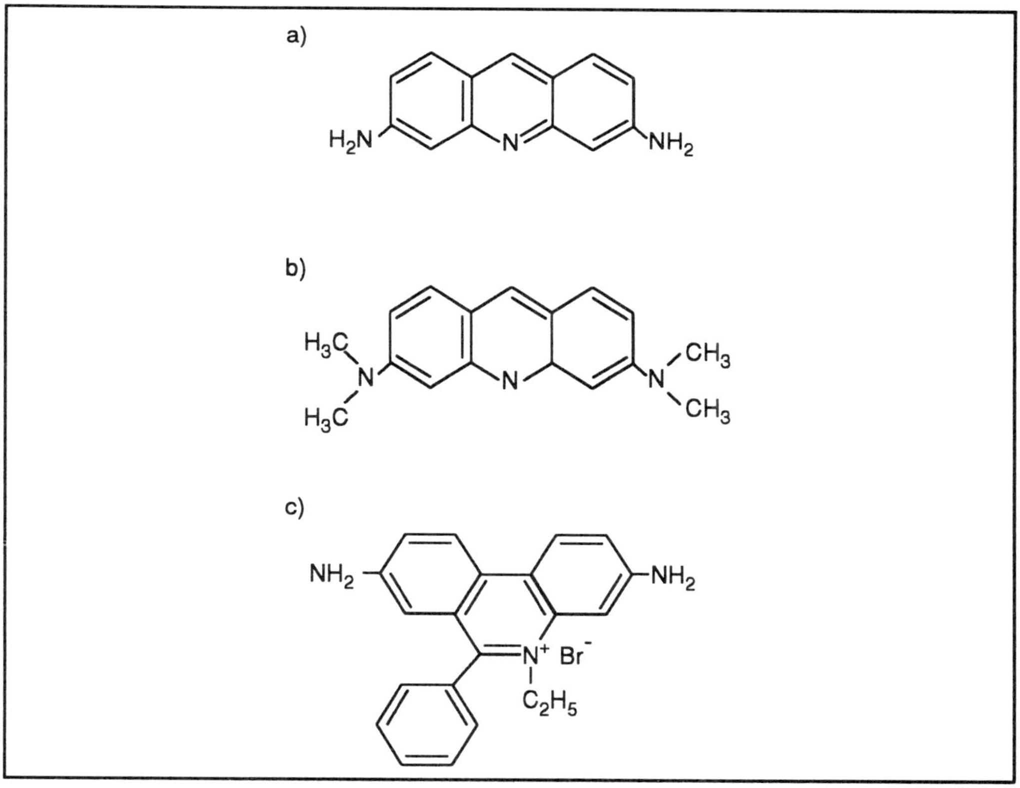

Figure 3.8 The structures of the intercalating agents a) proflavine b) acroflavine and c) ethidium bromide.

3.4.2 Radiation mutagenesis

Radiation may damage sugar-phosphate bonds in DNA and cause breaks in one or both strands, chemically alter bases or cause cross-linking between DNA strands or between DNA and chromosomal proteins. The major defect is, however, the formation of thymine dimers (Figure 3.9) which cause local distortion to the double helix.

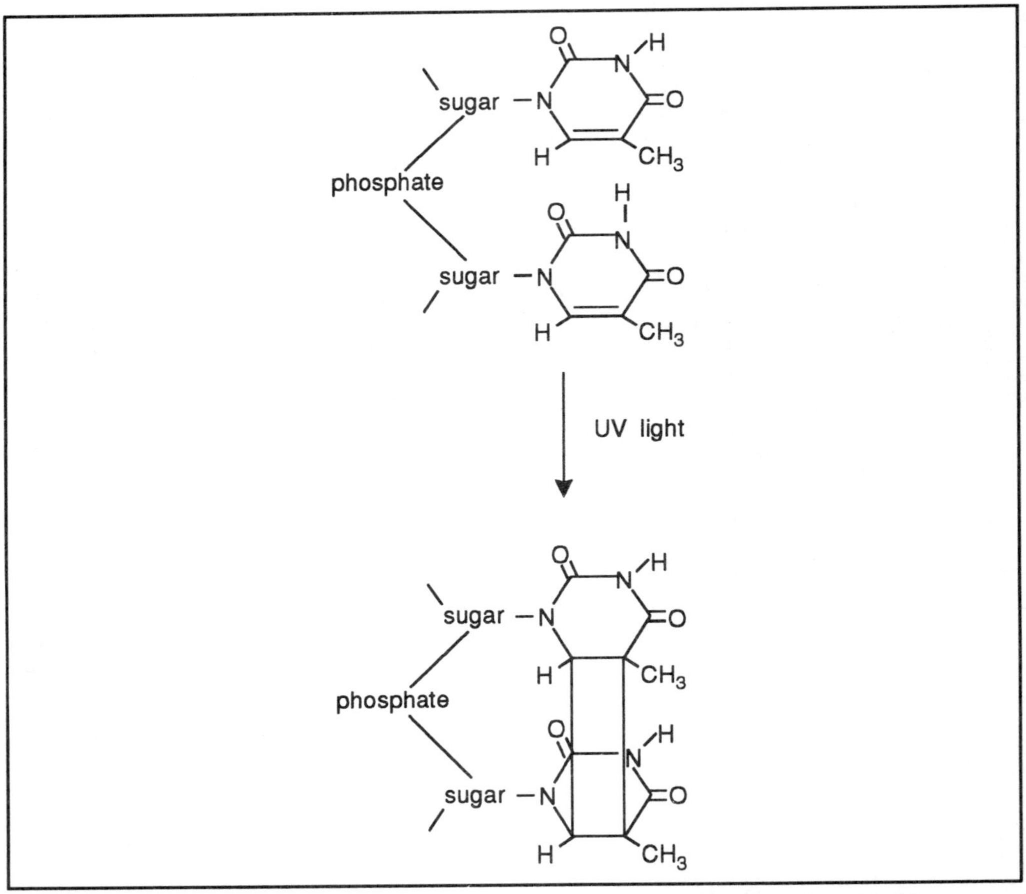

Figure 3.9 The action of ultra-violet light on adjacent thymine residues and the formation of cyclobutane dimers.

X-ray
α-particles
ultra-violet

The extent of damage to the DNA depends on the type of radiation (X-ray, α-particles and ultra-violet radiation) causing the damage and the amount of energy transmitted to the tissue. UV radiation in sunlight, with a wavelength of 260 nm, is probably the most common cause of radiation mutagenesis.

Π The graph below shows the incidence of leukaemia in Nagasaki and Hiroshima among survivors of the atomic bomb explosions. The bomb dropped on Hiroshima produced mainly neutrons which cause double stranded breaks in DNA. The Nagaski bomb produced X-rays which cause the production of thymine dimers. Explain the differences in the incidence of leukaemia in survivors in the two cities.

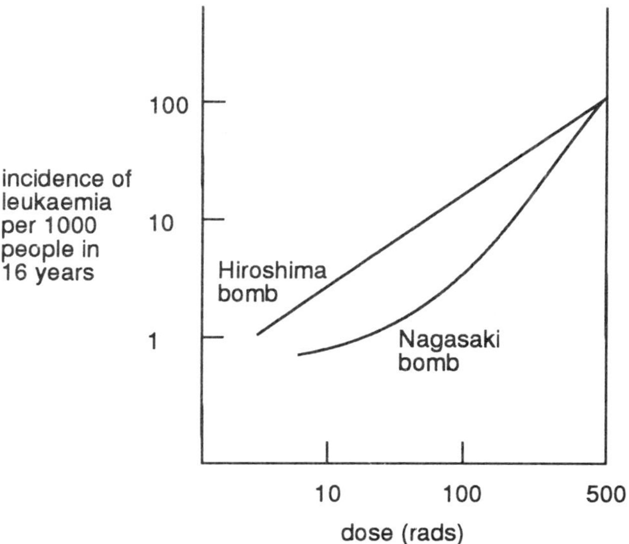

Since neutrons cause double stranded breaks in DNA the mutation rate is proportional to the dose of radiation giving the straight line graph. X-rays are less energetic and cause dimerisation of thymine in only one strand. Therefore the rate of mutation is not proportional to the dose (but to the dose^2).

3.5 Point mutations

mutagens

Point mutations may occur naturally as the result of the incorporation of an incorrect base during DNA replication or as a result of the interaction of DNA with mutagens, which alter the bases of DNA.

transition

transversion

A number of types of single base changes are possible. A transition is the change from one pyrimidine to another (C to T or T to C) or from one purine to another (A to G or G to A). A transversion changes a pyrimidine to a purine (T or C to A or G respectively) or *vice versa* (A or G to T or C respectively). These types of mutation are reversible if a second mutation restores the original base.

Before you attempt the next SAQ, we need to remind you of one of the conventions which is commonly used in describing nucleotide sequences in genes.

Consider the following sequences of bases found in a DNA molecule.

5′ AATCCGCAG 3′
3′ TTAGGCGTC 5′

Assuming that RNA is made by transcribing the bottom strand, the RNA produced will have the following nucleotide sequence.

5′ AAUCCGCAG 3′

(Remember that RNA is synthesised from the 5′ end and that U replaces T).

When this is translated it will lead to the incorporation of amino acid residues in the following order:

$^{+}NH_3$ Asn Pro Gln-COO-. (Check these with the codons given in Table 3.1).

convention used in describing nucleotide sequences of genes

Now let us return to the DNA sequence given above. Note that the strand of DNA that is not transcribed has the same nucleotide sequence as the mRNA that is produced except that U replaces T in the RNA molecule. In describing the nucleotide sequence of genes, it is usual to use the sequence of the strand that is not transcribed as this allows us to directly relate this to the amino acid sequence of the peptide that will be produced. If we used the base sequence of the transcribed sequence, we would first have to convert the base sequence to that found in the RNA product before converting it to the sequence of amino acids. Bearing this in mind, now attempt the following SAQ.

SAQ 3.2

A gene contains the bases 5'CGA AGT GGC GAT3' as part of the sequence of the non-transcribed strand.

1) What sequence of amino acids is specified by these codons?

2) If a mutation caused:

   a) an A to G transition;

   b) a G to A transition, what would the new sequences of amino acids be?

3.5.1 Frameshift mutations

In Figure 3.10 we have listed a variety of mutations. In this figure we have written the base sequence of the RNA. This is the same sequence as would be found on the non-coding strand of the DNA except that U would be replaced by T.

insertion deletion

frameshift

The insertion or deletion of a base alters the reading frame of the sequence in the gene. These mutations are collectively called frameshift mutations (Figure 3.10). Frameshift mutations may lead to major changes in the amino acid sequence of a protein since every amino acid residue after the mutation will be altered.

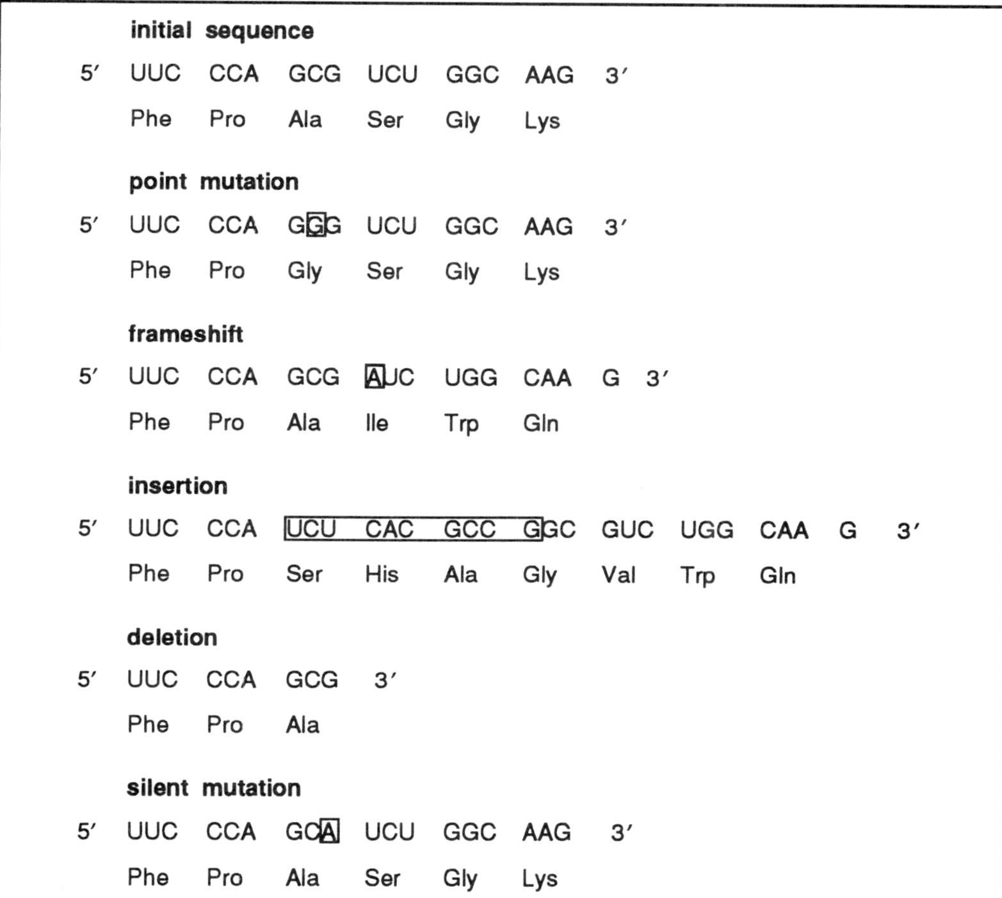

Figure 3.10 A sequence of bases codes for an amino acid sequence in part of a protein. Mutations alter the sequence of bases and therefore change the sequence of amino acid residues.

silent mutations

Since the genetic code is degenerate it is possible for a mutation to occur without altering the amino acid sequence of the protein (See Figure 3.10). Such silent mutations are only detectable by chemical analysis of the bases of the DNA. Table 3.4 shows the nine possible mutations arising from a single base change in the tyrosine codon. There are a total of 549 (61x9) possible substitutions in the codons coding for amino acids. Because of degeneracy about a quarter of these fail to produce amino acid substitution.

UAU (Tyr)		
CAU (His)	UGU (Cys)	UAC (Tyr)
AAU (Asn)	UUU (Phe)	UAA (Stop)
GAU (Asp)	UCU (Ser)	UAG (Stop)

Table 3.4 The effect of single base substitutions on the amino acid codon for tyrosine.

SAQ 3.3

1) Examine the codons for the amino acids proline, arginine and threonine in Table 3.1. In each case there are four codons for the amino acid. State the similarities in these codons.

2) Mutations may lead to the following types of amino acid change:

   a) Hydrophobic to hydrophobic

   b) Polar to polar

   c) Hydrophobic to polar or *vice versa*.

   Which of these would you expect to be most deleterious?

3.5.2 Deletions and insertions

Deletions and insertions are mutations caused by the movement of larger blocks of nucleic acid, that is two, three or more nucleotides long. Such mutations almost invariably lead to a gene which codes for a nonfunctional product.

3.6 Chromosomal mutations

chromosomal aberrations

Unlike point mutations, chromosomal mutations involve the movement of large pieces of the chromosome. Although these changes have historically been called mutations because they generate new phenotypes, a more modern designation is chromosomal aberrations. Chromosomal aberrations often lead to extreme abnormalities in an individual, although plants appear to tolerate such changes much better than animals.

3.6.1 Variation in chromosome number

aneuploidy

euploidy

Variation in chromosome number range from the addition of a single chromosome (aneuploidy) to the duplication of whole sets of chromosomes (euploidy). Table 3.5 lists the terminology used to designate the chromosome complement of a cell.

Term	Number of chromosomes present
Aneuploidy	2n plus or minus chromosomes
Monosomy	2n-1
Trisomy	2n+1
Tetrasomy	2n+2
Euploidy	Multiples of n
Monoploidy	n
Diploidy	2n
Triploidy, etc	3n, etc

Table 3.5 Terminology for describing chromosomal aberrations. n = haploid (monoploid) number of chromosomes, in which each chromosome is present in single copy.

What is the number of chromosomes present in tetraploidy?

You should have realised it is 4n.

The most common cause of aneuploidy is the gain or loss of a single chromosome caused by nondisjunction during the first or second meiotic divisions. Table 3.6 lists some examples of conditions which arise from aneuploidy.

Chromosome number	Syndrome	Consequences
47 XXY	Kleinfelter	Male genitalia, sperm not produce (additional X chromosome)
45XO	Turner	Female genitalia, rudimentary ovaries, abnormal physical development (only 1 X chromosome, no Y chromosome)
47XXX	-	May be normal, some underdeveloped secondary sexual characteristics
47 YY	-	Reduced intelligence, increased physical development (no X chromosome, additional Y chromosome)
46, $5p^-$	Cri-du-chat	Multiple anatomic malformations
47, 21^+	Down	Abnormal physical and mental development
47, 13^+	Patau	Additional chromosome 13, mental retardation, physical malformation

Table 3.6 Syndromes associated with the gain or loss of chromosomes. $5p^-$ indicated a deletion in the long arm of chromosome 5, superscript + indicates the presence of extra chromosomes.

polyploidy

Polyploidy is the presence of more than two sets of haploid chromosomes. Polyploids arise through a failure of all the chromosomes to segregate during meiosis. Alternatively, two spermatozoa may fertilise the same ovum producing a triploid zygote.

Polyploids are more common in plant than animal species. They often result in the production of larger individuals with flowers, ova and fruits increased in size. This increase is due to a larger cell size. Commercial seedless melons and several potato species are triploid while the commercial tomato is octaploid.

Triticum aestivum, the bread wheat, has evolved by a series of hybridisations and spontaneous doublings of chromosome numbers (polyploidy) as shown in the accompanying flow chart. Complete the figure (items marked 1) to 4)) to show how *T. aestivum* acquired its hexaploid genome (AABBDD). The specimen *T. monococcum* has a diploid complement of A type chromosomes and is therefore characterised as (AA).

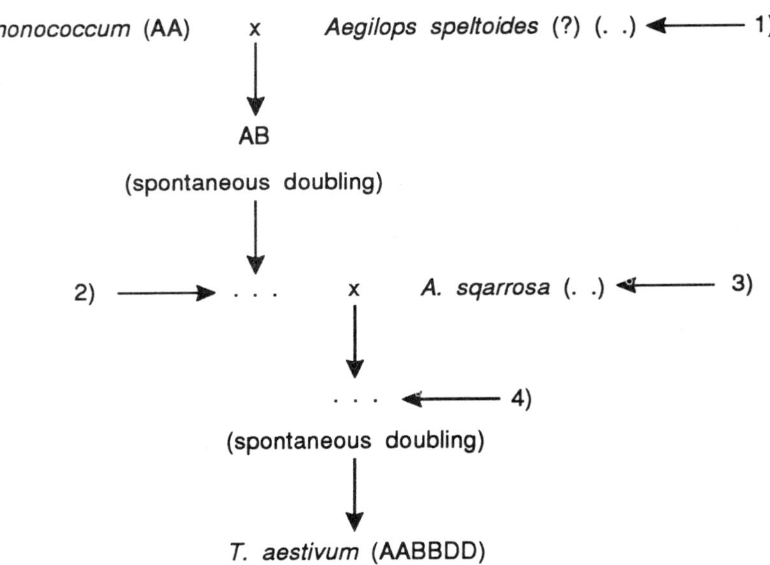

Your responses should have been 1) BB, 2) AABB, 3) DD and 4) ABD.

3.6.2 Variation in chromosome structure

Chromosomes may have pieces deleted from them, added to them or rearranged within them (Figure 3.11). Chromosomes can break spontaneously although the addition of chemicals or radiation can increase the rate of breakage. Deletions involve the loss of part of the chromosome. The missing part is the deletion. Duplications occur when a part of the chromosome is present more than once in the genome. They may arise through unequal crossing over during meiosis. Gene duplications may provide the material for the evolution of new genes. If a gene is essential, then any mutation in it may be lethal. If, however, the gene is duplicated the 'extra' copy is free to mutate and evolve into a new gene.

deletions

duplications

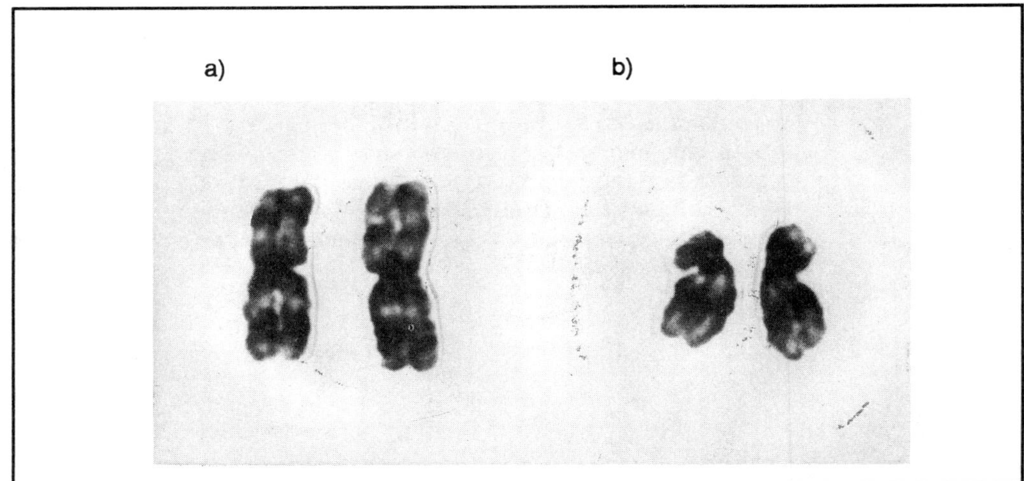

Figure 3.11 An example of a) an inversion and b) a deletion. Courtesy of Dr C. Harrison, The Patterson Laboratory, Christie Hospital, Manchester.

Π The two proteins, α-lactalbumin and lysozyme have a number of structural and chemical features in common. They have similar molecular mass, disulphide bonds at comparable positions and show about 40-50% homology in their primary structure ie 40-50% of the amino acid residues are identical in each sequence. However, the two proteins have different distributions among organisms and very different functions. α-Lactalbumin is restricted to mammals and forms a component of the enzyme lactose synthase. It is also a Ca^{2+} binding protein in milk. Lysozyme is an antibacterial enzyme found in the tissues and secretions of a number of vertebrate types. Both proteins are secreted from cells by similar mechanisms. Comment on the possible evolutionary origins/relationships of these two proteins.

The two have probably evolved from a common ancestral gene which probably coded for lysozyme (since it is the more widely distributed). Duplication of this gene allowed one of them to continue to code for lysozyme, the other to change to code for α-lactalbumin during the evolution of milk production in mammals.

inversion

An inversion does not involve a change in the quantity of genetic material but in its rearrangement. A piece of DNA is detached, turned through 180° and reinserted back into the chromosome.

translocation

A translocation is the movement of a piece of a chromosome to a new position. It may be a reciprocal translocation where material is exchanged between two non-homologous chromosomes or exchange may be within a single chromosome.

Down Syndrome commonly arises from the presence of an additional copy of chromosome 21 in the germ cell of the mother. There is, however, another inherited form called familial Down Syndrome. This arises from the translocation of part of chromosome 21 to chromosome 14. During meiosis some cells will be produced with duplicate copies of chromosome 21 which produce trisomy 21 when they fuse with the ovum.

3.7 Detection of mutagens

A variety of tests are available to test for mutagenicity. Some use bacterial and animal and plant systems, although cell cultures can also be used (Table 3.7). Normally, several of the tests would be used in testing a chemical for mutagenicity.

Ames test

The Ames test uses cultures of *Salmonella* which cannot synthesise histidine (ie a histidine auxotroph) and therefore require it in their growth medium. The Ames test is based on the principle that the mutagen causes the bacteria to revert from a form that cannot synthesis histidine to one that can. The bacterium is treated with a test compound and plated on a medium lacking histidine. The number of colonies grown, ie the number of revertants, is a measure of the mutagenicity of the compound.

Potential mutagens may require metabolism in the liver before they become mutagenic and so it is usual to incubate the mutagen with a crude liver extract before performing the test, the idea being that enzymes in the extract can convert the potential mutagen into the active form.

The systems listed in Table 3.7 can also be used to obtain data for chromosomal mutations.

Test Category	Organism	Chromosomal aberration: Translocation	Chromosomal aberration: Deletion/ Duplication	Chromosomal aberration: Nondisjunction	Mutation
Bacterial	*Salmonella typhimurium*				
Fungal	*Neurospora crassa*			+	+
Plant	*Vicia faba*	+	+	+	
Insect	*Drosophila melanogaster*				+
	Bombyx mori	+	+	+	+
Mammal	Mouse	+	+	+	
	Rat	+	+	+	
	Human	+	+	+	+
Cells	CHO	+	+	+	+

Table 3.7 Some commonly used tests for mutagens. Adapted from Klug, W.S. and Cummings, M.R. (1989) Concepts of Genetics, 2nd edition Pub. Merrill Publishing Company. CHO = chinese hamster ovary cells.

SAQ 3.4

A suspected mutagen, X, isolated from cigarette smoke was assessed by the Ames test. Use the following information to explain the action of X.

1) When the bacteria were directly plated out on agar medium lacking histidine they failed to produce colonies.

2) If X was pretreated with a liver extract a substantial number of colonies was produced.

3) The dimensions of X were approximately those of a basepair.

4) Analysis of a stretch of DNA after treatment with X showed it was deficient in four bases.

5) The ratio of purines to pyrimidines decreased.

6) The ratio of A/T decreased but G/C was unaffected.

SAQ 3.5

The accompanying graph show the mutagenicity of ethylenimine (●) and triethylenemelamine (■) using the standard Ames test. (Data from De Serres FJ and Shelby, MD, 1981, Comparative Chemical Mutagenesis, Plenum Press, New York and London).

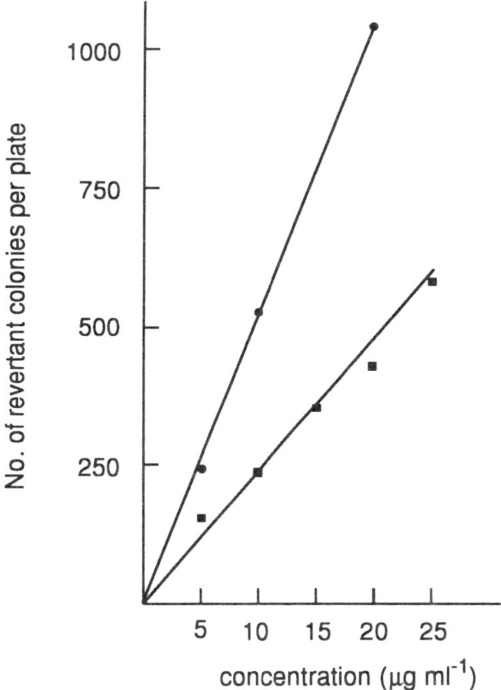

Determine:

1) The mutagenic potency of both compounds in terms of revertants/µg.

2) State which compound is the most mutagenic as assessed by this method.

3.8 The repair of mutations

Chemical changes may occur in all biological molecules. Changes in polymers such as RNA, proteins and lipids are less disastrous than in DNA as they are being constantly degraded and replaced. DNA must, however, be transmitted as little changed as possible from generation to generation. There are two methods of ensuring this accurate replication of the genetic material:

- the repair which occurs during replication which removes incorrectly incorporated bases;
- repair of damaged (mutated) DNA.

This latter occurs by one of a number of different strategies. DNA repair has been extensively studied in prokaryotes whereas much less is known about DNA repair in eukaryotes. The following section therefore describes the general situation in prokaryotes. We can assume that a similar situation may be found in eukaryotes.

3.8.1 Excision repair

Excision repair of a chemically altered base involves its removal, usually with several adjacent unmutated bases. Endonucleases cleave the sugar-phosphate backbone of DNA on the 5′ and 3′ sides of the defective base and remove up to eight bases including the damaged one. DNA polymerase I selects the appropriate bases to fill the gap using the intact strand of DNA as a template. DNA ligase joins the free ends to give a continuous strand (Figure 3.12).

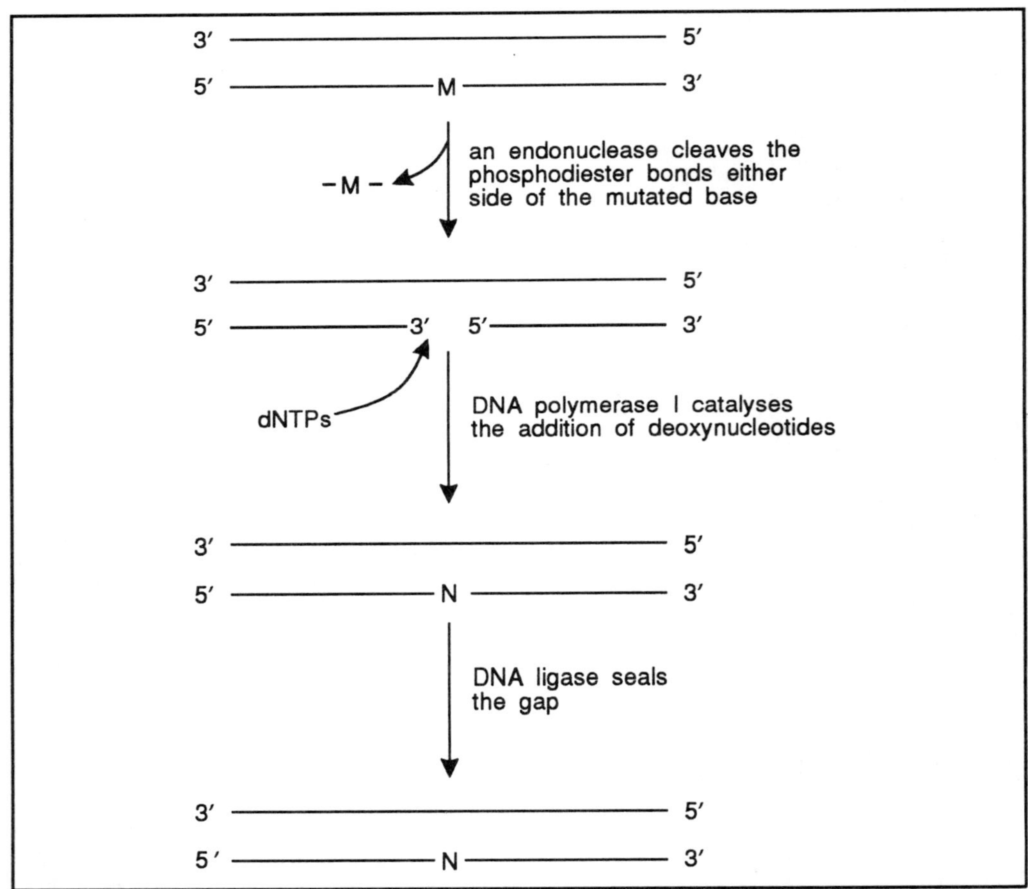

Figure 3.12 A summary of excision repair of mutated DNA. M represents a mutated base in one strand of DNA, N represents the normal base.

repair of thymine dimers

Thymine dimers in *E. coli* are repaired by a specific endonuclease consisting of three subunits. These are coded for by three genes called *uvr*A, *uvr*B and *uvr*C. The excised portion of the DNA strand is unwound by helicase II and the gap filled by DNA polymerase I (Figure 3.13). DNA which has been cross linked can also be repaired by this system, each strand being repaired separately and in turn.

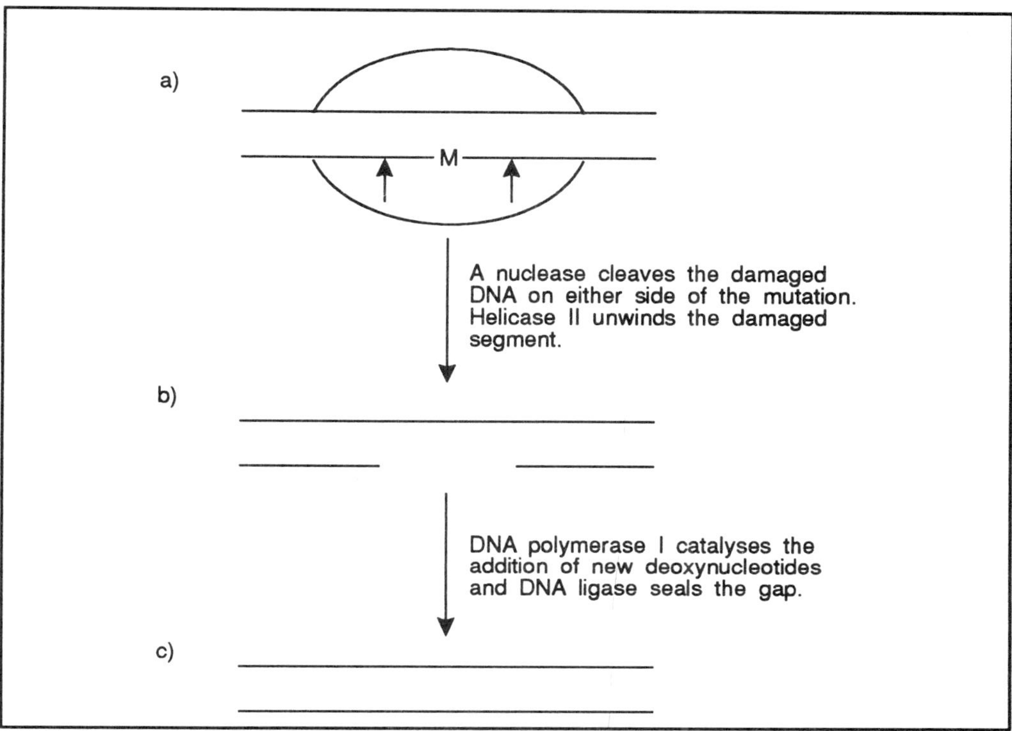

Figure 3.13 Excision repair involving the *uvr*ABC enzyme. a) The enzyme binds to DNA and detects distortions caused by mutations. b) A nuclease cleaves damaged DNA on either side of the mutation. Helicase II unwinds the damaged section. c) DNA pol I fills the gap and DNA ligase seals the gap.

Glycosylase enzymes

These enzymes remove the unnatural bases, hypoxanthine or uracil from DNA. Uracil may be incorporated into DNA as DNA polymerases can catalyse nucleotide addition using dUTP in place of dTTP. An enzyme, uracil-DNA N-glycosylase, removes dUMP from DNA by cleaving the glycosidic bond between uracil and deoxyribose. Uracil is released and a so-called apyrimidinic (AP) site is formed. This consists of a sugar without an attached base. A second enzyme, apyrimidinic endonuclease cleaves the phosphodiester bond on the 5′ side of the sugar. DNA polymerase I removes the defective deoxyribose phosphate along with several other deoxyribonucleotides and replaces them with nucleotides complementary to the undamaged strand. DNA ligase seals the free ends.

apyrimidinic sites

Uracil, formed by the deamination of thymidine, is removed by this mechanism.

Apurinic sites may also be formed by the removal of altered purines (eg alkylated guanine) and a similar repair process to that just described is used to repair the defect (Figure 3.14). At least six distinct glycosylases with different base specificities have been isolated from *E. coli*.

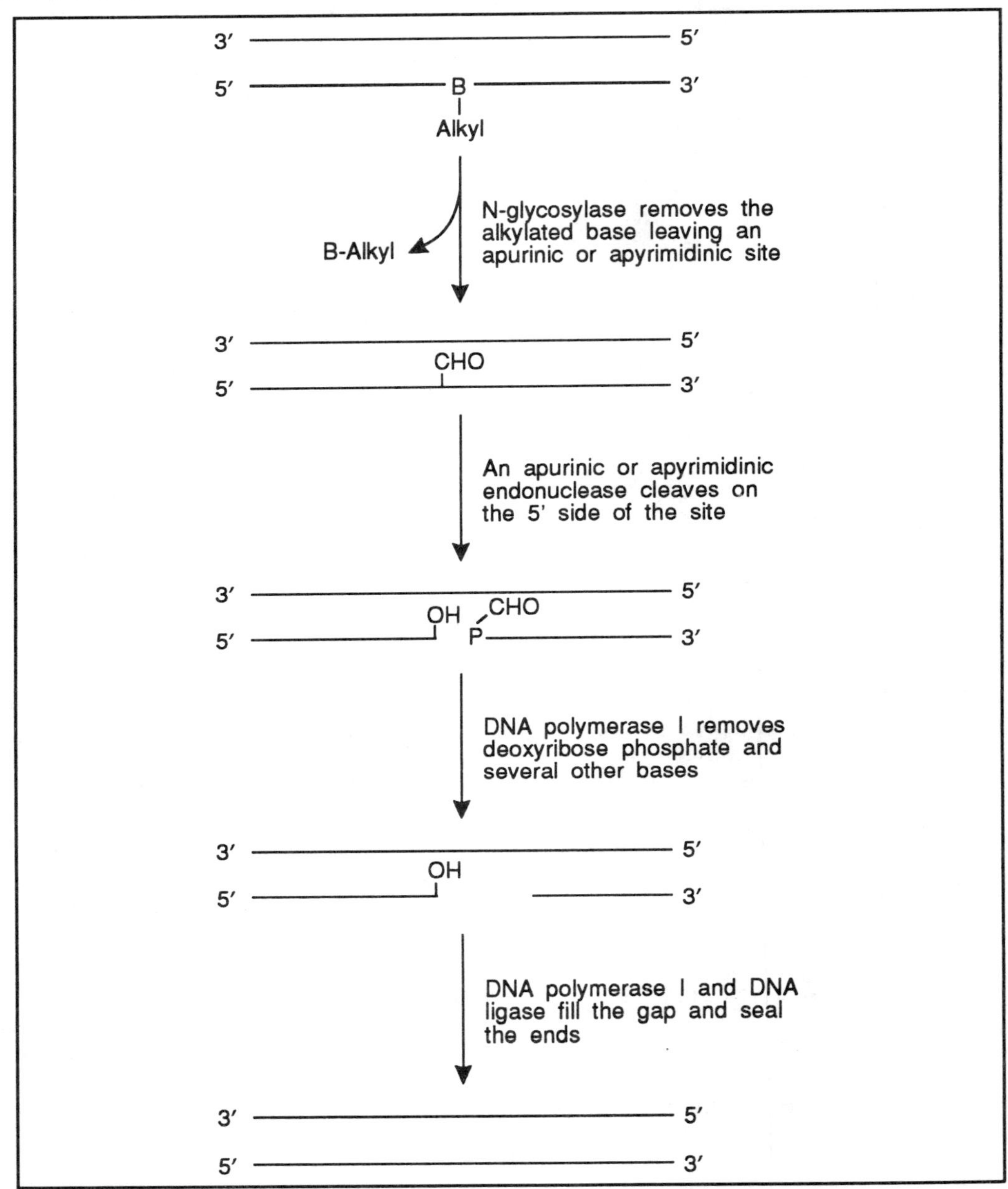

Figure 3.14 The repair of DNA containing an alkylated base by the DNA N-glycosylase enzyme. The N-glycosylase removes the alkylated base leaving an apurinic or apyrimidinic site. Further enzyme-catalysed steps restore the normal base composition (see text).

3.8.2 Post-replication repair

If high levels of DNA damage occur, then excision repair systems may become saturated. Thus replication of DNA may occur before its repair is completed. However, when the DNA polymerase comes into contact with a damaged base or a thymine dimer in the parental DNA strand, the enzyme stalls. Two processes may then come into play: recombination repair and SOS repair.

recombination repair

SOS repair

Recombination repair

Use Figure 3.15 to follow the description given.

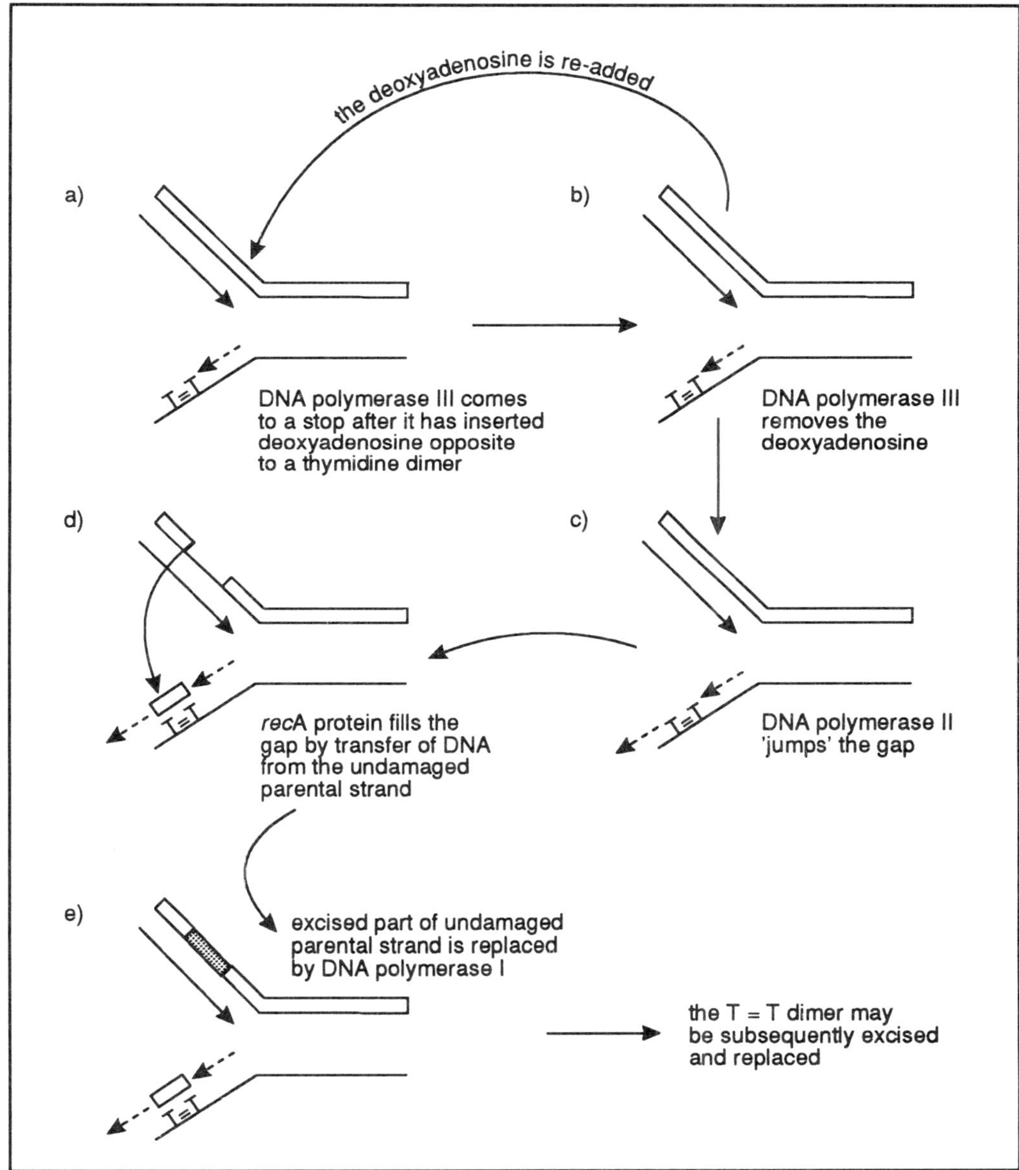

Figure 3.15 Recombination repair involving the *rec*A protein (see text for description).

When DNA polymerase III is halted during replication by a thymine dimer, it inserts a deoxyadenylate nucleotide in the growing DNA strand opposite to the defect. However, the distortion of the helix by the thymidine dimer induces the polymerase III to immediately remove the deoxyadenosine. Another deoxyadenylate is then inserted, thus the enzyme remains at the site inserting and removing deoxyadenosines (Figures

3.15a and b). Eventually the enzyme jumps the gap and synthesis of DNA can recommence at the next Okazaki fragment, leaving a short gap (Figure 3.15c).

The gap is filled by taking the corresponding section of DNA from the undamaged parental strand and inserting it in the newly synthesised daughter strand complementary to the lesion. The gap formed in the undamaged DNA strand can now be filled since there is a template to direct DNA polymerase I (Figure 3.15d and e).

A protein encoded by the *rec*A gene in bacteria is essential for the transfer of DNA sections between strands.

SOS repair

error-prone and error-free repair

This type of repair is used as a last resort by the bacterium when all other repair systems are inactivated or overwhelmed. The repair involves the introduction of mis-matched bases. In other words the DNA polymerase adds any base to overcome the damaged area. This type of repair is called error-prone to distinguish it from the error-free mechanisms of repair described above. In SOS repair the gaps opposite thymine dimers are filled by replication rather than transfer of DNA sections. The random incorporation of mismatched bases may be due to the proof-reading capacity of the DNA polymerase being inactivated.

There is some evidence that DNA polymerase II may be involved in the addition of the mis-matched bases.

3.8.3 Other methods of repair

In addition to post-replication repair, there are also several specific enzyme-mediated repair processes.

Repair of methylated bases

alkyltransferases

A series of enzymes, the alkyltransferases, can repair alkylated bases. O-6-methyltransferase catalyses the transfer of a methyl group from position O-6 of guanine to a cysteine residue on the enzyme. The enzyme is inactivated be being methylated and so may only be used once.

Photoreactivation

photolyase

The enzyme, photolyase can repair the defect of thymine dimers. The enzyme acts on the cyclobutane ring by binding to the dimer in the presence of visible light of wavelength 300-600 nm. The enzyme uncouples the dimer and the pyrimidines are converted back to their original states (Figure 3.16). Bacteria have higher levels of photolyase than eukaryotic cells.

Π See if you can speculate why this might be so.

Obviously the most common factor which generates thymine dimers in nature is UV light present in sunlight. It is likely that single celled organisms (typically prokaryotes) exposed to sunlight will be completely penetrated by UV light and therefore damaged. It makes sense to have a repair mechanism that is switched on by light (sunlight). With multicellular systems (typically eukaryotes), UV and visible light do not penetrate far into the tissues and thus it is not so imperative (or as effective) to have such a light activated repair system.

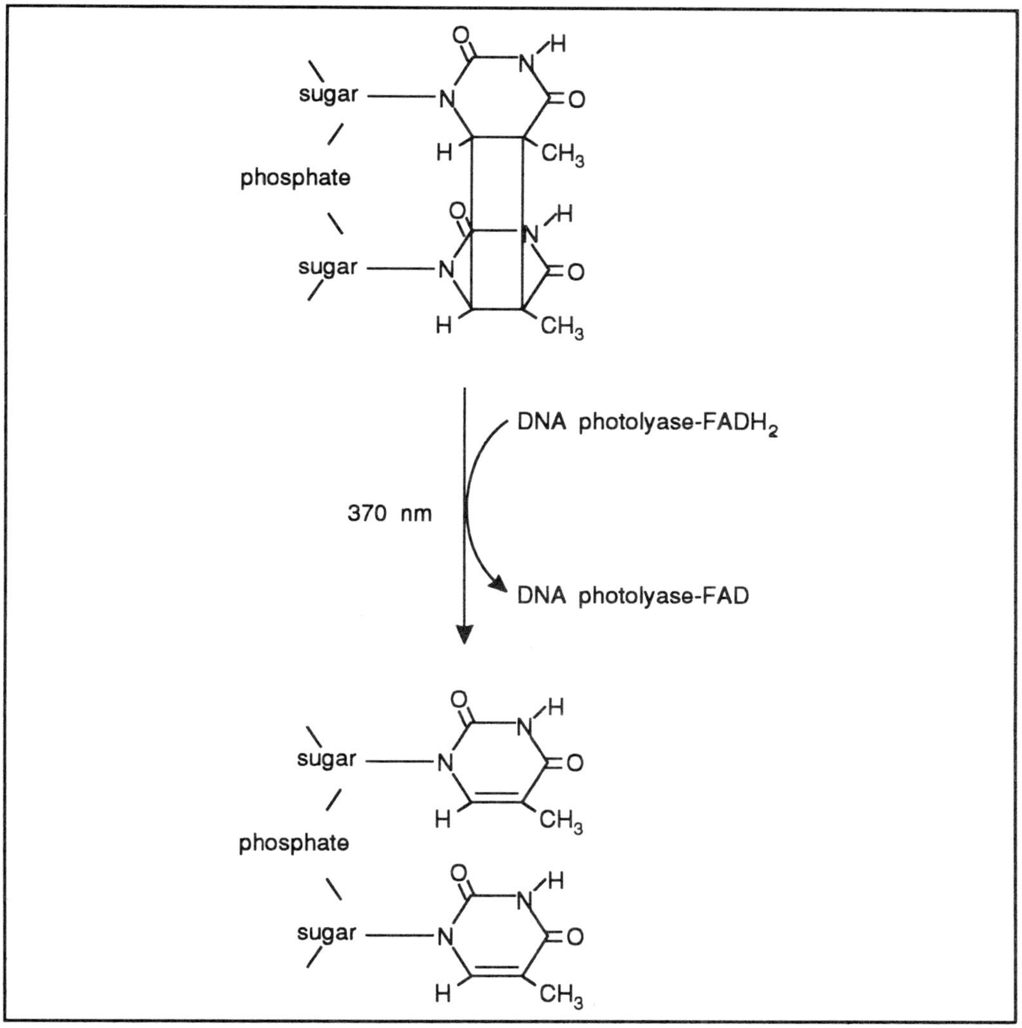

sugar
phosphate
sugar

370 nm
DNA photolyase-$FADH_2$
DNA photolyase-FAD

sugar
phosphate
sugar

Figure 3.16 The enzyme DNA photolyase catalyses the breakdown of thymine dimers in the presence of light of wavelength 370 nm. The enzyme contains an FAD prosthetic group.

SAQ 3.6

Fill in the missing words using the words provided below.

[ ] repair involves base removal by enzymes called [ ]. [ ] repair also requires the enzymes [ ] and DNA [ ]. Alkylated bases are removed by [ ] enzymes and [ ] dimers are excised by proteins coded for by the [ ], [ ] and [ ] genes. The removal of alkylated bases may leave an [ ] or [ ] site.

[ ] repair occurs after substantial DNA damage and involves movement of sections of DNA from the undamaged to a damaged strand. The [ ] protein is essential for the transfer. Both [ ] and [ ] repair are [ ]-free, and restore the correct sequence of bases in the DNA. If [ ] are introduced during repair then the process is [ ]. An example is [ ].

Alkylated bases can be repaired by [ ]. These enzymes catalyse the transfer of an [ ] group from the base to a [ ] residue on the enzyme. The enzyme may only be used [ ] as it is then inactivated. Thymine dimers may also be repaired by an enzyme [ ]. The enzyme uses [ ] to cleave the [ ] ring.

Word list

photolyase, cysteine, post-replication (twice), *uvr*A, SOS-repair, DNA *pol* I, apyrimidinic, excision (three times), alkyltransferases, *uvr*B, mismatched bases, ligase, error-prone, DNA, *rec*A, bases (twice), alkyl, error, glycolyase, *uvr*C, thymine, cyclobutane, visible light, endonucleases, apurinic, once.

SAQ 3.7

The graph below shows the per cent survival of bacterial cultures deficient in certain types of repair. The *pol*A and *uvr* strains are deficient in excision repair. The *rec*A mutants are deficient in gap repair. One strain lacks both repair systems. Explain the different responses of the bacteria to UV light.

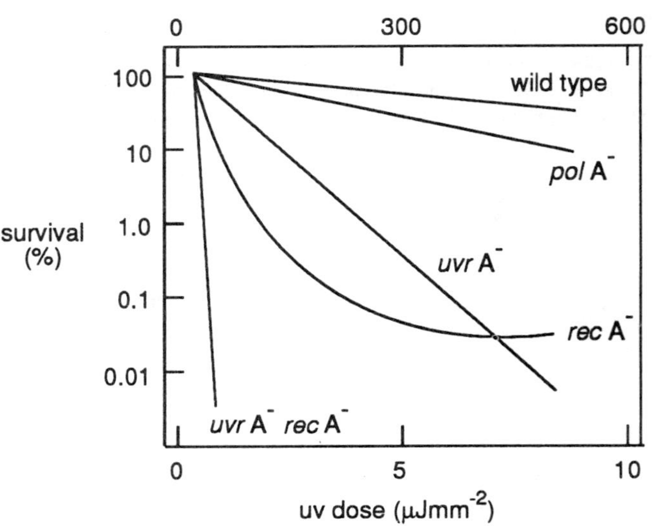

3.9 Failure-to-repair mutations

Although the DNA repair systems are not well characterised in eukaryotes, the failure to repair mutated or damaged DNA may result in the appearance of one of a group of autosomal recessive diseases.

Xeroderma pigmentosum

Xeroderma pigmentosum (XP) is a group of diseases in which one or more of the enzymes of the excision repair pathway are lacking. In these conditions, radiation - induced mutations cannot be repaired. Excision repair of thymine dimers does not occur leading to an extreme sensitivity to ultraviolet radiation and a high incidence of skin cancer. Patients suffering from XP die at an early age from multiple skin cancers.

Ataxia telangectasia

Ataxia telangectasia (AT) is also associated with an inability to repair damaged DNA. Fibroblasts from such patients are more sensitive to α-radiation than cells from normal patients.

Franconis anaemia Blooms syndrome

Franconis anaemia and Blooms syndrome are characterised by reduced DNA repair and increased chromosomal instability. There is an increased incidence of lymphoreticular cancer.

SAQ 3.8

The following graph shows the survival of cultured fibroblasts after treatment with increasing doses of ^{60}Co gamma rays. A second set of cells from an AT patient is also shown. Explain why the cell survival curves differ.

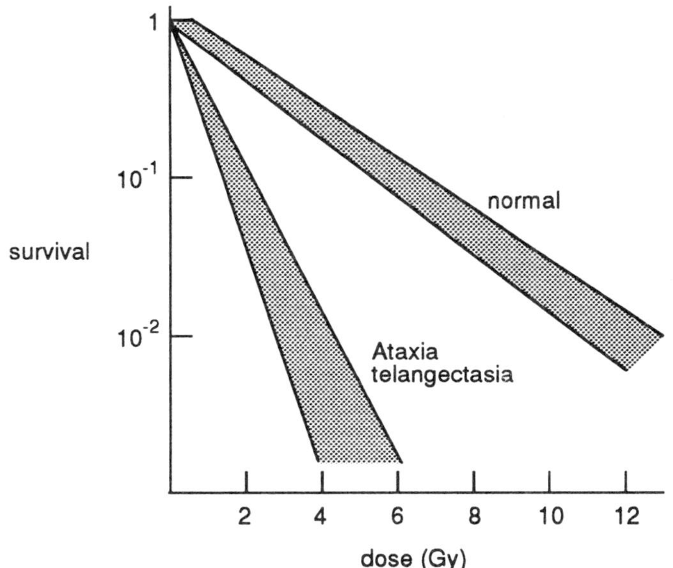

3.10 Uses of mutations

chromosomal mapping

Mutations have been widely used in basic research and also have uses in applied sciences and industry. For example, the inducement of mutations in experimental organisms have been widely used to identify individual steps in complex biochemical pathways (see Garrod, inborn errors of metabolism, Section 3.2). In organisms such as *Drosophila* or *Zea mays*, the extensive mapping of chromosomes has been possible by measuring the relative distances between markers, eg mutations, by determining the extent to which cross-overs are produced by recombination. The investigations of mutations in organisms like *Drosophila* has also given valuable insights into how differentiation and development is controlled at the molecular level.

The active selection of desirable traits has been used for centuries to improve strains in agriculture and horticulture. Chance mutations have also led to desirable characteristics such as hornless cattle and sheep. Induced mutations have been less successful in commercial animals, but the development of new colour strains in plants is one positive outcome to such treatments.

Micro-organisms are used to produce a number of commercially important materials. However, the fermentation-based procedures do not usually rely on the wild-type organisms because they produce too low a yield of product. Rather, the organism is subjected to successive rounds of mutagenesis until a high yielding strain is produced. Table 3.8 shows the development of a modern high producing strain of a pencillin-producing organism from an original, rather poor yielding strain using a number of different mutagenic treatments and some (lucky) spontaneous mutations.

Strain of *Penicillium chrysogenum*	Yield (mg l^{-1})	Treatment (if any)	Institution
NRRL-1951	60	Spontaneous	Northern Regional Laboratory
NRRL-1951.B25	150	X-rays	Carnegie Institute
X-1612	300	Ultraviolet light	University of Wisconsin
WIS Q-176	550	Ultraviolet light	University of Wisconsin
WIS B 13-D 10	-	Spontaneous	University of Wisconsin
WIS 47-638	-	Spontaneous	University of Wisconsin
WIS 47-1564	-	Spontaneous	University of Wisconsin
WIS 48-701	-	Nitrogen mustard	University of Wisconsin
WIS 49-133	-	Spontaneous	University of Wisconsin
WIS 51-20	-	Ultraviolet light	Eli Lilly & Co
E-1	-	Nitrogen mustard	Eli Lilly & Co
E-3	-	Nitrogen mustard	Eli Lilly & Co
E-4	-	Nitrogen mustard	Eli Lilly & Co
E-6	-	Nitrogen mustard	Eli Lilly & Co
E-8	-	Nitrogen mustard	Eli Lilly & Co
E-9	-	Nitrogen mustard	Eli Lilly & Co
E-10	-	Nitrogen mustard	Eli Lilly & Co
E-12	-	Nitrogen mustard	Eli Lilly & Co
E-13	-	Nitrogen mustard	Eli Lilly & Co
E-14	-	Nitrogen mustard	Eli Lilly & Co
E-15	-	Spontaneous	Eli Lilly & Co
E-15.1	7 g l^{-1}		

Table 3.8 The use of mutations in the development of high penicillin producing strains of *Penicillium chrysogenum*. Adapted from Primrose, S.B. (1987) Modern Biotechnology, Blackwell Scientific, Oxford.

SAQ 3.9

Calculate the percentage increase in yield between the initial and final strains of the penicillin-producing organism using the data in Table 3.8.

3.11 Uses of restriction endonucleases and mutations

restriction enzymes and detection of polymorphisms

It is possible to detect variations (polymorphisms) in particular stretches of DNA using restriction endonucleases. These enzymes hydrolyse both strands of DNA at specific sequences of two, four or six bases depending upon the particular enzyme. These sites are called palindromes because they have the same base sequence on each strand reading in the same direction (5'-3') (Figure 3.17). The object is to use a restriction endonuclease which cuts DNA at or very near a point where a critical polymorphism occurs. Since the polymorphism alters the base sequence, it will destroy an existing or possibly form a new palindrome site. Thus if the DNA containing the polymorphism is incubated with the enzyme, different lengths of DNA will be produced compared with the digestion of the 'normal' DNA by the restriction endonuclease. The different lengths of DNA may be separated gel electrophoresis and analysed by transferring ('blotting') the fragments onto a nitrocellulose sheet. It is then possible to detect the fragments of interest by allowing a radioactive piece of DNA complementary to the sequences of interest (ie a 'probe') to bind to the appropriate fragments as shown in Figure 3.18. The presence of the radioactivity then shows the presence of the relative bands. This process is called Southern blotting after the researcher who devised the technique.

Southern blotting

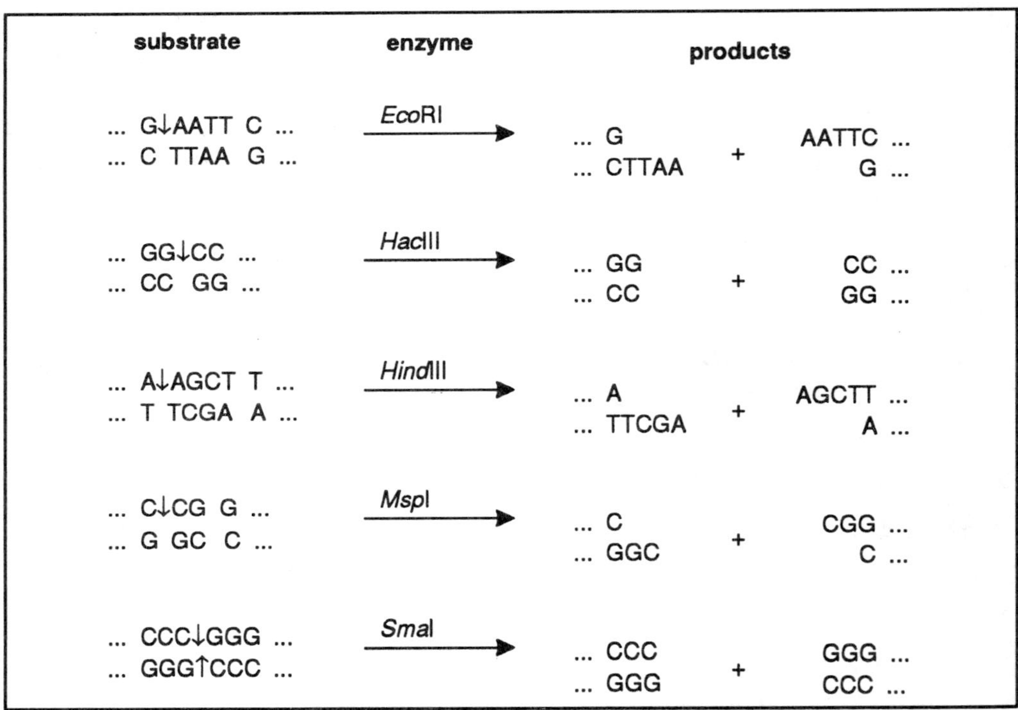

substrate
enzyme
products
... G↓AATT C ...
... C TTAA G ...
EcoRI
... G
... CTTAA
+
AATTC ...
G ...
... GG↓CC ...
... CC GG ...
HacIII
... GG
... CC
+
CC ...
GG ...
... A↓AGCT T ...
... T TCGA A ...
HindIII
... A
... TTCGA
+
AGCTT ...
A ...
... C↓CG G ...
... G GC C ...
MspI
... C
... GGC
+
CGG ...
C ...
... CCC↓GGG ...
... GGG↑CCC ...
SmaI
... CCC
... GGG
+
GGG ...
CCC ...

Figure 3.17 The specificities and products of some restriction endonucleases. The arrows indicates the sites of hydrolysis.

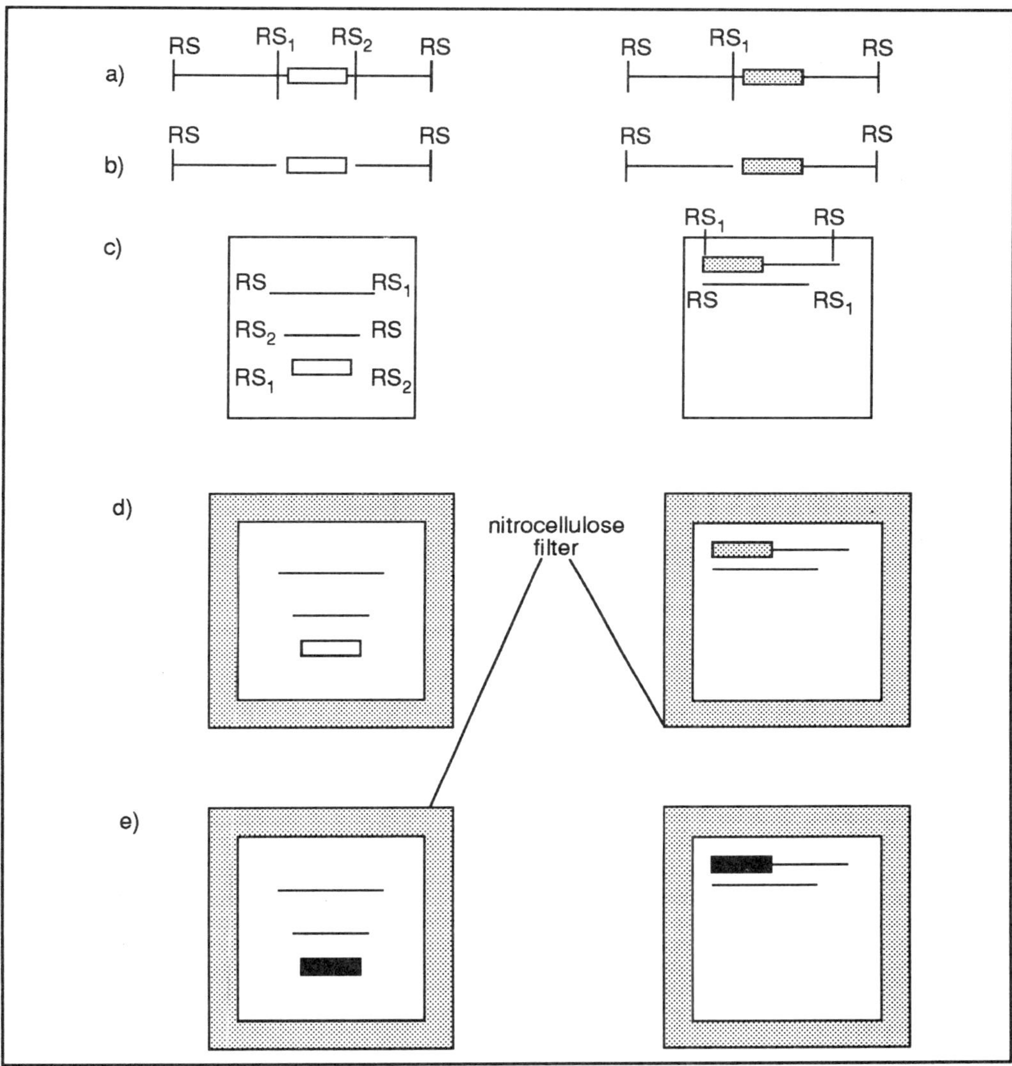

Figure 3.18 The detection of polymorphisms in DNA using restriction endonucleases. a) the 'normal' genes (unshaded) is associated with two restriction sites, RS_1 and RS_2, whereas RS_2 is absent from the section of DNA containing the mutated gene (shaded). b) Hydrolysis with the appropriate restriction endonuclease produces different lengths of DNA. c) The DNA fragments are separated by electrophoresis and d) transferred to a nitrocellulose sheet ('blotting'). e) Detection of the polymorphism on the nitrocellulose sheet by autoradiography using a DNA probe which binds to both the normal and mutated genes.

3.11.1 Restriction fragment length polymorphisms

Eukaryotic DNA contains large amounts of non-coding regions between, and even within, the genes. These non-coding stretches, like all DNA, accumulate random base changes which produce or destroy palindrome sites.

Π Would you expect coding or noncoding sections of DNA to accumulate random base changes at the higher rate? Explain your answer.

The noncoding, since mutations in the coding strand will be expressed and, being likely to be deleterious, will be selected against.

restriction fragment length polymorphisms

RFLPs

Two homologous members of a chromosome pair have broadly the same palindromic sites. However, a random point mutation in one member can occur giving a difference in restriction sites (Figure 3.19). Thus that chromosome will show a different pattern of restriction fragments following digestion with the restriction endonuclease in question. These differences in restriction patterns are inherited according to simple Mendelian patterns and are known as restriction fragment length polymorphisms or RFLPs.

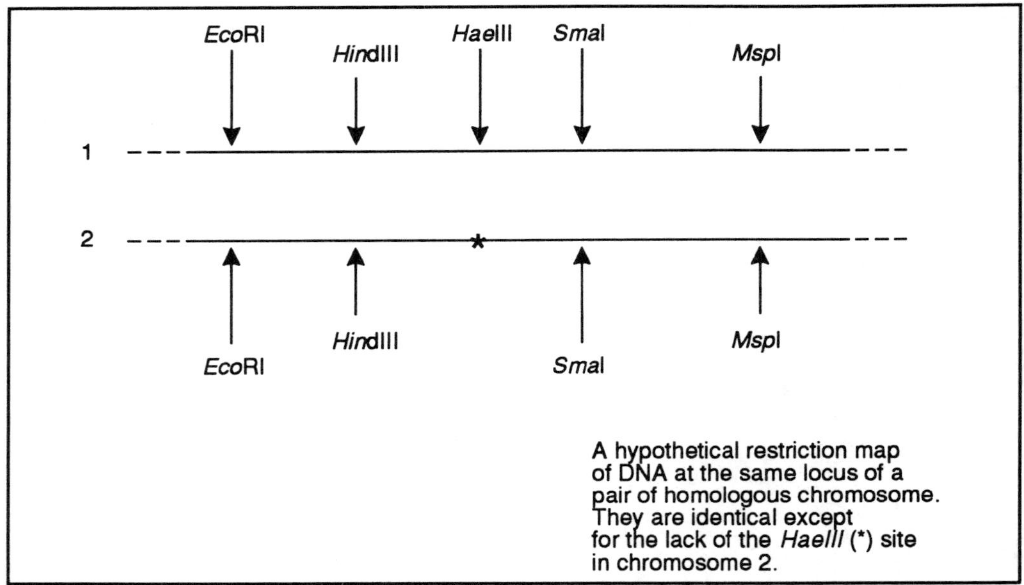
EcoRI
HindIII
HaeIII
SmaI
MspI
1
2
HindIII
MspI
EcoRI
SmaI
A hypothetical restriction map of DNA at the same locus of a pair of homologous chromosome. They are identical except for the lack of the HaeIII (*) site in chromosome 2.

Figure 3.19 Hypothetical restriction maps of DNA on two homologous chromosomes at the same loci. They are identical, except that the HaeIII (*) site is absent in chromosome 2.

RFLPs and inherited diseases

forensic uses of RFLP analysis

It is possible to study and diagnose inborn errors if the change in restriction pattern (ie the polymorphism for the palindrome site) is closely linked to the gene which, when defective, causes the disease. Generally, inherited diseases are caused by recessive genes, thus both parents may be non-affected carriers. This may be shown by a RFLP analysis of DNA from blood samples. Defects in foetuses may also be diagnosed using aminiocentesis or chorionic villi samples to obtain foetal DNA. The technique has also received widespread publicity in paternity testing, identifying rapists and establishing the relationships between immigrants and close relatives wishing to join them.

SAQ 3.10

The detection of sickle cell anaemia does not always have to rely on RFLPs. Hydrolysis of the normal β-globin gene with the restriction endonuclease *Mst*II gives a band of different length to that produced with Hb^{s} DNA (Figure a). The specific changes giving rise to the difference in length (in kilobases, kb) is shown. Complete Figure b), by 1) drawing in the banding patterns for *AA* (normal homozygote), *SS* (homozygote for sickle cell disease) and the heterozygote, *AS*. 2) What do you conclude about the foetus (F) whose banding pattern is shown?

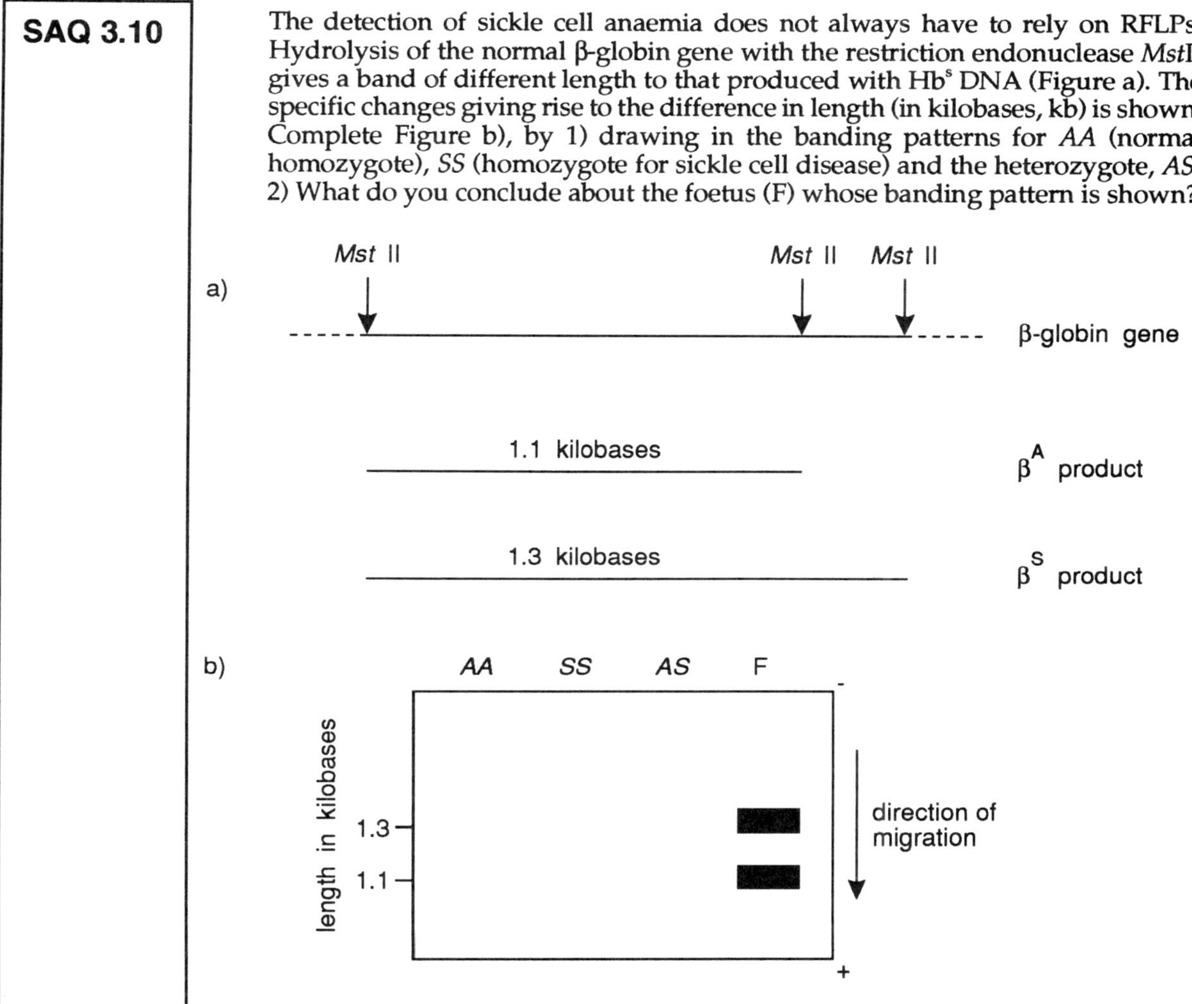

3.12 *In vitro* mutagenesis

protein engineering

In vitro mutagenesis is a method of specifically mutating a gene by introducing a specific base change. The protein which is coded by the gene will be altered with an amino acid substitution corresponding to the mutation. Such mutations provide information about the role of specific amino acid residues in the structure and function of proteins. They also allow protein engineering, that is the production of proteins which are more efficient in some desired aspect than their natural counterparts. Several techniques for site specific mutagenesis are available. We will not provide a lot of details of these here but we will provide a brief overview. Details of these techniques are covered in the BIOTOL text 'Strategies for Engineering Genes'.

SAQ 3.11

A hypothetical enzyme, catabolase has possible commercial uses. It catalyses the production of a desirable product, its tertiary structure is known and its catalytic mechanism is well understood. Unfortunately its industrial impact is limited because at temperatures which give viable rates of reaction catabolase is rapidly thermally denatured.

The enzyme possess two regions of extended pairs of β-sheets. In one pair, a serine residue is found in each strand with their side groups pointing directly at one another and in close proximity. A similar situation exists in the other β-sheet region, except the residues in question are alanines. Neither the serine or alanine residues are involved in catalysis.

Describe how site-directed mutagenesis ('protein engineering') may possibly produce a more thermally stable catabolase. Experimental details are not required.

3.12.1 Chemical mutagenesis

bisulphite deaminates cytosine residue

Bisulphite (HSO_3-) deaminates cytosine residues in single stranded DNA to produce uracil residues. To generate sections of single stranded DNA, the double stranded DNA is treated with a restriction endonuclease which nicks the DNA at a specific site, followed by an exonuclease which degrades the DNA from the nick leaving a section of single stranded DNA. Subsequent treatment with bisulphate then changes all cytosine residues to uracils in the single stranded portion. If DNA polymerase I, DNA ligase and a mixture of the four deoxyribonucleotides are incubated with the mutated DNA under appropriate conditions, the mutated strand forms a template which is copied. During copying, transitions from a G, C to an A, T pair will take place (Figure 3.20).

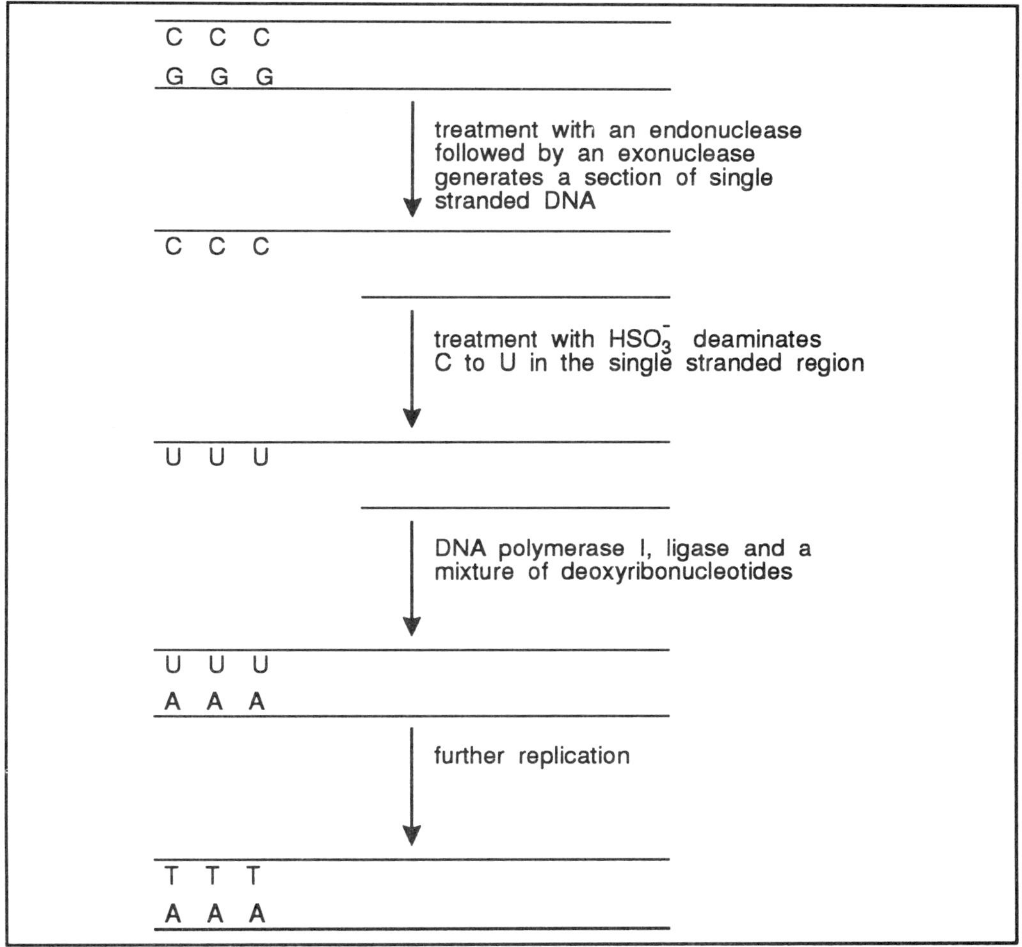

Figure 3.20 Treatment with HSO_3^- causes a GC to AT transition by deaminating cytosine forming uracil.

Chemical mutation is relatively nonspecific. All cytosine residues in the section of DNA will be mutated, although some specificity can be introduced by controlling the length of the single strand, the concentration of bisulphite and the time of exposure.

SAQ 3.12

A protein contains an amino acid sequence Pro-Ile at the active site. Only one of these residues is involved in catalysis. How can you use the processes of *in vitro* mutagenesis using HSO_3^- to decide which of these is involved?

3.12.2 Base mispairing

A second method of introducing a mutation is shown in Figure 3.21. A section of single stranded DNA is generated from double stranded DNA using a restriction endonuclease followed by an exonuclease. The DNA is then repaired using DNA polymerase I and a mixture of three of the four deoxyribonucleotides. For example, if dGTP is omitted an incorrectly paired base will be inserted opposite the C of the template strand. Again, this mechanism is rather imprecise and does not allow a specific base to be mutated.

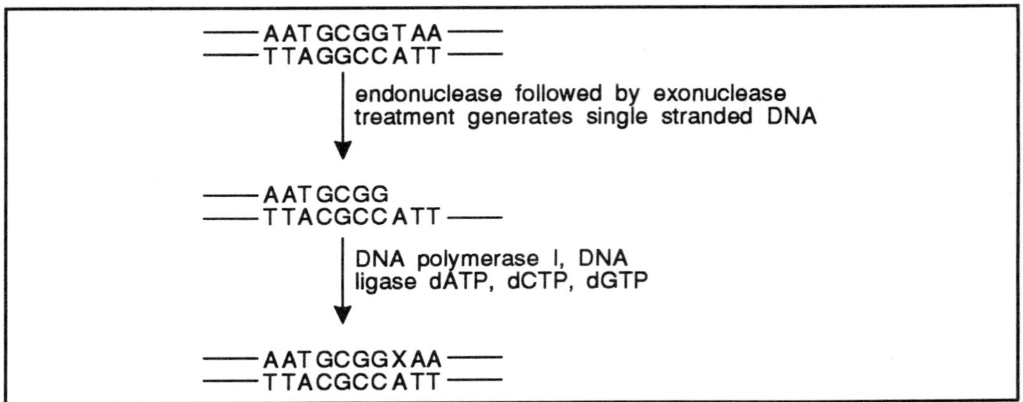

Figure 3.21 Specific mutagenesis by base mispairing. Double stranded DNA is enzymically converted to the single stranded form which is allowed to replicate in the absence of dTTP. Another base is randomly incorporated opposite adenine residues of the template strand.

3.12.3 Mutagenic copying

mutagenic analogues (HO-dCTP)

Mutagenic copying uses a highly mutagenic analogue of a normal base. An example is N^6-hydroxydeoxycytosine 5′-triphosphate (HO-dCTP, Figure 3.22) an analogue of cytosine. DNA polymerase I will incorporate this analogue opposite G or A residues of the template strand depending on whether the HO-dCTP is in the amino or imino form. The amino form pairs with G while the imino form pairs with A. The effect is therefore to cause A to G or G to A transition.

HN—OH
N
O
N

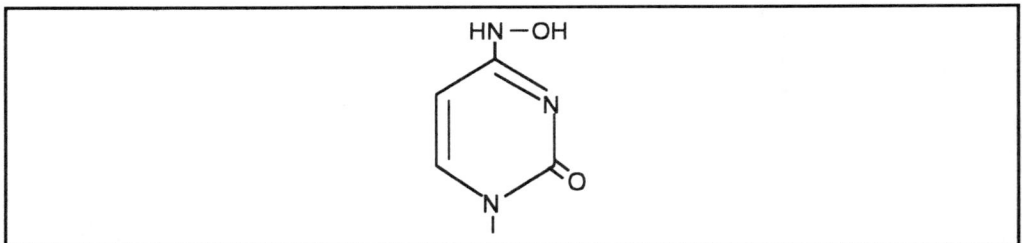

Figure 3.22 The structure of N^6-hydroxydeoxycytosine.

3.12.4 Synthetic oligonucleotide method

use of single stranded M13 vectors and synthetic oligonucleotides

This is the most precise method for introducing a base mutation. The gene under study is introduced into a single stranded vector, usually the single stranded phage M13 (Figure 3.23). A short synthetic oligonucleotide of 15-20 bases identical to a short section of the gene but containing a specific mutation is first produced. This is allowed to bind to a length of single stranded DNA containing the gene. The oligonucleotide is then extended in length using DNA polymerase and ligase, the single stranded DNA forming the template. The double stranded DNA is introduced into *E. coli* where it replicates producing a mixture of both normal and mutant genes. When the genes is expressed it, therefore, produces a polypeptide with a specific amino acid change.

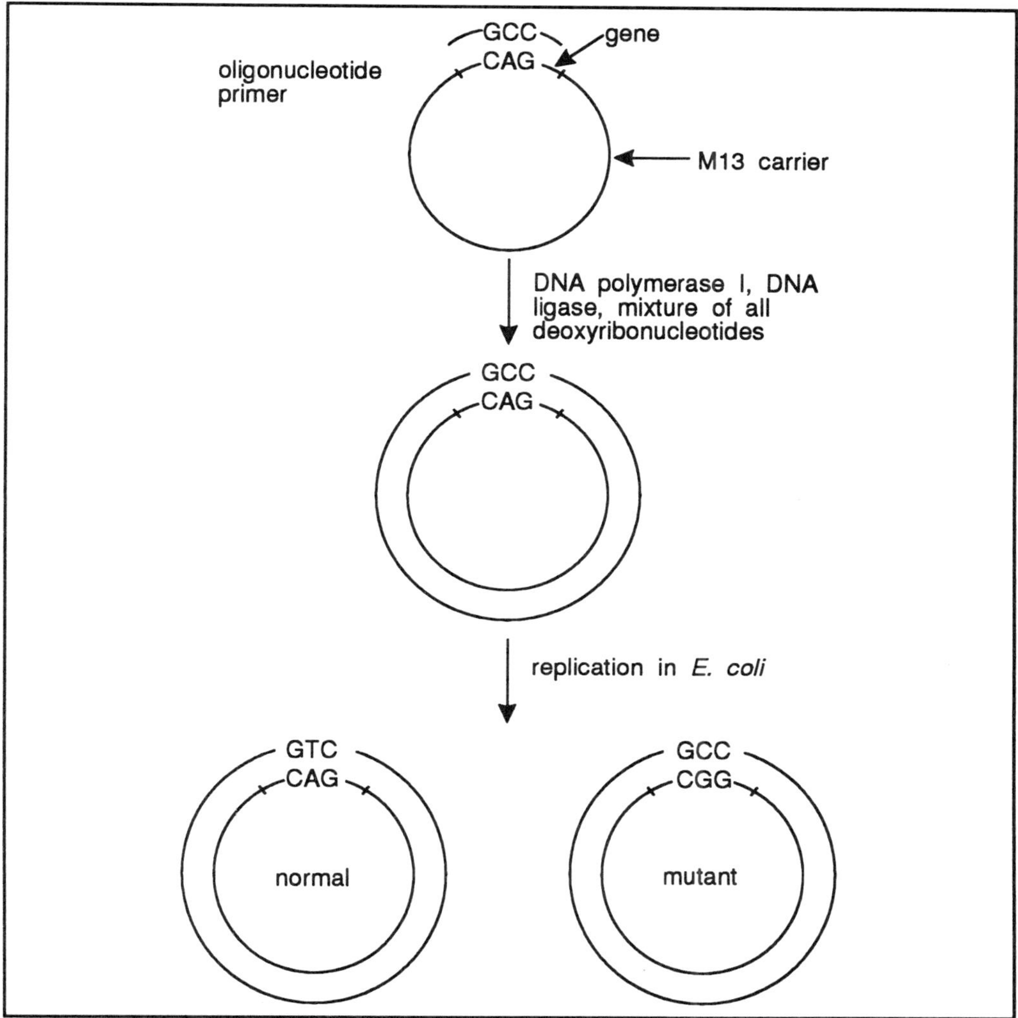

Figure 3.23 The gene to be mutated is incorporated into a carrier, the phage M13. An oligonucleotide primer with a mispaired base is prepared. This is extended enzymically to produce a double stranded DNA molecule, which is replicated in *E. coli* to produce equal numbers of the mutated and wild type genes.

The process can be repeated using a variety of oligonucleotides which differ only in the base substituted. The only limitations are the need to know the base sequence of the gene and to be able to chemically synthesise the oligonucleotides. You should note there are many different ways of carrying out this process. If you wish to learn more, we recommend the BIOTOL text 'Techniques for Engineering Genes'.

3.13 Cancer and mutations

Cancer is a generic name for a large group of diseases in which cells continue to divide irrespective of the tissues demand for new cells. The cancerous cells may migrate in the body and grow in other tissues where they have impaired differentiation (metastases).

Cancers arise from mutations in a single cell. Table 3.9 lists some of the evidence which associates mutations with cancer.

- defects in DNA repair increase the probability of cancer developing
- chromosome instability is found in many types of cancer
- some cancers are inherited suggesting a mutation is being passed to offspring
- some cancers contain mutated oncogenes
- susceptability to some carcinogens is linked to the amount of cellular enzymes which convert it to a metabolically active form

Table 3.9 The evidence that cancer results from mutations.

oncogenes

Mutations lead to an increased risk of cancer. Indeed, over two hundred mutations which lead to an increased risk of cancer have been identified. There is evidence that certain genes called oncogenes are capable of causing cancer. In normal cells these genes are expressed without causing cancer and play an essential role in the development and growth of the organism. Oncogenes may code for proteins which are growth factors or growth factor receptors or which act in the nucleus affecting the transcription of DNA. The role of oncogenes which code for transcription factors will be considered in a later chapter. Alteration of these genes or a change in their expression leads to cancer.

3.13.1 Activation of oncogenes by mutation

bladder tumours

mutations in *ras*

Bladder tumours have been shown to arise by a specific point mutation in an oncogene called *ras*. This codes for a phosphorylated protein of molecule mass 21 000 Daltons. In normal cells, amino acid residue 12 of this protein is valine (GTC); in cancer cells this is mutated to code for glycine (GGC). This particular mutation can be introduced by mutagens such as nitrosomethylurea. The mutated protein has a decreased ability to hydrolyse GTP which is a common intracellular messenger. Other oncogenes may mutate in a similar fashion producing other types of cancer.

3.13.2 Chromosomal translocations and oncogenes

chronic myelogenous leukaemia

Chromosomal translocations occur at a high frequency in many types of cancer. In chronic myelogenous leukaemia (CML) there is a translocation between chromosomes 9 and 22 producing the so-called Philadelphia chromosome. Many oncogenes are situated at the point of translocation and are activated by the translocation. For example, in CML an oncogene called *abl* is on chromosome 9 and the oncogene *sis* is on 22. After translocation these genes are switched with *abl* now found on chromosome on 22 and *sis* on 9. This type of genetic analysis is increasing our understanding of cancer.

3.13.3 Viruses and cancer

Viruses may enter cells and become incorporated into the DNA of the host cell. At some stage of its life cycle the virus will leave the cell and in the process may take a small section of the host DNA with it. Commonly, oncogenes are acquired by the virus in this way. When the virus enters a new cell and becomes incorporated into the host DNA it may insert the oncogene in such a position that there is inappropriate expression of the gene.

SAQ 3.13

Identify which of the following are true and which are false. Explain your answers.

1) Cancers may arise through mutations in somatic cells.

2) Oncogenes are found only in cancer cells.

3) Oncogenes are activated only by mutations.

4) Specific chromosomal translocations accompany the development of specific cancers.

5) Some oncogenes code for proteins which are involved in regulating cell growth.

Summary and objectives

Mutations are changes in the genome of a cell. If the change occurs in a gamete then it is inheritable, and may lead to an inborn error of metabolism. Many such errors are known to lead to recognised clinical conditions. These diseases may arise from changes in single recessive genes, from chromosomal abberations or be the result of a combination of genetic factors.

Mutations in somatic cells may also cause diseases. For example, mutations in oncogenes can lead to cancer, while mutations in the genes concerned with the repair of DNA may result in numerous clinical problems.

The study of mutations has proved beneficial in several branches of science, particularly genetics. The induction of mutations in organisms for scientific and commercial reasons is sometimes of great utility.

Now you have completed this chapter you should be able to:

- explain the types of mutation that can occur in DNA;
- describe the consequences of mutation in terms of protein structure and phenotypic characteristics;
- describe the role of mutagens in inducing mutations and the techniques of *in vivo* mutagenesis;
- outline the role of mutants in research, medicine and industry;
- describe the different mechanisms of repairing mutated DNA and identify the role of the proteins involved;
- illustrate the genetic diseases resulting from deficiencies of repair enzymes;
- describe the consequences of failure to repair DNA in particular the development of cancer.

The organisation of the eukaryotic genome within the nucleus

The organisation of the eukaryotic genome within the nucleus

4.1 Introduction

The most prominent structure within eukaryotic cells is the nucleus (Figure 4.1). This organelle was first observed by Brown as long ago as 1833. However, the application of electron microscopy and modern biochemical techniques have been successful in establishing a number of its structural features and details of its functions. Most eukaryotic cells contain a single nucleus, although it is absent in the functionally mature forms of some cell types, for example erythrocytes and plant sieve tubes. However, a number of cell types (eg mammalian liver cells) may contain two or even three nuclei. Striated muscle consists of extended fibres formed by the fusion of the cell membranes of a number of cells. This fibre is therefore a syncytium and contains many nuclei.

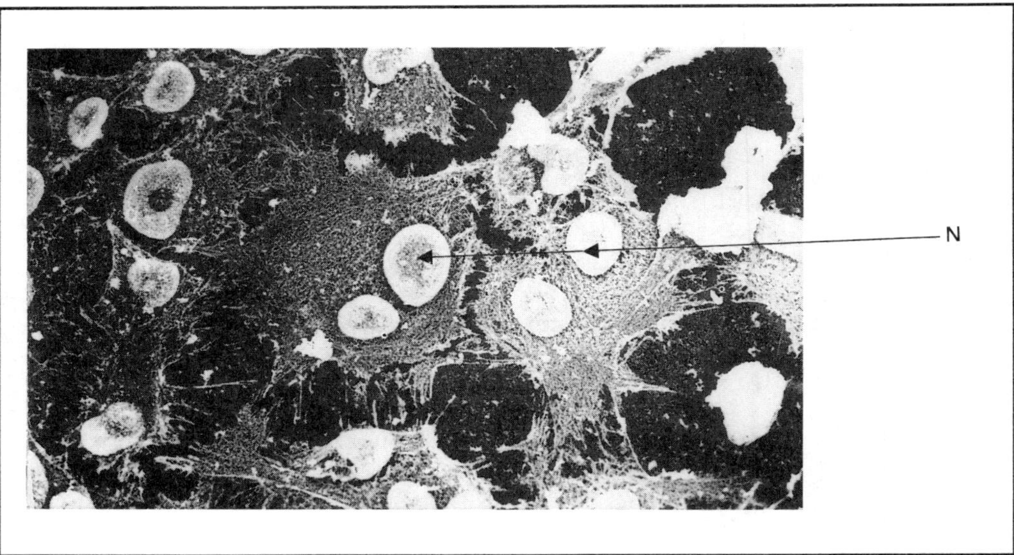

Figure 4.1 Scanning electron micrograph of *Xenopus* cells grown in tissue culture. The cells where initially fixed to preserve nuclear fine structure then extracted with detergent to remove most of the protein before treatment to leave a clear view of the nuclei (an example is denoted by 'N'). x770. Courtesy Dr T. Allen, Department of Structural Cell Biology, Paterson Institute for Cancer Research, Christie Hospital, Manchester. Copyright Dr T. Allen, Department of Structural Cell Biology, Paterson Institute for Cancer Research, Christie Hospital, Manchester.

typical nuclear diameters are 5-25 μm

Nuclei are usually spherical, although their size and shape varies, as does their size relative to the total cell volume. They are typically 5-25 μm in diameter. This comparatively small volume contains the enormously long molecules of DNA, and associated proteins, which constitute the chromosomes. Thus nuclei contain and protect the genetic information and are responsible for its expression. All nuclei contain at least one densely staining region called a nucleolus.

Π The cells of a rat contain the following amounts of DNA:

Cell type	DNA content (pg per nucleus)
thymus	6.3
liver	6 - 13
kidney	6.7
spleen	6.5
lung	6.7
pancreas	7.3
leucocyte	6.6
spermatozoa	3.3

Account for the differing levels of DNA in these cells.

Most variations are due merely to experimental error. However, spermatozoa are haploid (contain half the chromosomal complement) hence contain only half the DNA of other cells. Liver cells may have two nuclei and therefore may contain twice the amount of DNA of other cells.

In this chapter, we will first describe the physical structure of nuclei by examining the nuclear envelope and the matrix of the nucleus. We will, however, predominantly consider the packaging of DNA in nuclei. We will also examine the denaturation and the kinetics of renaturation of DNA which provides data on the base composition and degree of repetition of particular nucleotide sequences. We will conclude the chapter by briefly describing how restriction enzymes and restriction fragment length polymorphism enables us to physical map eukaryotic genomes and provides a way of diagnosing particular diseases.

4.2 The nuclear envelope

nuclear envelopes and nucleoplasm

Nuclei are surrounded by a nuclear envelope which separates its contents (sometimes referred to as nucleoplasm) from the rest of the cell, that is the cytoplasm.

Π Cylindrical cells of 10 μm diameter are 20 μm long and contain a spherical nucleus of 5 μm diameter. Determine: 1) the total volume of the cell; 2) the volume of its nucleus; 3) the volume of the cytoplasm and 4) the percentage of the total volume contributed by the nucleus.

1) Volume of the cell = $\pi r^2 h = \pi \times 5^2 \times 20 = 1570.8\ \mu m^3$

2) Volume of the nucleus = $4/3\ \pi \times r^3 = 4/3\ \pi \times 2.5^3 = 65.4\ \mu m^3$

3) Volume of the cytoplasm = $1570.8 - 65.4 = 1505.4\ \mu m^3$

4) % = 65.4/1570.8 x 100 = 4.2%

Note how misleading the impression given from a 2 dimensional section or drawing is, when the nucleus appears to constitute a much greater proportion of the cell than 5%.

outer nuclear membrane continuous with ER

The envelope consists of two nuclear membranes. The inner membrane is in contact with the nucleoplasm while the outer appears to be continuous with the endoplasmic reticulum (ER) and often has ribosomes on its outer surface (Figure 4.2). Each membrane is about 10-15 nm thick. They are separated by a perinuclear space about 10-15 nm across. This space is seemingly continuous with the lumen of the ER.

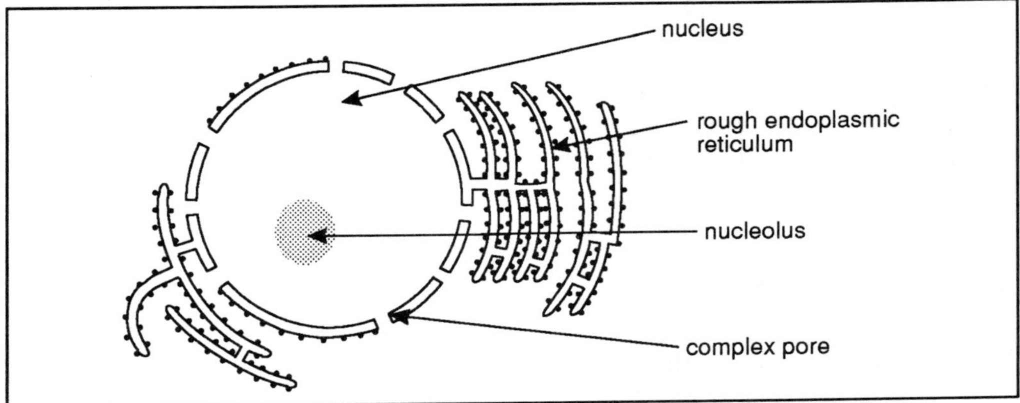

Figure 4.2 Representation of portion of a nucleus.

Π What is the 'normal' range of thicknesses of the nuclear envelope?

The sum of the inner and outer membranes plus the perinuclear space is 30-45 nm.

Nuclear pores

The inner and outer nuclear membranes fuse at a number of places forming circular nuclear pores (Figures 4.2 and 4.3).

nuclear pores (porosomes)

These pores puncture the envelope and constitute connections between the nucleoplasm and cytoplasm. The size and number of pores varies; however, diameters of 80-100 nm are common. They may be extensive in number and can occupy up to 30% of the surface area of the nucleus. Careful examination of nuclear pores by electron microscopy shows they have a complex structure, often called a porosome (Figure 4.4). Each pore is surrounded by eight proteins, which appear as granules 10-20 nm in diameter, at both outer and inner surfaces. Filamentous 'spokes' extend inwards from the granules, while fibres, about 3 nm in diameter and over 20 nm long, extend from the pore complex into both the nucleus and cytoplasm. It is thought that the nuclear pores allow the selective movement of materials into and out of the nucleus. This material is often visible as a 'plug' in the middle of the pore.

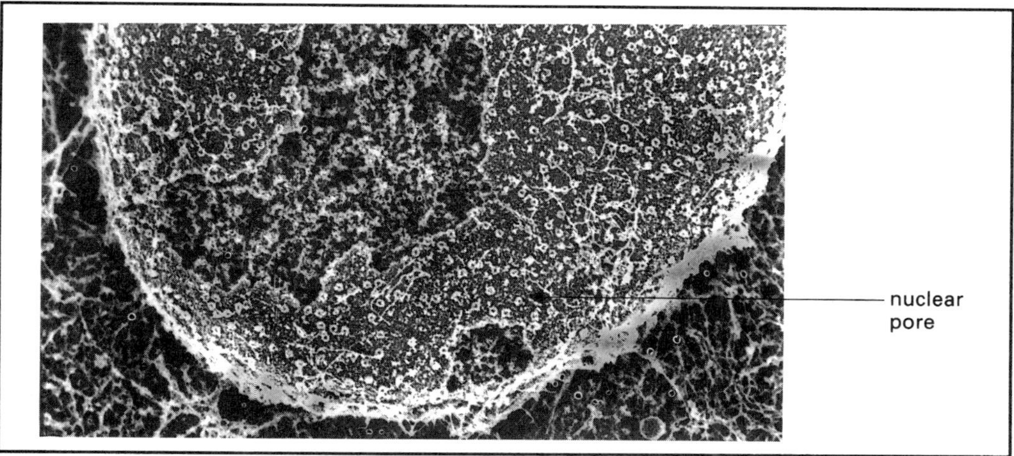

Figure 4.3 Scanning electron micrograph of portion of the nucleus of a *Xenopus* cell, treated as in Figure 4.1. Note the numerous nuclear pores (example arrowed). x 13000 Courtesy Dr T Allen, Department of Structural Cell Biology, Paterson Institute for Cancer Research, Christie Hospital, Manchester. Copyright Dr T Allen, Department of Structural Cell Biology, Paterson Institute for Cancer Research, Christie Hospital, Manchester.

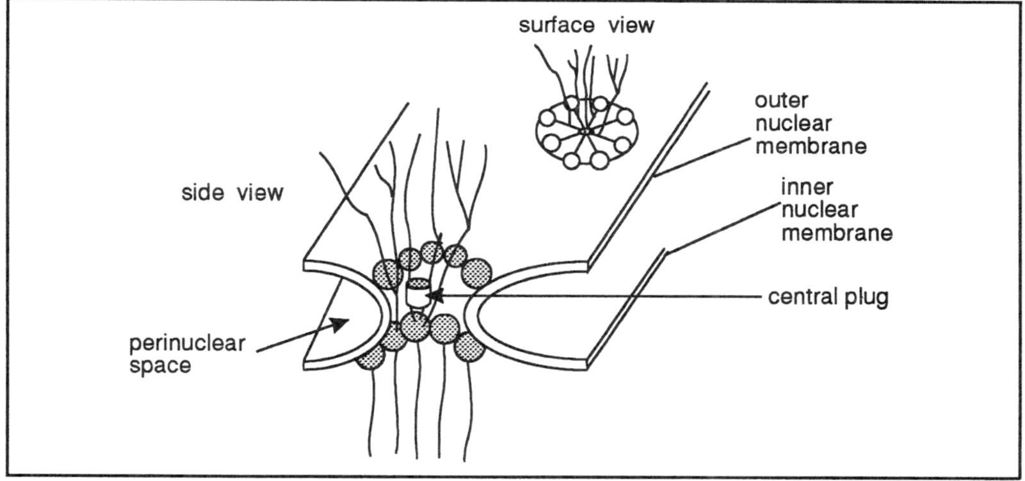

Figure 4.4 Schematic representation of nuclear pore structure. Redrawn from Starr, C.M and Hanover, J.A (1990) Structure and function of the nuclear pore complex: new perspectives. BioEssays, 12(7). 323-329 (see text for a description).

4.3 Nucleolus

All nuclei possess one or more densely staining portions called nucleoli (singular nucleolus). The nucleolus is rich in RNA and proteins and is largely concerned with the biosynthesis of ribosomal RNA (rRNA) and the production of ribosomal subunits (see Chapter 5). Nucleoli are formed by the juxtaposing of regions, called nucleolar-organising regions or NORs, of specific nucleolar-organising chromosomes or NOCs. All eukaryotic cells possess at least one such chromosome. In humans,

nucleolar organising regions

chromosomes 13, 14, 15, 21 and 22 are nucleolar organisers, however, their NORs associate to give a single nucleolus.

Active cells, (those producing large quantities of protein) have large nucleoli, commensurate with their need for numerous ribosomes. Metabolically sluggish cells, as might be expected, have smaller nucleoli.

4.4 The nuclear skeleton

The nuclear skeleton consists of three distinct structures:

the nuclear lamina;

the internal nuclear matrix;

the nucleolar skeleton.

All three consist of networks composed of extremely insoluble fibrous proteins. Mitotic chromosomes also have a fibrous protein support called the scaffold, which assists in their highly ordered tight packing.

The nuclear lamina

three types (A, B, C) of nuclear lamins

The nuclear lamina lines the inner surface of the inner nuclear membrane and appears to be essential for the structural integrity of the nucleus. Thus dissolution of the nucleus during mitosis is linked to the breakdown of the nuclear lamina. The lamina is a meshwork of protein fibres, 30-100 nm thick depending upon the cell type. It is composed of three major types of proteins called lamins A, B and C. All have molecular masses of 60-75 000 Daltons. These proteins resemble each other in primary structure and have extended regions, which are largely helical, flanked by globular ends (Figure 4.5). In many structural respects lamins resemble the proteins which form the intermediate filaments of the cytoskeleton.

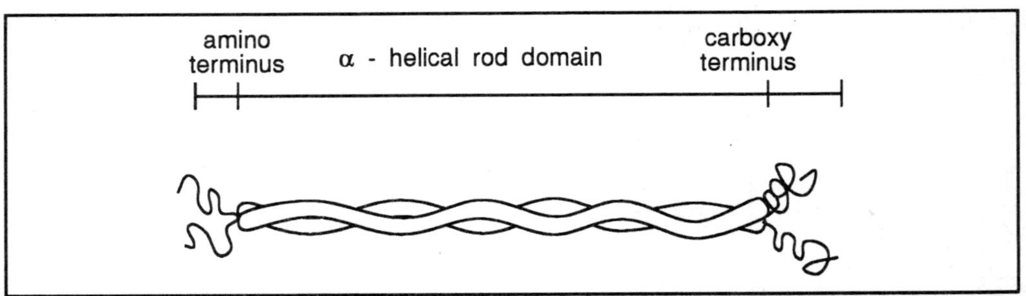

Figure 4.5 Two schematic representations of nuclear lamin protein structure. Redrawn from Krachmarov, C. and Zlatanova, J (1988) Nuclear skeletal structures. Biochemical Education, 16(3), 122-127.

Π What can plausibly be concluded about the evolutionary history of nuclear lamin proteins and those of intermediate fibres of the cytoskeleton given the resemblances between these proteins?

They are both likely to have evolved from a common protein ancestor.

lamin B

Lamin B appears to be closely associated with integral proteins of the inner nuclear membrane. Fibres of chromosomes (chromatin) also seem to be linked to the nuclear lamina (Figure 4.6). It is also thought that nuclear pores are closely associated with the lamina.

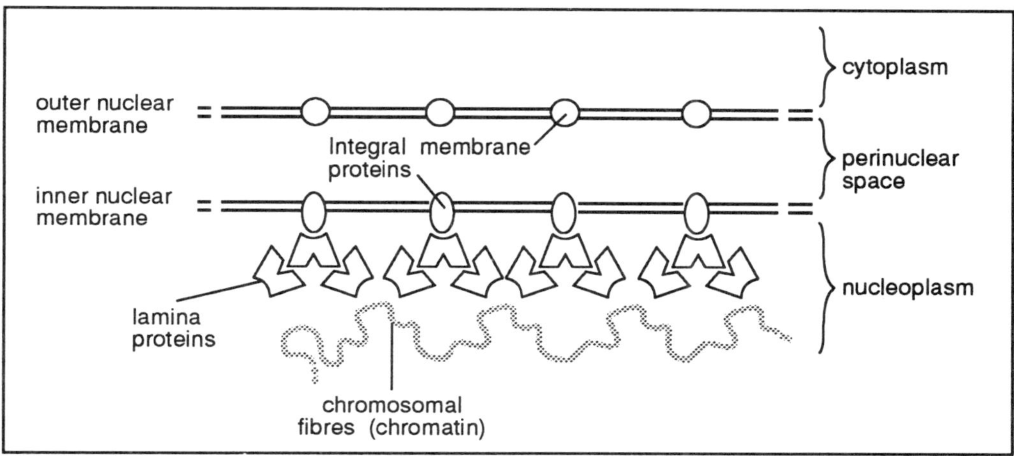

Figure 4.6 Highly schematic arrangement of the nuclear lamina and associated nuclear structures. Redrawn from Alberts, B. *et al* (1983) Molecular Biology of the Cell, 1st edition, page 431. Garland Publishing, Inc. New York and London.

The internal nuclear matrix

The internal matrix is the protein framework found within the interior of the nucleus. The loops of chromatin (Section 4.5) found in the interphase nucleus are thought to be attached to this matrix. Each loop possibly constitutes a discrete replication and transcriptional 'unit'. The DNA replication 'machinery' is found in the matrix.

The nucleolar skeleton

proteinaceous skeleton of nucleolus different from the rest of the nucleus

The nucleolus contains regions of the genome which dictate the production of pre-ribosomal RNA. It is also the site of the production of ribosomal nucleoproteins (see Chapter 5). The nucleolus is 'supported' by a proteinaceous skeleton which is composed of proteins which differ from those of the internal nuclear matrix, although it is probably attached to the matrix. However, little is known of the detailed structure of the nucleolar skeleton.

4.5 Chromatin

chromatin contains DNA and proteins

The nucleus contains the chromosomes of the cell. Each chromosome consists of a single DNA molecule and a number of different types of proteins. This complex of DNA and protein is called chromatin. Genes consist of DNA and control the growth, differentiation and activities of the cell.

Π *Acetabularia* species are single-celled green algae 3-5 cm long! Different species are distinguishable by their differing morphologies, particularly the shapes of their 'caps' (see Figure 4.7). The morphology and activity of the different species is determined by their respective nuclei which is found near the base of the stalk. What feature of *Acetabularia* makes them suitable for putting their nuclei in the 'wrong' cytoplasm?

Clearly, the morphology of these organisms makes it possible to surgically remove part of the cell containing the nucleus and to graft it onto another cell.

experiments with *Acetabularia*

Acetabularia are useful species in which to study the functions of nuclei because of the possibilities of transferring nuclei between different cells and also because growth and regeneration can occur after removal of portions of the cell. Figure 4.7 shows the results of a series of experiments on two species: *A. acetabulum* and *A. crenulata,* which differ in the shape of their caps. The caps were removed and the nuclei then reciprocally exchanged as shown. Regeneration produced hybrid forms with caps of intermediate morphology. Removal of these caps then led to the production of new caps. However, the morphologies of these new caps were a reversal of the original types.

Π Account for these results.

Cap formation must involve the synthesis of new proteins, which are dictated by the DNA via mRNA. Following the nuclear transplants, intermediate caps where formed because although the 'new' nucleus is specifying the production of new mRNA and proteins, the cytoplasm contains mRNA and proteins which where produced by the dictate of the original nucleus. However, these will have been 'used up' in the production of the intermediate cap. Hence its removal will lead to the production of a cap which is specified solely by the transplanted nucleus.

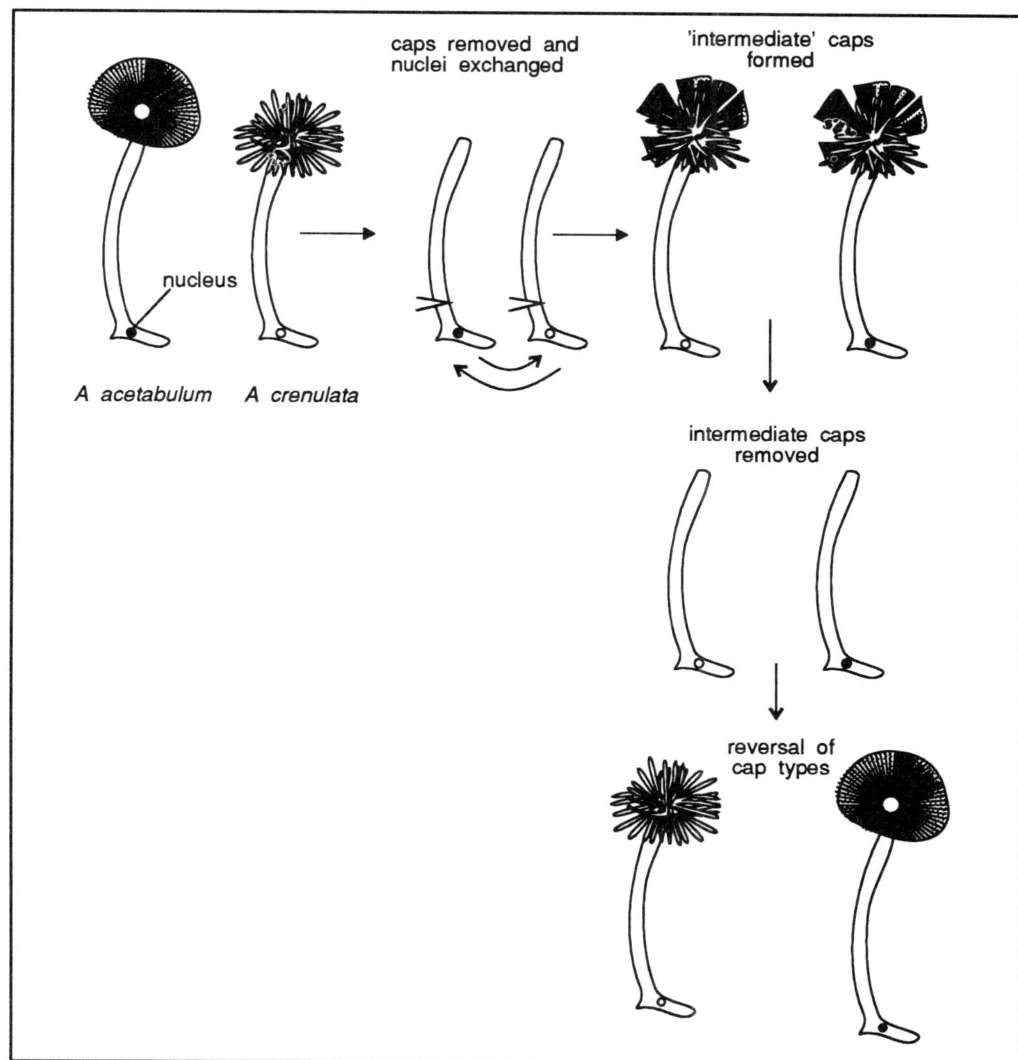

Figure 4.7 Experiments with *A acetabulum* and *A crenulata* in which caps and nuclei were removed and exchanged (see text).

euchromatin

heterochromatin

In the nucleus of interphase, (non-dividing cells), chromatin exists as long, extended fibres (Figure 4.8). In most cells, however, only 5-10% of the genes are active (see Section 4.10). This active chromatin, called euchromatin, is rather loosely coiled and stains relatively lightly (Figure 4.9). The remaining, non-active, chromatin is stored in a densely staining form called heterochromatin (Figure 4.9). The packaging of chromatin is described in the following section. In addition, the nucleus also contains a number of non-chromosomal proteins which are probably concerned with activities such as the control of gene expression and DNA replication.

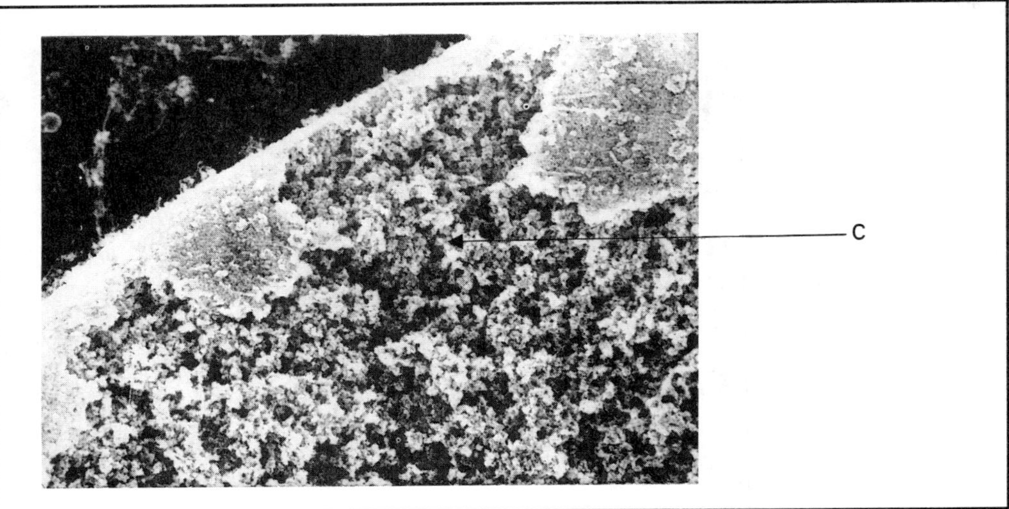

Figure 4.8 Scanning electron micrograph showing portion of the nucleus of a *Xenopus* cell prepared as in Figure 4.1. x26 000. The nuclear membrane is partially disrupted clearly revealing the fibres of chromatin (C). Courtesy Dr T. Allen, Department of Structural Cell Biology, Paterson Institute for Cancer Research, Christie Hospital, Manchester. Copyright Dr T. Allen, Department of Structural Cell Biology, Paterson Institute for Cancer Research, Christie Hospital, Manchester.

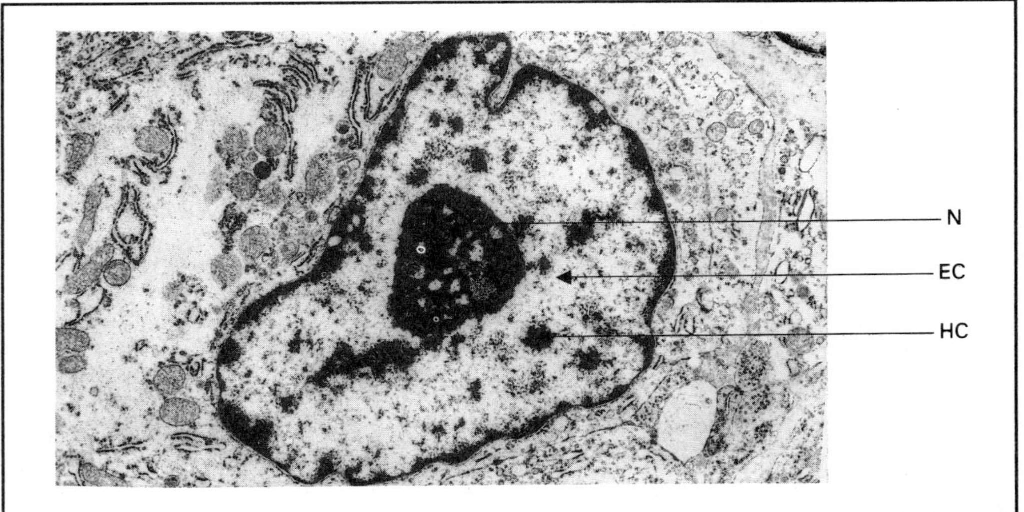

Figure 4.9 Transmission electron micrograph of a thin section of a parotid cell nucleus, showing heterochromatin (HC), euchromatin (EC) and the nucleolus (N). x10 000. Courtesy Dr. A. Curry, Public Health Laboratory, Withington Hospital, Manchester.

4.6 Packaging of DNA into the nucleus

DNA molecules are enormously long. For example, the DNA molecules in the diploid human chromosome complement of 46 have a total length of about two metres! Thus to pack them within the limited confines of the nucleus calls for great care.

Π Calculate the ratio of 1) the total length of human DNA to that of the diameter of the nucleus (10 μm); 2) the percentage of the volume of the nucleus occupied by the DNA (DNA molecules are 2 nm in diameter).

1) The length of the human DNA is about 2m and each nucleus is about 10 μm in diameter. The ratio of these lengths is about 2×10^5:1. This is a clear indication that the DNA must be carefully packaged in order to fit into the nucleus.

2) The volume of the nucleus is $\frac{4}{3}\pi r^3 = \frac{4}{3}\pi 5^3 = 524\ \mu m^3$.
The volume of the DNA is $\pi r^2 h = \pi \times (1 \times 10^{-3})^2 \times 2 \times 10^6\ \mu m^3 = 6.3\ \mu m^3$.

Thus the % of the volume occupied by DNA $= \frac{6.3}{524} \times 100 = 1.2\ \%$.

In other words, although the DNA molecules are much larger than the nucleus, if they are properly packaged, they only need to occupy a very small fraction of the volume of the nucleus. (Think of a ball of string analogy. If the string is unwound, it is much longer than a room. If rolled up into a ball, it can be safely accommodated in a drawer in a cupboard within the room!).

packaging of DNA into nucleosomes and solenoids

Eukaryotic DNA possess several levels of structural organisation which are illustrated in Figure 4.10. The DNA is packaged consecutively into nucleosomes, fibres (solenoid), loops and even more tightly folded states in mitotic chromosomes.

Nucleosomes consist of octomers of eight histone proteins which form a bead around which is wound approximately two and a half turns of DNA (146 basepairs) (Figure 4.11).

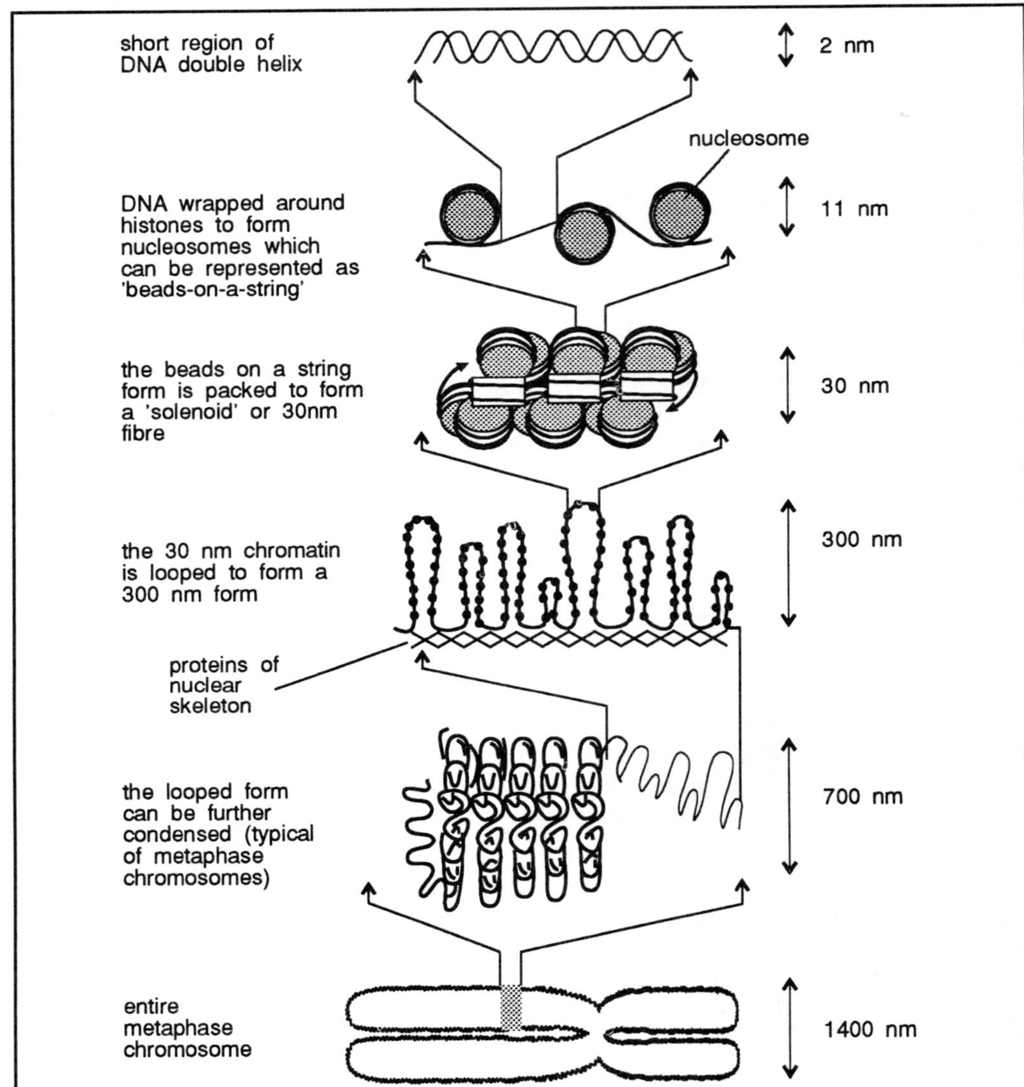
short region of DNA double helix
2 nm
nucleosome
DNA wrapped around histones to form nucleosomes which can be represented as 'beads-on-a-string'
11 nm
the beads on a string form is packed to form a 'solenoid' or 30nm fibre
30 nm
the 30 nm chromatin is looped to form a 300 nm form
300 nm
proteins of nuclear skeleton
the looped form can be further condensed (typical of metaphase chromosomes)
700 nm
entire metaphase chromosome
1400 nm

Figure 4.10 Schematic illustration of stages in the folding of chromatin. Based on Alberts, B. et al (1989) Molecular Biology of the Cell, 2nd edition, page 504. Garland Publishing, Inc, New York and London.

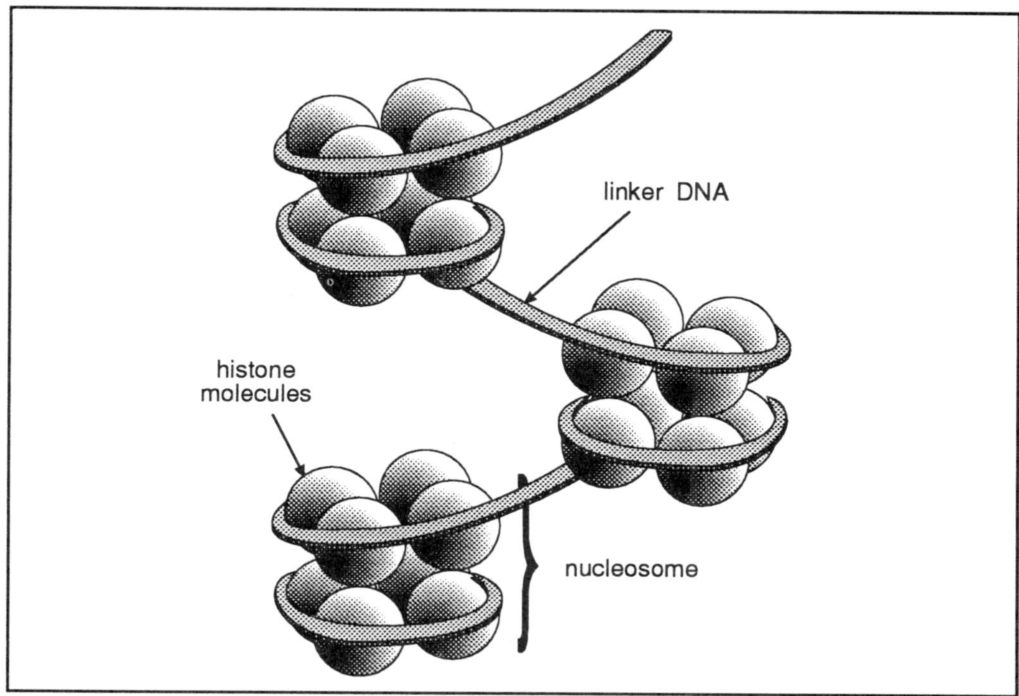

Figure 4.11 Diagrammatic representation of nucleosomes. Each nucleosome consists of DNA wrapped around eight histone molecules (see text for details).

Π Histones are small proteins rich in the amino acid residues lysine and arginine (Figure 4.12). Are histones acidic or basic proteins? What sort of molecular interactions are likely to occur between histones and DNA.

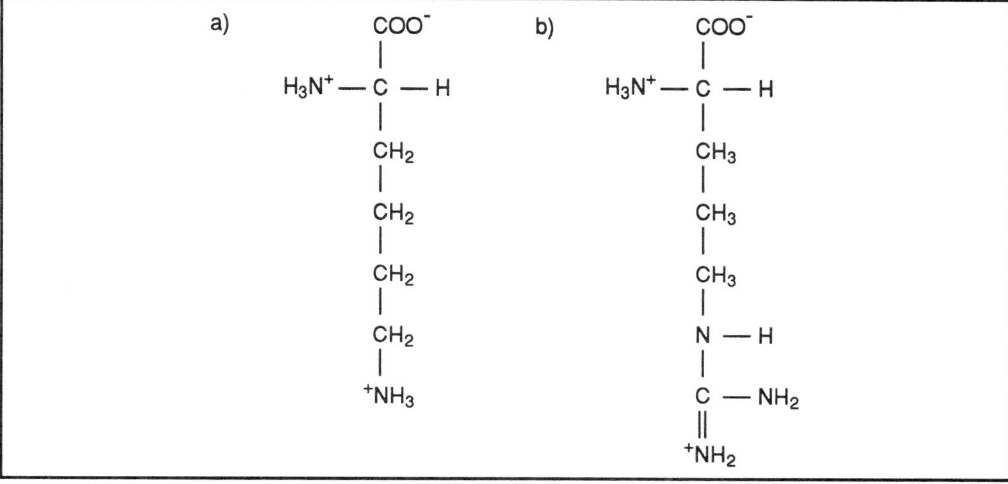

Figure 4.12 a) lysine and b) arginine.

Histones are basic proteins. About 80% of the amino acid residues occur in α-helices, many of which associate with the major groove of the DNA double helix. This nucleoprotein complex is stabilised by ionic interactions between the negatively charged phosphate groups of the DNA and the positively charged side groups of the lysine and arginine residues.

nucleosomes and linker DNA

solenoid and loops

The DNA between the nucleosomes is called linker DNA. For obvious reasons this level of organisation is often called 'beads on a string'. The helical winding of the nucleosomes generates the next structural form: the 30 nm fibre or solenoid. This, in turn, is folded into loops. In the interphase nucleus chromatin occurs in the loop form. Loops are 250-400 nm long and, on average consists of about 60 000 bp of DNA. The loops are anchored and stabilised by binding to proteins of the nuclear skeleton (see Section 4.4).

SAQ 4.1

Given that each bp is 0.34 nm long, how many loops of 60 000 bp would be expected to form in the nucleus of a 'typical' human cell if all the chromatin occurred as euchromatin?

Successive folding of the 30 nm fibre generates the highly condensed chromatin of the mitotic chromosome. Each round of folding produces a thicker fibre, with progressive shortening of the chromosome. The effectiveness of the packaging at each stage can be estimated by calculating the packing ratio, that is the length of the chromosome before the folding step divided by its length after folding (Figure 4.10).

SAQ 4.2

The total length of DNA in a human diploid cell is about two metres. This is compacted into mitotic chromosomes with a packing ratio of 10^4. What is the total length of the mitotic chromosomes?

4.7 Denaturation of DNA

separation of DNA strands is accompanied by changes in absorbance, optical rotation and viscosity

The two polydeoxyribonucleotide strands of the double helix of DNA are held together in a relatively rigid structure by hydrogen bonding and hydrophobic forces. Although these forces are weak, and therefore easily disrupted, their large numbers maintain the structure of the molecule. Disruption of these bonds allows the double helix to unwind, producing a more flexible, single stranded molecule. This process is called denaturation. Denaturation can be measured in a number of ways. The aromatic purine and pyrimidine bases absorb light maximally at a wavelength of about 260 nm. If native, double stranded DNA is denatured then the solution of DNA shows increased absorbance at 260 nm (Figure 4.13). This increase is called the hyperchromic effect. Changes in optical rotation and viscosity also accompany denaturation and can be used to quantify it.

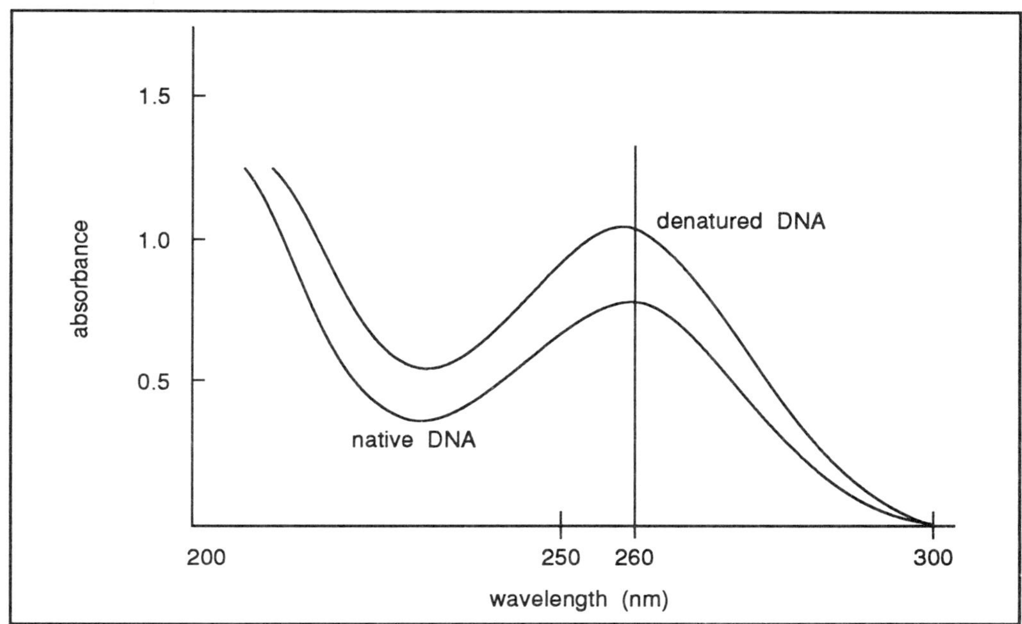

Figure 4.13 The hyperchromoic effect. When DNA is heated it changes from its native double stranded form to a denatured single stranded structure, which absorbs more light of 260 nm.

The effect of temperature on denaturation

melting of DNA

The unwinding of the DNA double helix on heating is called melting. The melting temperature or T_m, is the temperature at which the DNA is midway through its transition from a double stranded to a single stranded structure, ie fifty per-cent of the DNA will have melted at its T_m. Figure 4.14 shows the effect of increasing temperature on the melting of DNA isolated from a number of organisms. The temperature at which the DNA melts is dependent on the proportion of GC to AT basepairs. The richer the DNA in GC basepairs, the higher the T_m. This effect occurs because each GC basepair is stabilised by three hydrogen bonds while each AT pair by only two. The T_m of DNA varies in a linear manner with GC content (Figure 4.15). This relationship means that AT-rich regions of DNA will melt before GC-rich regions.

A useful relationship is given by:

$\%\ GC = 2.44\ (T_m - 69.3)$ in 0.2 mol l^{-1} NaCl.

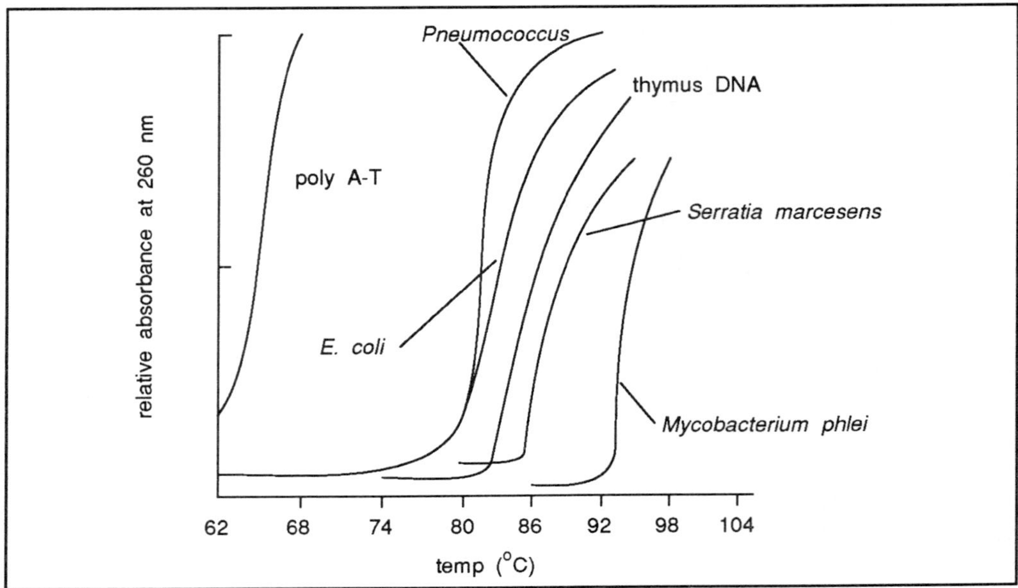

Figure 4.14 The stylised changes in absorbance of DNA isolated from a number of sources as the temperature is increased. As the proportion of GC base pairs in the DNA increases, the curve is shifted to the right. Data from Smith *et.al* (1983) Principles of Biochemistry, 7th edition, page 143. Published by McGraw Hill, London. Note that the actual temperature at which these transitions in absorbance occurs depends on the ionic strength, pH and other factors (see text).

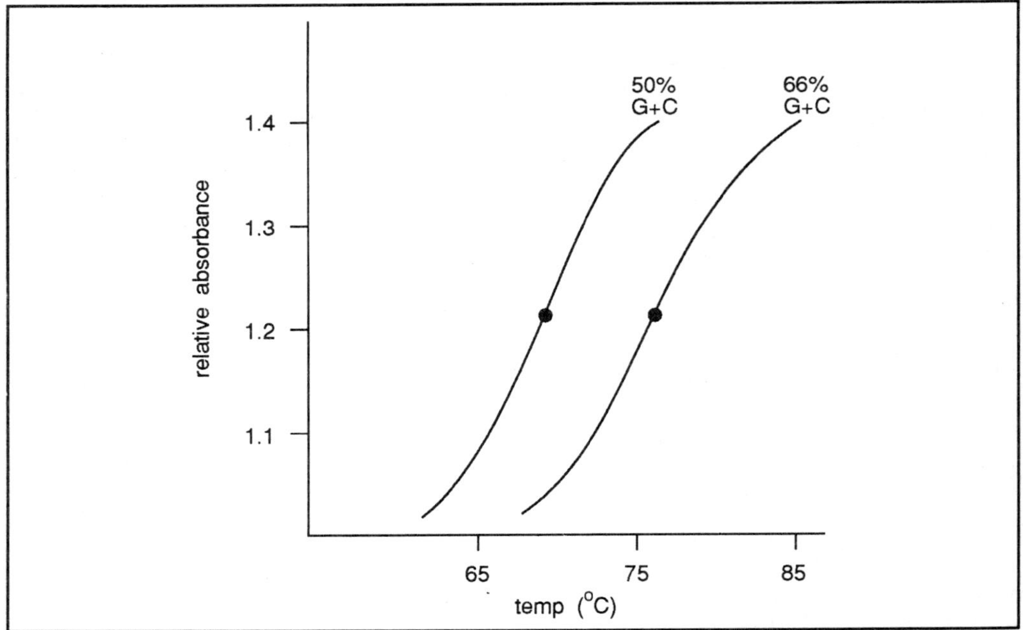

Figure 4.15 Melting curves for DNA molecules with increasing proportions of GC base pairs. The T_m for *E. coli* DNA which contains 50% GC base pairs is 69°C. For *P. aeruginosa* DNA, which has 66% GC base pairs, the melting temperature is 76°. Note that the actual temperature at which the DNA melts depends on the presence or absence of factors which influence the strength of hydrogen bonds.

SAQ 4.3

If a solution of double stranded DNA is heated and then cooled over a short period and the absorbance at 260 nm followed what would you expect to see if:

1) the solution is heated to 5 °C below the T_m;

2) the solution is heated to 20 °C above T_m before it is rapidly cooled?

An empirical relationship relates GC content of DNA and T_m at a fixed ionic strength. This is % G+C = 2.44 (T_m - 69.3) in 0.2 mol l^{-1} NaC1

Urea and formamide lower the T_m of DNA. Urea disrupts hydrogen bonds and formamide aids the interaction of hydrophobic bases with water. In 8 mol l^{-1} urea the T_m is lowered by about 20°C.

SAQ 4.4

1) Calculate the T_m of a DNA molecule in which the GC base pair content is a) 20% and b) 40%.

2) If the melting temperature of a DNA molecule in 0.2 mol l^{-1} NaC1 is 110 °C, calculate the per cent GC base pairs.

4.8 Renaturation of DNA

If a solution of DNA is denatured by heating and then cooled below its T_m the double stranded structure will reform. This renaturation is dependent on the complementary strands forming the appropriate basepairings. If the initial concentration of denatured DNA is C_0 (expressed in moles of bases per litre) the change in concentration, dC, of the single stranded DNA can be represented by the equation:

$$dC/dt = kC_0$$

C is the concentration of nonrenatured DNA at time t and the equation integrates to give:

$$C/C_0 = 1/(1+kC_0t)$$

$C_0t_{1/2}$

If a graph of C/C_0 against log C_0t is drawn (Figure 4.16) then the time at which C/C_0 = 0.5 can be estimated. This value is called the $C_0t_{1/2}$ (pronounced 'cot-a-half'). The reciprocal of $C_0t_{1/2}$ is the rate constant k which is related to the number of basepairs (N) in the DNA when the DNA lacks repeated sequences.

DNA complexity

N is sometimes called the complexity (designated by X) of the DNA and is directly proportional to $C_0t_{1/2}$ in that it is a measure of the number of unique sequences in the DNA. C_0t analysis may be used to determine how often particular sequences of bases occur in DNA. Figure 4.16 shows plots of C_0t against the percentage reassociation of eukaryotic DNAs. To carry out this determination, the DNA is first degraded into small pieces of about 300 basepairs.

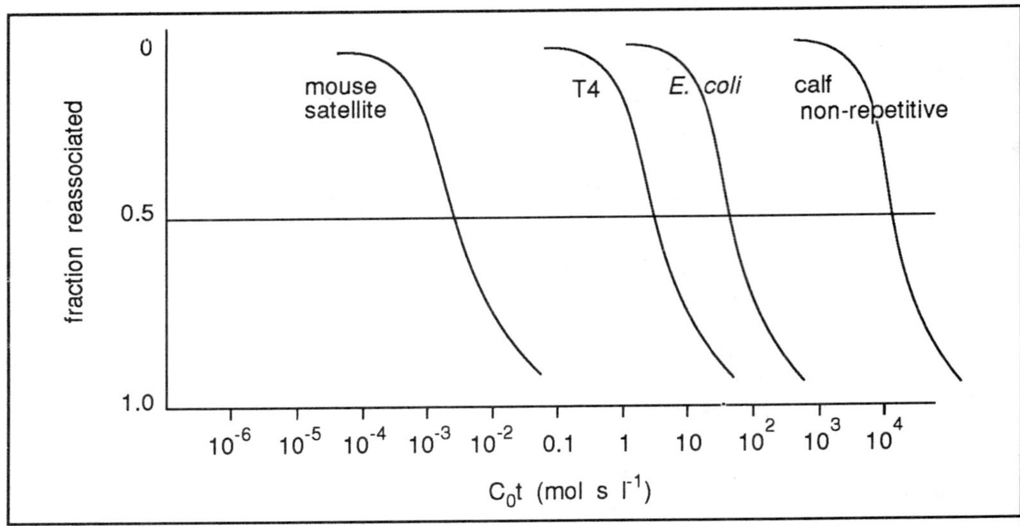

Figure 4.16 Stylised graphs of log Cot plotted against the fraction of DNA reassociated (C/Co). From such a plot, the Cot value at which C/Co is 0.5 can be estimated. Data from White *et. al* (1983) Principles of Biochemistry, 7th edition, page 145. Published by McGraw Hill, London.

Let us examine what is happening during the renaturation process. Consider two samples of DNA, one from an organism with a large genome (organism A) and one with a much shorter genome (organism B).

Thus:

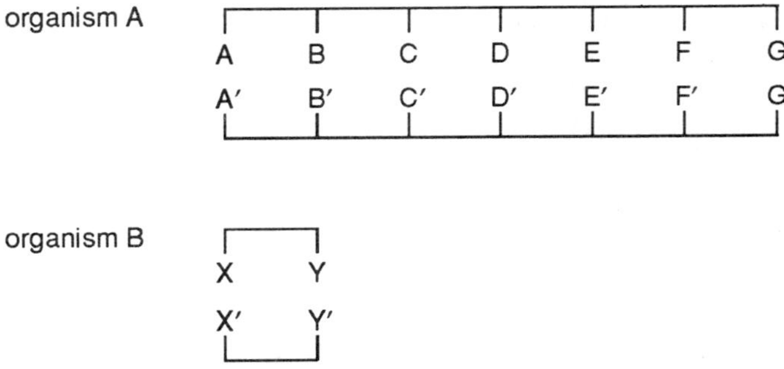

If we take the same total amount (C_0) of DNA from each organism, break into short fragments and melt prior to renaturation, then the situation with organism A will be:

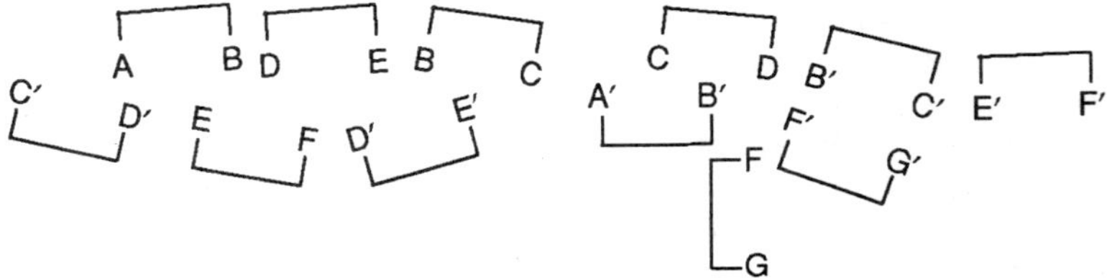

The chances of each strand meeting its complementary strand is small and thus the rate of renaturation is slow.

The situation with organism B is however different. Using the same total amount (C_0) of DNA then in solution we will have:

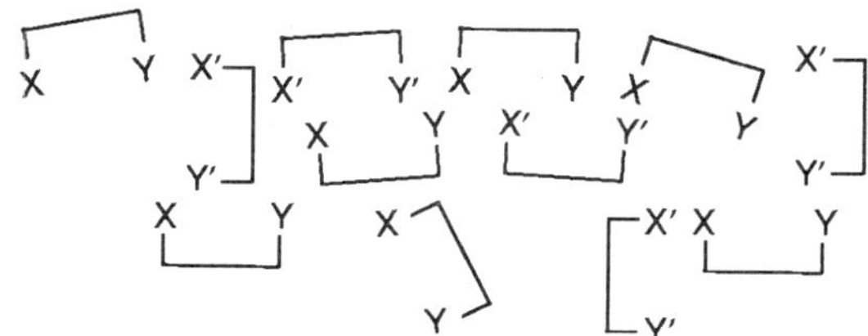

The chances that complementary sequences will meet are high and thus renaturation is much faster. In other words, we can use the rates of renaturation to get some idea of the size of the genomes of organisms. If the renaturation conditions are the same (eg C_0, temperature, base composition) then a genome of size X should renature four times faster then a genome of size 2X, and sixteen times faster than a genome of size 4X. The rate will depend on the concentration of both strands. In practice, the relationship is not as simple as it is influenced by the base composition (remember the GC basepair has three hydrogen bonds, AT has two and this influences renaturation kinetics). The relationship is also only valid if each of the nucleotide sequences in the DNA are unique (present in single copy).

Consider the following portion of a genome in which sequence A occurs several times.

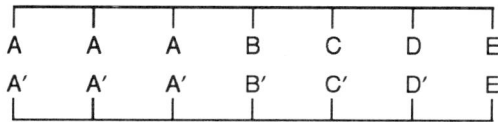

A A A B C D E

A' A' A' B' C' D' E'

If we break this into fragments and melt it, then we would have the following pieces.

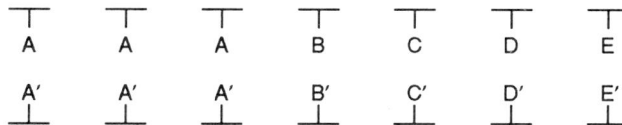

A A A B C D E

A' A' A' B' C' D' E'

If we now allow these to renature, then A is more likely to collide with fragments bearing A' than B is to collide with fragments bearing B'. In other words, repetitive sequences will renature faster than unique sequences.

three types of sequences in eukaryotic DNA

Three types of sequences are found in eukaryotic DNA (see Figure 4.17). That which reassociates rapidly represents a small fraction of the total but is present as millions of copies and is called highly repetitive DNA. These sequences are largely found in the centromeric regions of chromosomes. The next fraction to reassociate is present in thousands of copies. This fraction is called moderately repetitive DNA.

The sequences which reassociate last are present as single unique copies and are the structural genes of the organism.

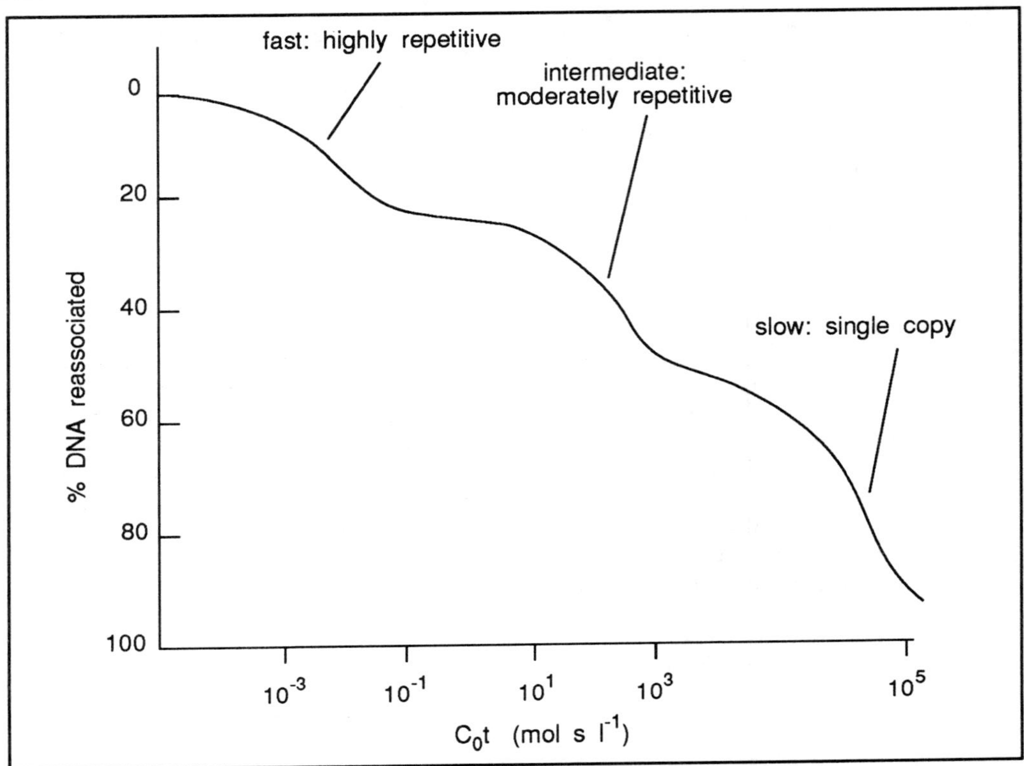

Figure 4.17 Stylised renaturation of DNA from a typical eukaryote. The fraction of DNA reassociated is plotted against C_0t. Three classes of DNA are distinguishable as indicated.

Figure 4.18 shows a comparison of the reassociations of prokaryotic and eukaryotic DNA and clearly demonstrates the differences in association between the two types of organisms.

The reassociation curve for prokaryotic DNA (Figure 4.18) indicates that there are a few repeating units (these include for example ribosomal genes and the genes for transfer RNA) but that the majority of the DNA consists of unique sequences. In contrast, the eukaryotic DNA contains significant amounts of repeated sequences.

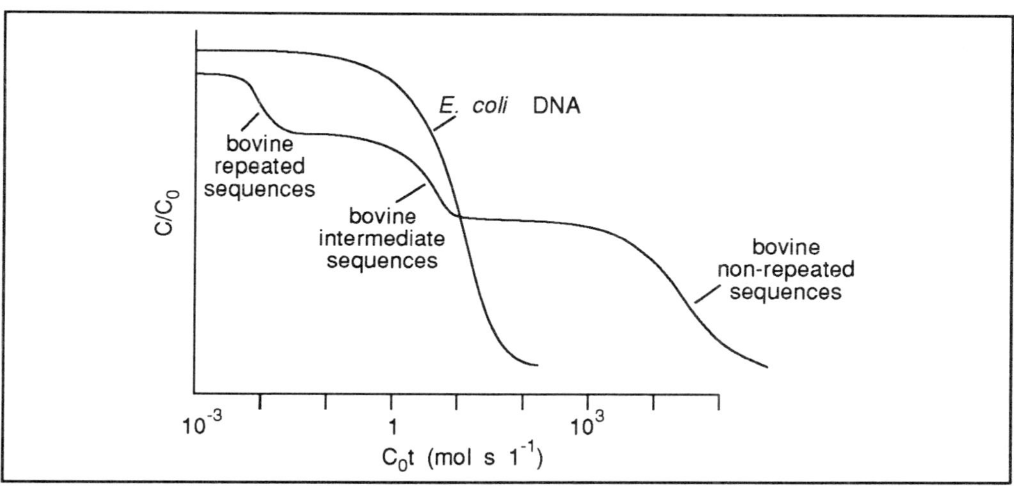

Figure 4.18 A comparison of the reassociation of eukaryotic and prokaryotic DNA. The curve for *E. coli* is consistent with the DNA consisting of nonrepeated sequences, while the eukaryotic consists of a mixture of repeated and unique sequences.

complexity of DNA

The complexity of DNA (X) is related to $C_0t_{1/2}$ as follows:

$$X = k\ C_0t_{1/2\ (pure)}$$

k is a constant and has a value of 5×10^5 (units = nucleotides l mol^{-1} s^{-1}). $C_0t_{1/2(pure)}$ is the $C_0t_{1/2}$ which each class of DNA would have if it was purified and renatured separately from the other classes, as opposed to $C_0t_{1/2}$ mixture which is measured when the total genome is renatured.

$$C_0t_{1/2(pure)} = C_0t_{1/2\ (mixture)}\ GF$$

The genome fraction (GF) is the proportion of the DNA represented by each class of DNA and can be obtained from the C_0t curve by extrapolating horizontally to the vertical axis at the end of each phase, and calculating the proportion of DNA which renatured during that phase.

SAQ 4.5

Study the reassociation curve for the DNA sample shown below.

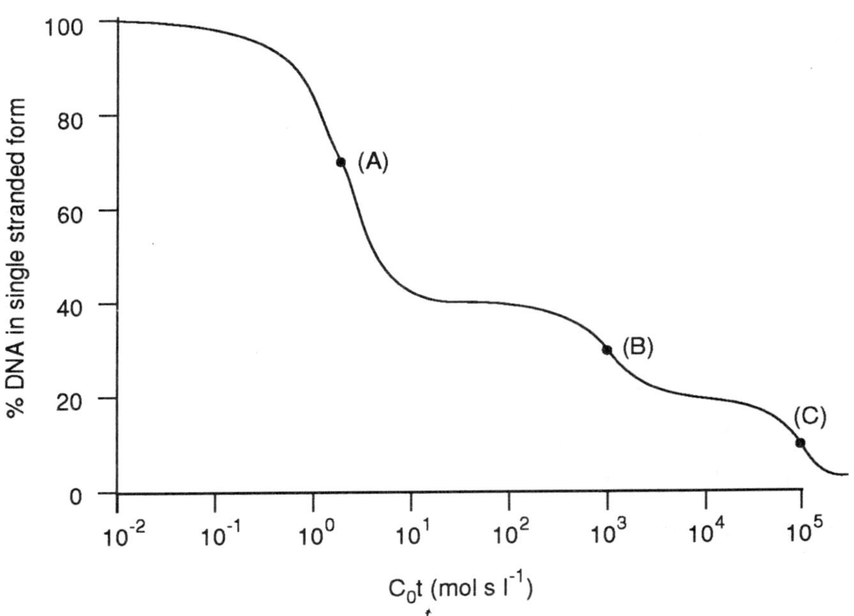

1) How many different classes of DNA are present?
2) What fraction of the total genome does each class represent?
3) What is the $C_0t_{1/2\ (mixture)}$ of each class and therefore the $C_0t_{1/2}$ pure of each class?
4) What is the complexity (X) of each class?

Highly repetitive DNA

satellites often represent repetitive DNA

Highly repetitive sequences can be separated easily from the bulk of DNA by density-gradient centrifugation in CsC1. These sequences often form 'satellites' around the main band DNA and are thus called satellite DNA. In some eukaryotes, satellite DNA may account for as much as 20% of the total DNA. Some of these repeated sequences are concentrated at the centromere, the point of attachment of sister chromatids. They may also serve as the point of attachment of spindle fibres during mitosis. Centromeric DNA is highly reiterated. Three centromeric satellites are found in the fruit fly *Drosophila virilis*: 5'-ATAAACT-3'; 5'-ACAAACT-3' and 5'-ACAAATT-3'. Centromeric sequences are often unique to a particular species and may differ even in closely related species.

repetitive sequences often associated with centromeres

Higher organisms also contain sequences not connected with the centromere but which occur in large numbers. These sequences are evenly distributed throughout the genome. In the human genome, the majority of these sequences belong to the *Alu* family. *Alu* elements are about 300 bp long and there are approximately 10^6 copies present. The function of *Alu* sequences is uncertain although they may contain origins of replication. *Alu* sequences are homologous in part to the 7SL RNA. This RNA molecule combines with six proteins to form a small RNA-protein complex (the signal recognition particle) that is required for the translocation of proteins across the membranes of the rough endoplasmic reticulum (see Chapter 8).

Alu elements are repetitive sequences

Moderately repetitive DNA

Moderately repetitive DNA sequences of less than 500 bp are present in the eukaryotic chromosome in more than 10^4 copies. Among these are sequences with homology to the tRNA genes.

SAQ 4.6

Fill in the missing words using the list of words at the bottom of the question.

The change in structure of DNA from a double to a [ ] form is called [ ]. This results in an increased [ ] at 260 nm, which is called the [ ] effect. The unwinding of the double helix is called [ ]. The [ ] is the temperature at which [ ] of the DNA is single stranded. The richer the DNA in [ ] base pairs, the higher the [ ].

If [ ] DNA is slowly cooled below its [ ] it [ ]. The product of concentration x time when 50% of the DNA has renatured is called [ ]. The [ ] of DNA is a measure of [ ] in DNA and is directly proportional to its complexity. $C_0t_{1/2}$ analysis indicates there are [ ] types of sequence in eukaryotic DNA. These are [ ] which are present in millions of copies, [ ] which are present as thousands of copies and [ ] which are present as single copies.

Use the following words: unique sequences; absorbance; melting; renatures; GC; denatured; $C_0t_{1/2}$ (two times); highly repetitive; denaturation; T_m (three times); non-repetitive; 50%; moderately repetitive; hyperchromic; three; single stranded.

4.9 Analysis of gene expression in eukaryotes

differential gene expression

Studying the renaturation of DNA by C_0t analysis provides information about the number of unique sequences in the genome. Virtually all proteins are coded for by gene sequences which are non-repetitive ie are found in the unique class of DNA sequences. It has been estimated that eukaryotes have 30-50 000 genes. Not all of these genes will be active at any one time since eukaryotic cells become differentiated by selectively suppressing some and activating other genes. Thus a liver cell will express a different set of genes and contain a different set of proteins to, for example, a kidney cell. However, some genes must be expressed in all cells. For example, genes which encode the enzymes of carbohydrate metabolism, and the structural proteins of the cytoskeleton. These genes are sometimes referred to as 'housekeeping genes'.

4.10 Estimation of the number of genes expressed in a cell

The genes expressed in a cell are derived from the non-repetitive DNA of the genome. Expression of a protein-coding gene must involve the synthesis of mRNA which is complementary to the base sequence of the gene.

Π Hybridisation of mRNA to DNA is carried out by binding single stranded DNA to a paper filter and exposing it to radioactively labelled mRNA. The mRNA will bind to complementary sequences on the DNA. In a typical experiment, the cloned DNA encoding a protein, α-fetoprotein, was hybridised to total mRNA extracted from foetal liver and adult liver. The cloned DNA hybridised to the foetal mRNA but not to the adult liver mRNA. What information does this provide about the expression of this gene?

Since the DNA only binds to mRNA extracted from foetal liver the α-fetoprotein gene is expressed only in these cells.

R_0t analysis

If the total mRNA of the cell is extracted and hybridised to the DNA then the amount of DNA bound to mRNA will be a measure of genes active in the cell. This procedure is called R_0t analysis. The results are usually expressed as a graph of DNA hybridised to mRNA against the product of the initial mRNA concentration and the time of hybridisation (Figure 4.19).

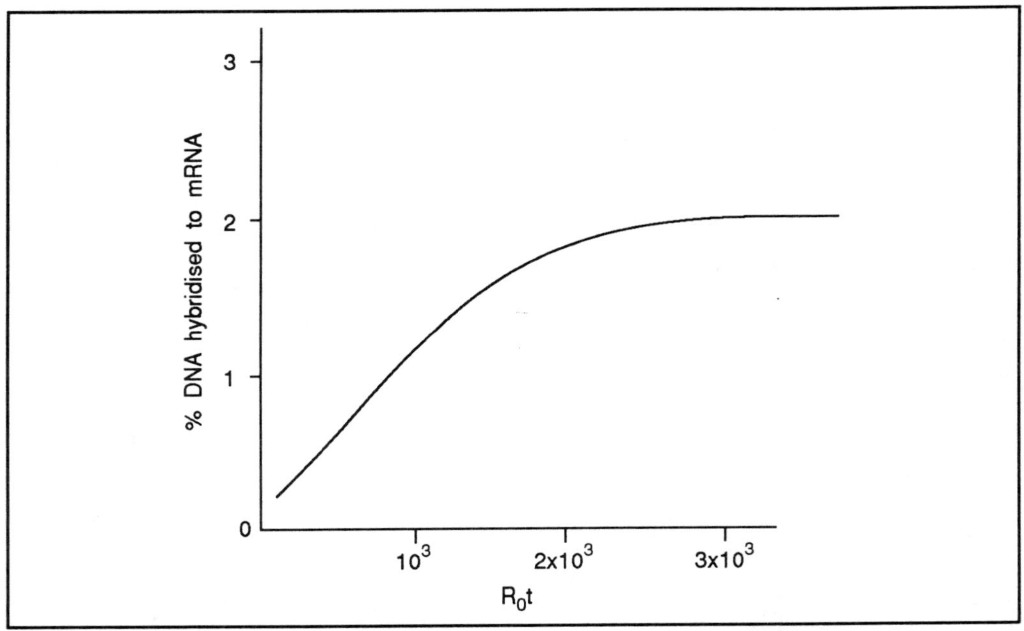

Figure 4.19 R_0t analysis. Excess mRNA from a eukaryotic cell is hybridised to DNA. Thus in the example illustrated about 2% of the genes in the single stranded DNA are being expressed, ie 4% of the double stranded form.

R_0t analysis of a range of eukaryotic cells indicates that from 7500 to 20 000 genes are usually expressed. Obviously the specific genes expressed depends on the cell type.

SAQ 4.7

In a typical R_0t analysis 2.05% of non-repetitive DNA is hybridised to mRNA. For a double stranded DNA molecule this would be equivalent to 4.1% of the DNA. If the genome contains 9.4×10^8 base pairs of non-repetitive DNA and the average length of a mRNA molecule is 2000 nucleotides, calculate the number of structural genes expressed in the cell. (Assume this is equivalent to the number of different mRNA molecules).

4.11 Do all somatic cells contain DNA in exactly the same form?

4.11.1 Euchromatin and heterochromatin

Earlier (Section 4.5) we described the existence of chromatin in two quite distinct forms. The loosely coiled form (euchromatin) stains lightly whilst the densely packed form (heterochromatin) stains rather darkly. The regions of euchromatin contains genes which are being actively expressed whilst the heterochromatin contains genes which are 'switched off'. Bearing in mind that different cell types in multicellular organisms express different genes, we must anticipate that the euchromatin regions in the different cell types will be different from each other.

polytene chromosomes and chromosome puffs

The classic experiments that showed that the loosely coiled form of chromatin was associated with transcription came from the study of the polytene chromosomes of the giant cells of *Drosophila* larval salivary glands. Incubation of these cells with ^{3}H-uridine, showed that the radioactivity became associated with the loosely packed regions of the chromosomes (so called chromosome puffs). The interpretation of these observations is that RNA synthesis (as measured by ^{3}H-uridine) took place in the loosely packed chromosomal regions.

Further studies with the same organism showed that treatment of cells with the insect hormone ecdysone lead to new chromosomal puffs being generated. These too were associated with RNA synthesis. Ecdysone regulates development in insects and controls their development. As the organism progresses through different developmental stages, new puffs arise and old puffs recede. Thus the selective unfolding of looped domains in chromosomes allows transcriptions of different genes. If, therefore, we examine the chromosomes in different cells from a multicellular organism we would find that they had different arrangements of euchromatin and heterochromatin regions.

lampbrush chromosomes

Further evidence for the unpacking of chromatin associated with active transcription comes from the study of oocytes. Microscopic studies of these reveal so called lampbrush chromosomes. These consist of tightly packed heterochromatin interspersed by regions of uncoiled DNA. We can represent these in the following way:

Again, using ^{3}H-uridine, the loosely packed regions can be shown by autoradiography to be regions of active transcriptions.

4.11.2 Gene deletion, amplification and rearrangement

gene rearrangement in invertebrates

It is generally assumed that all somatic cells retain all of their DNA. There is, however, no fundamental reason why they should. If they retain only those genes that are needed to fulfil their function then the loss of non-essential genes would be rather inconsequential. Despite this, however, it seems that most somatic cells retain all (or most) of their DNA during development and differentiation. There are, however, examples where DNA loss does occur. The evidence is strongest from lower eukaryotes (eg the worm *Ascaris* and crustacean *Cyclops*. Early in development, some cells eliminate specific heterochromatin regions. Up to 50% of the DNA may be lost).

Π Is heterochromatin eliminated from germline cells which give rise to ova or spermatozoa?

Loss of genetic material from the germline cells would lead to the production of gametes with incomplete genetic complements and subsequent development would be impaired or prevented. DNA is not eliminated from germline cells.

gene rearrangement in vertebrates

Using restriction enzyme technology, DNA rearrangements have been demonstrated in vertebrates. A good example is provided by B-cells which produce antibodies. During the development of these lymphocytes, gene segments which will give rise to different portions of the antibody molecules are rearranged relative to one another. The particular arrangement of these gene portions specifies the antibody that the cell will produce. Different B-cells carry different arrangements and therefore produce different antibodies. We can represent this process in the following way:

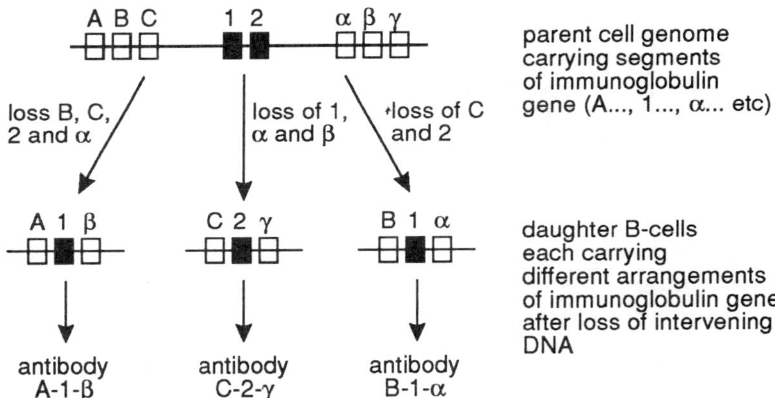

This is very much a simplification of the process, but it does illustrate that not all somatic cells contain exactly the same DNA. In this case only a very small fraction of the total genome is rearranged in the progeny B-cells. (A detailed description of the organisation and molecular biology of immunoglobulin biosynthesis is given in the BIOTOL test 'Cellular Interactions and Immunobiology').

gene amplification

In contract to the rearrangement of specific segments of DNA, in some cases specific regions may be amplified. The best documented example of this is in the developing oocytes of animals. These cells are required to produce very large amounts of proteins and therefore need many ribosomes. To achieve this, multiple copies of the rRNA genes are made. These can be demonstrated by hybridising radioactivity labelled rRNA to 'melted' DNA in oocytes and comparing the amount of hybrid formed to that produced using other types of cells. (In *Xenopus* oocytes the amplification can be as great as x1000). Other examples of specific gene amplification includes the genes which code for the proteins which form the hard coats of insect eggs.

The examples of gene deletion, rearrangement and amplification described above indicates that not all somatic cells carry exactly the same genetic complement. However, such irreversible changes in the genome appears to be limited in higher plants and animals and is largely confined to particular cell types or to particular developmental stages.

Π Many somatic cells from plants can be cultured as single cells *in vitro*. Using suitable hormonal and nutritional regimes, these individual cells may each produce whole plants. What does this tell you about the genetic complement of these somatic cells?

The conclusion must be that the somatic cells contain all of the genes that are necessary for the production of all of the organs of the plants and to carry out all of their metabolic and physiological functions. In other words, there has been no large scale (if any) deletion of genes in these somatic cells.

4.12 Mapping the arrangement and fine structure of genes

In Chapter 2, we described how the analysis of the phenotypes arising from conventional genetic experiments enabled us to identify linkage groups and to map genes within the genome. Before the advent of restriction enzyme technology, the genetic maps obtained for eukaryotes were rather coarse. Only in the case of the major RNA species (ribosomal and transfer RNA) did we have details of the arrangement of genes at, what we might call, a refined level. The cellular accumulation of these RNA species meant that we could isolate them in sufficient quantities to use them as 'probes' to map these genes by hybridisation. Using this technique, it was established that the major rRNA species were synthesised by genes arranged in tandem along the genome at the nuclear organiser sites. We can represent this in the following way:

tandem gene arrangement

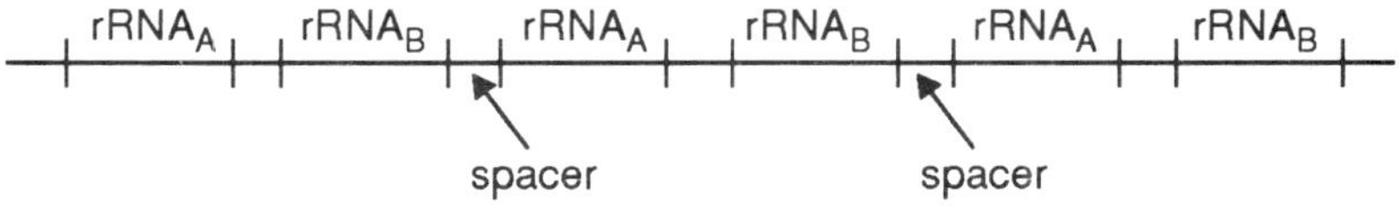

$rRNA_A$ $rRNA_B$ $rRNA_A$ $rRNA_B$ $rRNA_A$ $rRNA_B$

spacer spacer

These studies also showed that rRNAs were produced as long transcripts and subsequently 'tailored' to produce mature rRNAs. As you will learn in later chapters, this is a great simplification of these events and the tailoring of transcription products is a common feature of gene transcription in eukaryotes.

The advent of restriction enzyme technology and of gene cloning has, however, led to tremendous advances in our understanding of gene organisation and structure in eukaryotes. Much of the information in the remainder of this text has been derived from these techniques, so it is worthwhile at this stage considering their use in mapping and identifying genes. It is, however, beyond the scope of this text to examine the techniques used in detail. Detailed descriptions of these techniques are given in the BIOTOL text 'Techniques for Engineering Genes'.

4.12.1 Restriction endonucleases - a reminder

In Chapter 3 (see Section 3.11) we introduced the enzymes, restriction endonucleases since they are of use in mapping mutations. They are also useful for detecting mutations. In this chapter, we have been concerned with the organisation of eukaryotic genomes within the nucleus. Restriction enzymes have an important uses in mapping genes.

Here, we will examine the application of restriction endonucleases in gene mapping. We have reiterated some of the principle features of these enzymes. This should provide a useful brief revision.

We remind you that the restriction endonuclease are enzymes which recognise specific base sequences in DNA and hydrolyse one or both strands of the DNA molecule. The enzymes are found only in prokaryotes and their natural function is to degrade 'foreign' DNA which enters the bacterium and may be incorporated into the host DNA. The host DNA is not degraded by the endonuclease because the recognition sequences are also recognised by a corresponding methylase which adds a methyl group in a specific manner. Thus the site is now protected from hydrolysis by the restriction enzyme.

naming of restriction enzymes

Restriction enzymes are named as follows. The first three letters of the name denotes the bacterium from which the enzyme was extracted, *Eco* for *Escherichia coli*. A further letter may denote a strain for example *Eco*K or *Eco*R for the K and R strains respectively. If more than one enzyme is found in the organism the enzymes are designated by the addition of a Roman numeral. *Hind*III is one of the three enzymes from *Haemophillus influenzae* strain d.

Π Let us assume that an *Escherichia coli* strain R produces two restriction enzymes. What would they be called?

You should have deduced that they would be given the names *Eco*RI and *Eco*RII.

There are three distinct groups of restriction enzymes of which the type II are the best understood.

Type II restriction endonuclease

restriction enzymes recognise palindromes

Type II restriction endonuclease have two subunits which bind to complementary strands of DNA. They recognise sequences of from four to eight bases, which are palindromes and show 180° rotational symmetry. The product of the cleavage may have short single stranded sections as in the product of *Eco*RI or blunt ends as occurs with *Hind*II. In the latter case there is no single stranded section.

Π Do any of the other restriction enzymes listed in Table 4.1 produce blunt ended DNA when they cleave double stranded DNA.

You should have spotted that *Hae*III, *Hind*II and *Xba*1 also produce blunt ended hydrolysis products.

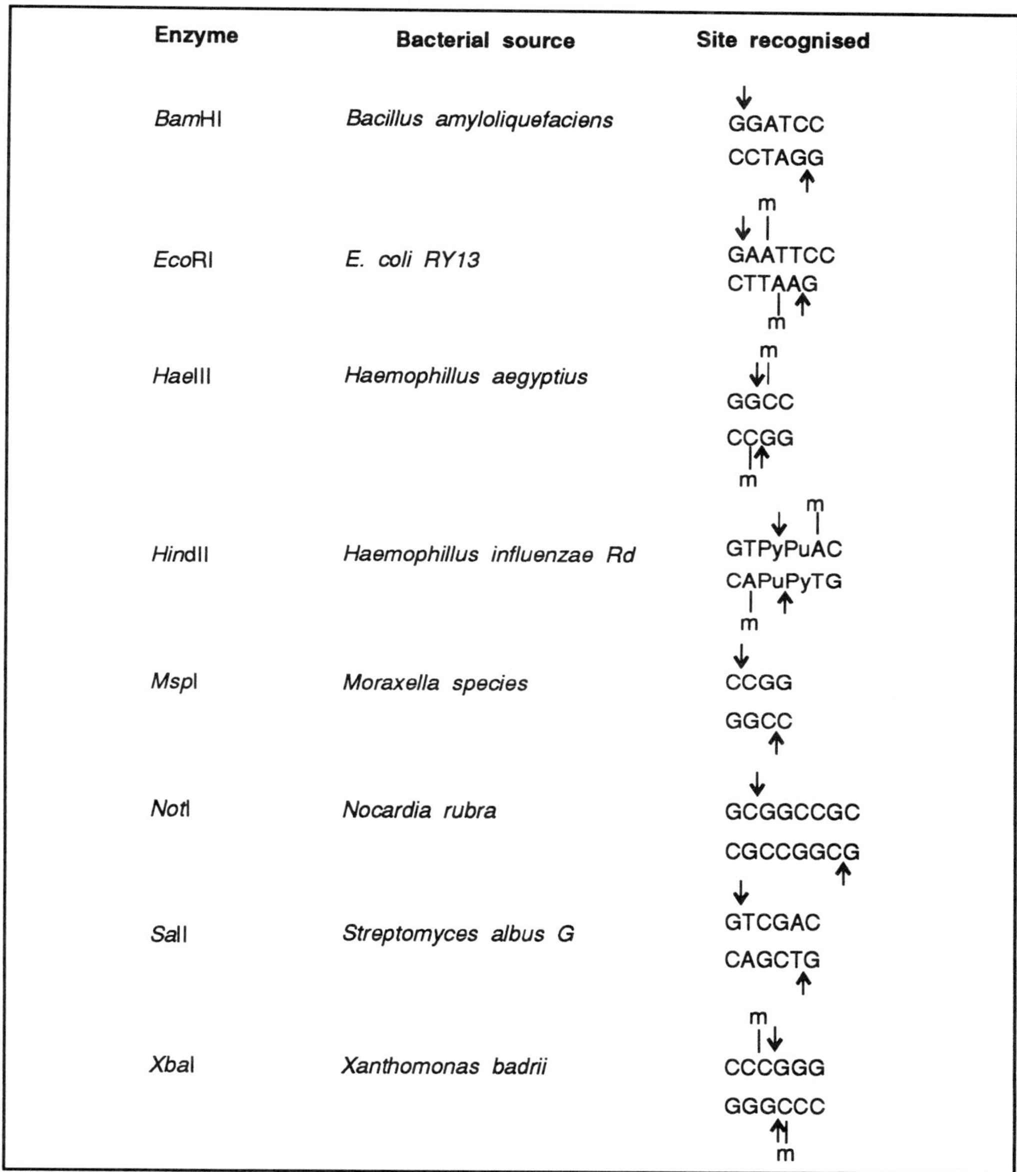

Enzyme	Bacterial source	Site recognised
*Bam*HI	*Bacillus amyloliquefaciens*	GGATCC CCTAGG
*Eco*RI	*E. coli RY13*	GAATTCC CTTAAG
*Hae*III	*Haemophillus aegyptius*	GGCC CCGG
*Hind*II	*Haemophillus influenzae Rd*	GTPyPuAC CAPuPyTG
*Msp*I	*Moraxella species*	CCGG GGCC
*Not*I	*Nocardia rubra*	GCGGCCGC CGCCGGCG
*Sal*I	*Streptomyces albus G*	GTCGAC CAGCTG
*Xba*I	*Xanthomonas badrii*	CCCGGG GGGCCC

Table 4.1 Site of action of some type II restriction endonucleases.
m indicates a methylated base, Pu, Purine; Py, Pyrimidine bases. The site of cleavage is indicated by an arrow.

4.12.2 The principles of using restriction enzymes for gene mapping

Type II endonuclease cut DNA at defined sequences. In any DNA molecule, the sequences recognised by an endonuclease are randomly scattered and occur infrequently. By using several enzymes and in different combinations, it is possible to map the sites of restriction sequences for each enzyme. This process is called restriction mapping. Restriction mapping is possible because gel electrophoresis may be used to separate the fragments of DNA of different sizes resulting from the digestion.

Figure 4.20 shows an example of restriction mapping. If a DNA molecule is treated with restriction endonuclease I (RE I) a large DNA fragment A is produced. If fragment A is treated with enzyme II (RE II) two further sequences B and D are generated. If enzyme II is used first then fragment C is generated. Exposure of C to enzyme I will generate D and E. Electrophoresis may be performed in the presence of DNA molecules of known base length to determine the sizes of the fragments A to E. This procedure may be used to determine the order of gene loci on the chromosome. The nucleotide sequence of fragments A to E and of genes located in the fragments can be determined. We will not go into details of this here but, if you wish to follow this up, we recommend the BIOTOL texts 'Genome Management in Prokaryotes' or 'Analysis of Amino Acids, Proteins and Nucleic Acids'. The sequences of nucleotides in a gene can be translated into the amino acid sequence of a polypeptide using the genetic code. Nucleotide sequences corresponding to genes can be used to make an RNA or DNA copy using a mixture of nucleotide triphosphates and the appropriate enzyme. If the nucleotides are labelled with ^{32}P then the RNA or DNA will also be labelled and can be used as a 'probe' to search for the gene in DNA molecule. The use of such probes will be described later in this text.

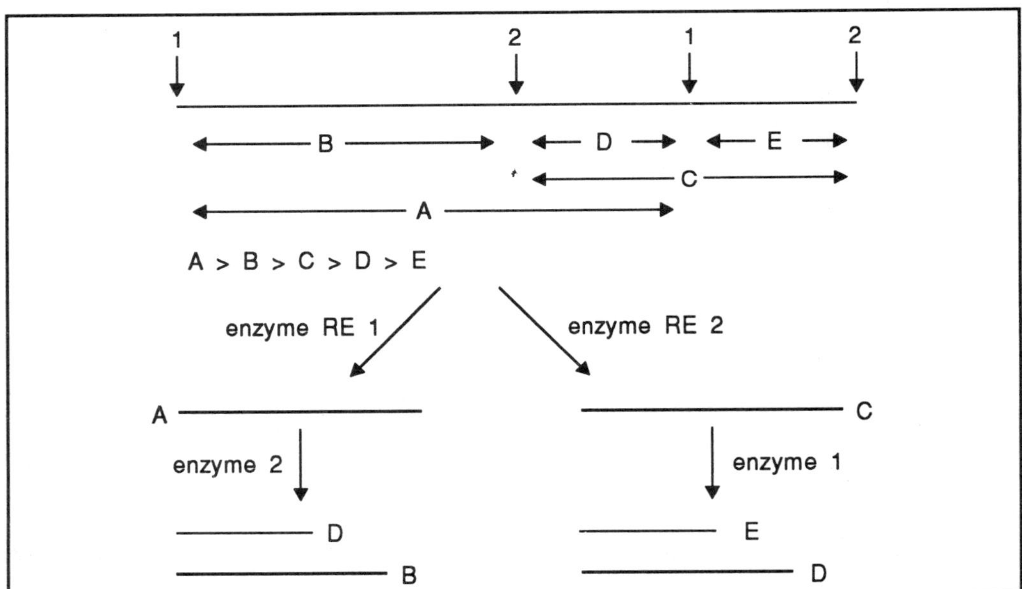

Figure 4.20 The mapping of restriction endonuclease sites using two endonucleases I and II. The DNA fragments produced from the enzymic digestion may then be separated by gel electrophoresis.

4.12.3 Chromosome walking and other techniques for mapping genes

By using restriction enzyme, we can cut DNA into fragments and clone these fragments into suitable vectors. Molecular biologists have a battery of techniques that enable them to identify those parts of the cloned DNA fragments which encode the gene. Here we will briefly examine some of these.

Nuclease protection

In this technique, a mRNA (or a piece of DNA complementary to this mRNA = cDNA) is hybridised to 'melted' cloned fragment. It is then treated with a nuclease which hydrolyses single stranded DNA.

Thus:

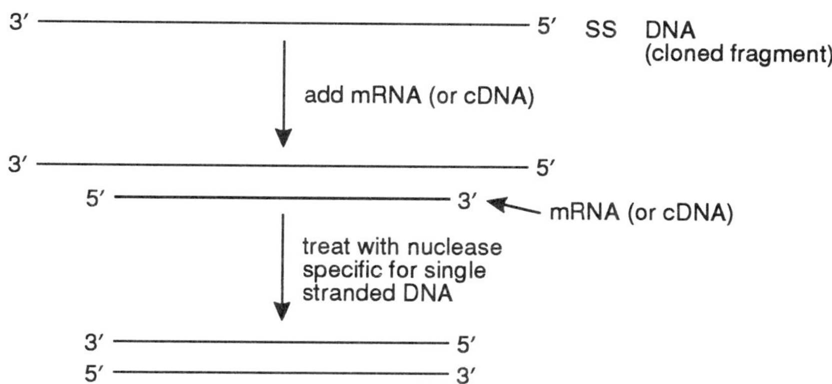

In this way, we are able to identify the exact position of the gene within the DNA fragment. It also enables us to identify the nucleotide sequences at each end of the gene and to explore whether or not this contains regulatory sequences.

Chromosome walking

Chromosome walking is a technique which enables us to 'walk' along a genome. Here we will describe the general principles.

Let us assume we have isolated the mRNA from gene A and have prepared DNA copies (cDNA) of this to use as a probe.

We prepare fragments of a genome using a restriction enzyme and clone these in a suitable vector. We then identify those clones which will hybridise with the gene A probe. We subsequently isolate the clones which have hybridised with the gene A probe and then sub-clone the sequences at the downstream end of the clone and use these to re-probe the library. Suppose that the sequence of the clone is actually in a second gene.

Thus:

gene A gene B

new probe

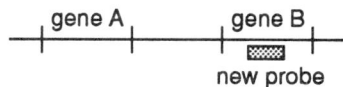

If we identify any clones which hybridise with probe B, we would expect that these clones may have the structure.

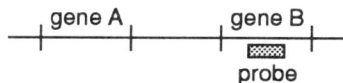

gene A gene B

probe

Other clones may have the structure.

gene B gene C

probe

In other words, we have now identified overlapping fragments of DNA and have managed to 'walk' from gene A to gene C.

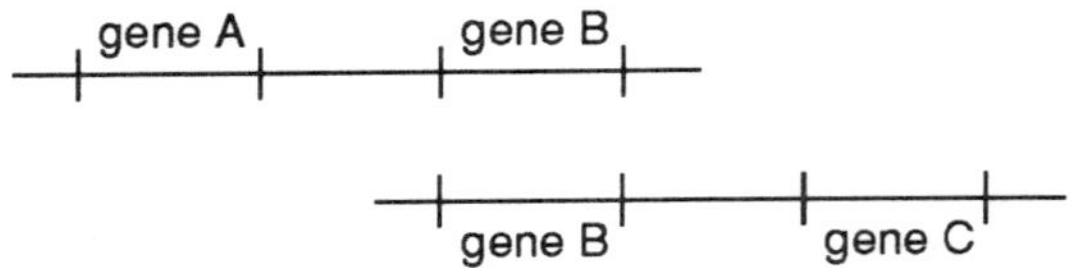

This is very much an oversimplification of the processes involved, but at least you will understand the principles involved. In practice, the size of the 'steps' that can be taken in 'walking' along the chromosome depends on the sizes of fragments generated. This in turn is governed by the restriction enzymes used to cut the DNA and the type of vector used to clone the DNA fragments.

Footprinting

Footprinting is a technique used to identify regulatory sequences. Again, cloned DNA fragments can be used. If regulatory proteins are added to this cloned DNA, they bind with their regulatory sites on the DNA. If this DNA is then incubated protein complex with nucleases, the exposed DNA is hydrolysed. The DNA in the complex is protected from hydrolysis. Thus:

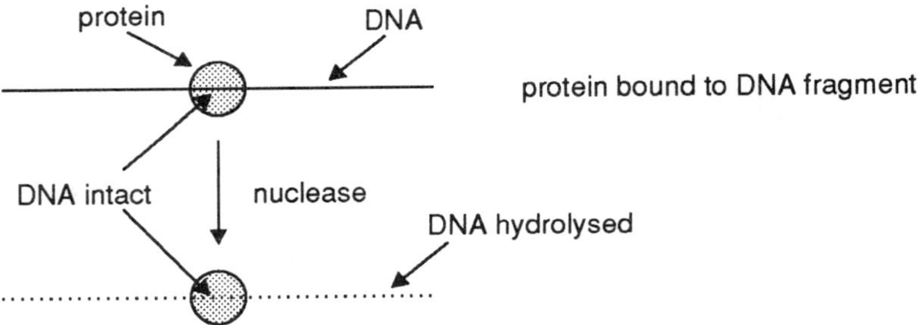

The DNA in the complex can be recovered and analysed. In this way DNA sequences recognised by regulatory proteins can be identified.

4.12.4 Genome mapping using restriction enzymes and gene probes

Classical genetic techniques of recombination, crossing over and deletion have been used to locate genes on specific chromosomes. These techniques produce a genetic map but do not give information on the base sequence of the gene or its physical position on the chromosome. This latter information constitutes the physical map of the chromosome. The ultimate aim of genome mapping is to match the genetic and physical maps. A number of methods have been used to achieve this. In *in situ* hybridisation the DNA in mitotic chromosomes is partially denatured and radiolabelled, complementary mRNA or DNA is allowed to bind to the DNA. Probes (ie the labelled DNA or RNA) complementary to the base sequence of a gene can thus be used to locate genes on chromosomes (Table 4.2). However, a more detailed physical map of the genome can be obtained using restriction enzymes.

locating genes on chromosomes

Gene	*Chromosomal location
Collagen, type 1 and 2	2p12
α-Interferon	7q12
β-Globin	11p15
Insulin	11p15
Growth hormone	17q12
Immunoglobulin κ light chain	2p12
Immunoglobulin λ light chain	22

Table 4.2 Some human genes which have been located by *in situ* hybridisation.
*The first number indicates the chromosome, p and q are the short and long arms respectively of the chromosome. The second number refers to the chromosome band.

mapping eukaryotic genomes is difficult

The genome of *E. coli* contains 4.5 x 10^6 basepairs. If the bacterial DNA is treated with the restriction enzyme *Not*1 which cleaves at a rare sequence, 21 fragments are produced. The position of these fragments on the *E. coli* genome may be determined by one of several methods. For example, using previously cloned genes as probes and by using known deletions. Mapping the complex map of eukaryotes is more difficult but similar techniques can be used. The large size of most eukaryotic genomes means too many fragments are produced using restriction enzymes to be resolvable by current techniques. The small genome of the nematode *Caenorhabditis elegans* has been analysed by treating with the enzyme *Not*1 and cloning the extremely large fragments into a suitable vector (usually yeast artificial chromosomes). Large amounts of the fragment are generated and these can be further sequenced using a battery of restriction enzymes. Overlap between the larger fragments has been predicted by computer programmes. About 95% of the *C. elegans* genome has been covered by overlapping fragments. This process of mapping to the human genome is being conducted in many laboratories.

4.12.5 Restriction fragment length polymorphisms and gene identification

Generally the two chromosomes of a pair will show the same restriction pattern when treated with a restriction enzyme. However, if one chromosome acquires a random base-pair change it may destroy or introduce a new restriction site. This difference will be inherited in a Mendelian fashion and is called a polymorphism. These polymorphisms may be observed as a difference in the length of fragments after treatment with endonuclease and are called restriction fragment length polymorphisms (RFLPs). As we learnt in Chapter 3 in sickle cell disease, the gene for β-globin has acquired a single mutation (A to T). This mutation abolishes a recognition site for the endonuclease *Dde*I (Figure 4.21). Normal DNA produces two bands while the mutated DNA shows a single band following digestion with *Dde*I. This banding pattern can be used in diagnosis of sickle cell disease using foetal tissue.

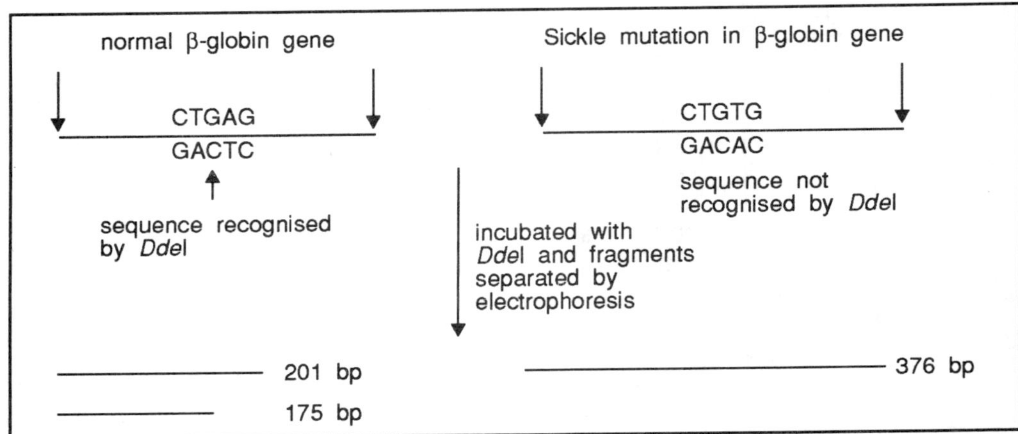

Figure 4.21 The result of treating DNA containing a 'normal' β-globin gene with the endonuclease *Dde*I results in the generation of two fragments of 201 and 175 bp respectively. If there has been an A to T transition, then one of the recognition sites is abolished and only a single band of 376 bp results. This is the case in sickle cell disease.

RFLPs provide an enormous number of markers for genetic diseases (Table 4.3). The RFLP may be cloned, radioactively labelled and used to find a complementary sequence on the chromosome.

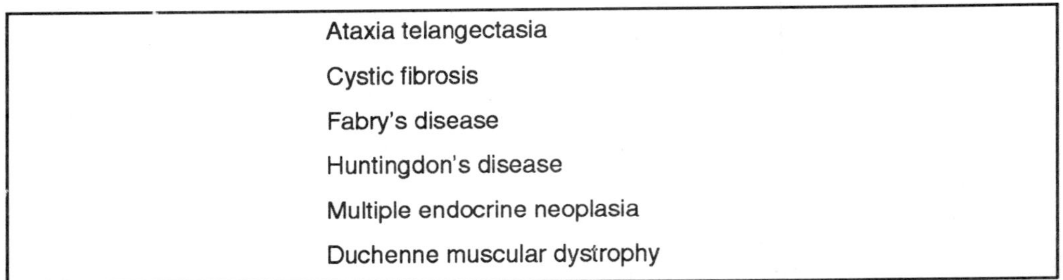

Diseases
Ataxia telangectasia
Cystic fibrosis
Fabry's disease
Huntingdon's disease
Multiple endocrine neoplasia
Duchenne muscular dystrophy

Table 4.3 Examples of some diseases which may be diagnosed by the use of RFLPs.

Cystic fibrosis (CF) is a common genetic disorder affecting 1 in 1600 of the population. One in 20 of the population are carriers. The disease is caused by a mutation in a gene on chromosome 7 which controls $C1^-$ movement across epithelial cells. The location of the mutated gene can be used as an illustration of the methods used to identify genes. A major advance in the identification of the CF gene has followed the use of random markers (including RFLPs) assigned to each autosome to look for genetic linkage to the disease.

Treatment of DNA from cystic fibrosis patients with the endonuclease *Taq*I produces DNA fragments of 7.0 and 4.0 kb. The 7.0 kb fragment appears to be a marker for the disease in some families. Patients who are homozygous for the marker suffer from the disease. Carriers are heterozygous. Figure 4.22 shows a gel electrophoresis pattern obtained from one family and used in the diagnosis of the disease cystic fibrosis.

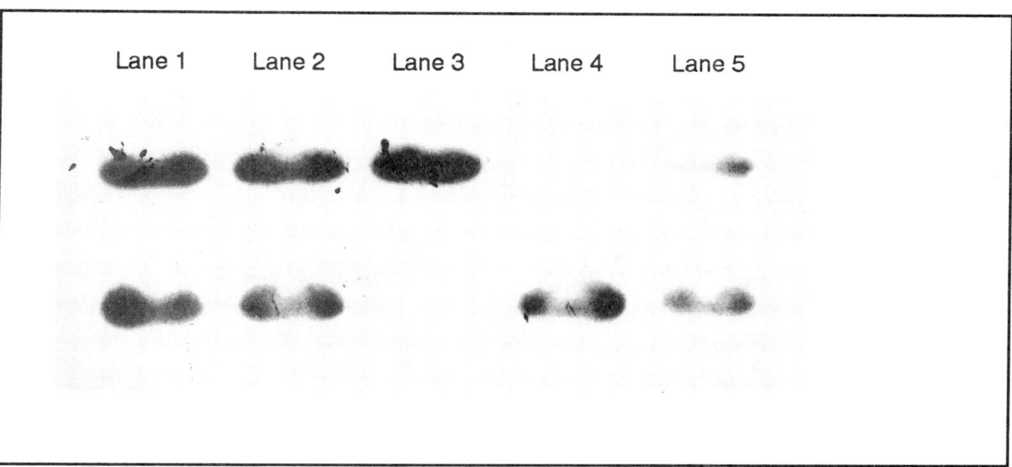

Figure 4.22 DNA analysis of a family suffering from cystic fibrosis. A RFLP of 7 kb generated by treating the DNA with the restriction endonuclease *Taq* I is indicative of the presence of the disease. The DNA fragments are separated by gel electrophoresis. Lane 3 is a sample from a child suffering from cystic fibrosis and homozygous for the 7 kb marker. Lanes 1 and 2 are samples from the mother and father respectively, who are both heterozygous. Lane 4 is another sibling from the same family but who is not a carrier and Lane 5 a further child who is a carrier. Photograph provided by Mr H. Hughes, Department of Biological Sciences, Manchester Metropolitan University.

SAQ 4.8

Identify which of the following are true and which are false.

1) Restriction endonuclease are enzymes which catalyse the hydrolysis of RNA at palindromes.

2) Restriction endonuclease are enzymes which catalyse the hydrolysis of methylated DNA.

3) The enzyme *Eco*R 1 catalyses the hydrolysis of DNA at the sites indicated:

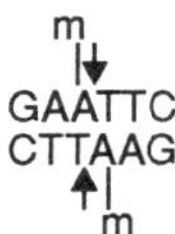

Summary and objectives

Each chromosome in a eukaryotic cell contains a single molecule of DNA. The chromosomes are contained within, and protected by, a double membrane-bound structure called the nuclear envelope.

Communications between the contents of the nucleus and the cytoplasm is facilitated by the nuclear pores which puncture the nuclear envelope. The structure of the nucleus is maintained by a system of fibrous proteins which form the nuclear skeleton. The DNA molecules within the chromosomes are elaborately folded and condensed with proteins to shorten their length and allow them to be accommodated within the limited space of the nucleus.

DNA may be denatured by heating. However, when slowly cooled it renatures. Studies of this renaturation have shown that eukaryotic DNA contains unique, moderately repetitive and highly repetitive sequences. Similar studies have been used to estimate the number of active genes in any one type of eukaryotic cell as only 5-10% of the total present.

Restriction endonucleases are able to recognise short palindrome sequences and catalyse the hydrolysis of DNA in a highly specific manner. When combined with denaturation and renaturation methods, these properties make restriction endonuclease valuable tools for investigating the distribution of genes along DNA molecules, allowing genetic maps to be formulated, and for the diagnose a number of genetic diseases.

Now that you have completed this chapter, you should be able to:

- estimate the relative volumes of cells, nuclei and cytoplasm from supplied data;
- describe the basic structure of the nuclear envelope and the pores within this envelope;
- describe the role of lamins in nuclei;
- explain how DNA is packaged in eukaryotic cells and calculate the length of mitotic chromosomes from supplied data concerning the dimensions of DNA and packing ratio;
- describe the processes of DNA denaturation and renaturation and calculate the base composition of DNA from data relating to DNA melting;
- identify and calculate the complexity of different classes of DNA found in eukaryotic genomes from renaturation kinetics;
- calculate the number of genes being expressed in a cell from R_0t analysis;
- give examples which illustrate that part of the DNA of eukaryotes may be rearranged, deleted or amplified during specific phases of development;
- explain how restriction enzymes and restriction fragment length polymorphism enables us to physically map genomes;
- explain with suitable examples, how restriction fragment length polymorphism enables us to diagnose particular diseases.

Eukaryotic transcription units

Eukaryotic transcription units

5.1 Introduction

In this chapter we will look at the ways in which DNA stored in the eukaryotic cell nucleus is copied into the different classes of RNA molecules which are found in the cell cytoplasm. You should already be familiar with the basic processes of RNA synthesis and the regulation of prokaryotic gene expression. The following questions should serve as a brief revision checklist before we carry on.

SAQ 5.1

Fill in the spaces with the most appropriate word from the list supplied at the bottom.

1) The process of synthesising a single stranded RNA molecule using a DNA template is called [ ].

2) The enzyme involved in RNA synthesis is called [ ].

3) Unlike DNA synthesis, the initiation of RNA synthesis does not require a [ ]. ie transcription may start 'from scratch' at an internal position within the DNA molecule.

4) Particular RNA products are normally of a fixed, discrete size, eg all the *E. coli* 16S [ ] molecules in a cell are the same number of base pairs long.

5) RNA synthesis is not a random copying process, ie there are specific start and stop sites for transcription, called [ ] and [ ] sites, respectively.

6) The regions of DNA between transcription start and stop sites are called [ ].

Word list
attenuation, gene, inducer, initiation, operator, polyribonucleotide synthetase, primer, promoter, RNA polymerase, mRNA, rRNA, tRNA, ribosome, template, termination, transcription, transcription units, translation.

transcription unit

One important idea to grasp is the difference between a gene and a transcription unit. You may already be familiar with this point from the study of transcription units such as the *lac* operon in prokaryotes. The transcription product of the *lac* operon is a single polycistronic mRNA which encodes 3 different gene products. We will see in eukaryotes that although mRNAs are usually monocistronic there is also not necessarily a close identity between a gene and a transcription unit. Indeed, in eukaryotes it is often difficult to directly relate a gene, as defined by classical genetic techniques, to a clearly defined region of a DNA molecule.

5.2 The major classes of RNA molecules

You should be familiar with the major classes of RNA molecule and their roles in gene expression.

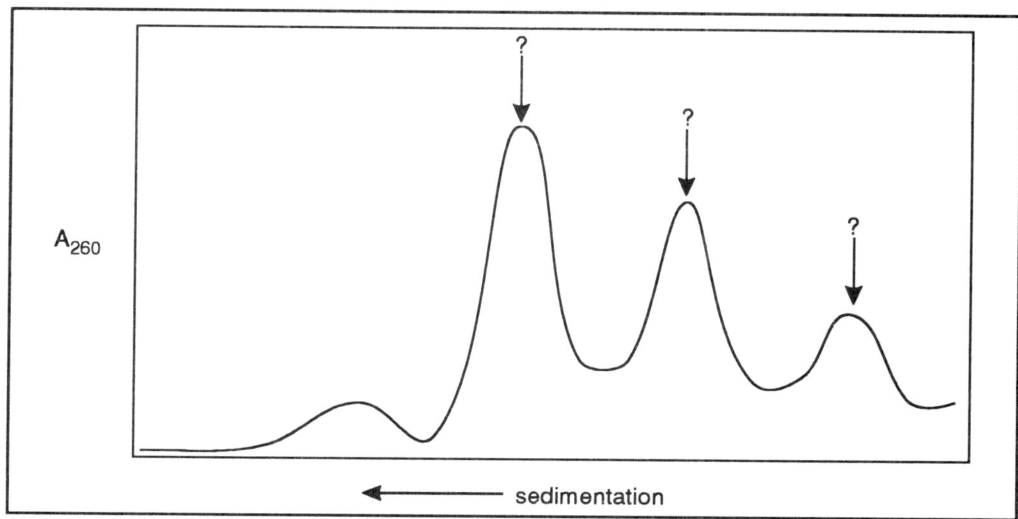

Figure 5.1 Velocity sedimentation of total cellular RNA through a sucrose density gradient.

Π Figure 5.1 shows the separation of eukaryotic cell RNA by sucrose density gradient centrifugation, one of the techniques first used to separate RNA molecules. The RNA is detected by UV absorbance at 260nm. From your knowledge of prokaryotic RNA identify each peak in the gradient. Table 5.1 should help you to work out the answer.

Svedberg units

S value

Table 5.1 shows the sizes of the major classes of both prokaryotic and eukaryotic RNAs, given in Svedberg units. (The S value is a measure of sedimentation rate during density gradient centrifugation). The table also shows the proportions of these classes of RNA in the cell. The three peaks in Figure 5.1, from left to right, are therefore 28S, 18S and (4S +5S) RNA.

Class	Size (Svedberg units)		Steady state amount %
	Prokaryotes	Eukaryotes	
rRNA	23S	28S	
	16S	18S	75
	5S	5S	
tRNA	4S	4S	15
mRNA	variable	variable	< 5

Table 5.1 Classes of RNA and their sizes from typical prokaryotes and eukaryotes. The approximate proportions of each found in cells are also reported.

Π From the table, why would you expect the size of mRNAs to be heterogenous, whereas rRNAs and tRNAs fall into discrete size classes? Also consider the proportions of the RNAs - can you explain why mRNA should be such a small proportion of total cellular RNA?

The mRNA population will be heterogeneous because different mRNAs code for different proteins of different sizes, and will be transcribed from different sized transcription units. However, within the mRNA total population there may be several copies of the same mRNA molecule which will each have an identical length. The number of copies of an mRNA species is referred to as its abundance.

abundance

The different proportions, by weight, of the different RNAs reflects their different roles in gene expression. At any one time, a single mRNA molecule may be translated by several ribosomes, requiring many tRNA molecules in the process. Notice that a small proportion of total cell RNA is unaccounted for. The remaining RNA includes hnRNA (heterogeneous nuclear RNA), snRNA (small nuclear RNA) and scRNA (small cytoplasmic RNA). The functions of some of these RNAs will be discussed in this and later chapters.

hnRNA

snRNA

scRNA

The distinction between the different classes of RNA is made in the eukaryotic cell even at the level of transcription, as we shall see in the next sections when we look at transcription units and the RNA polymerases which transcribe them.

RNA polymerases

5.3 Eukaryote RNA polymerases

Unlike prokaryotes, eukaryotes contain 3 different RNA polymerases which are functionally and structurally distinct. The major difference between them relates to the class of RNA which they synthesise. Thus RNA polymerase I synthesises 28S and 18S rRNA, RNA polymerase II produces mRNA and RNA polymerase III copies the 5S rRNA and tRNA genes.

RNA polymerase I, II and III

Π Where in the nucleus would you expect these RNA polymerases to be active?

RNA polymerase I is active in the nucleolus, the site of ribosomal RNA synthesis. The other two polymerases are active in the nucleoplasm. The three RNA polymerases can be separated by conventional protein purification methods, particularly ion exchange chromatography. They are all extremely large multisubunit enzymes which share some subunits in common. Table 5.2 shows the subunit structure of the three RNA polymerases from yeast.

nucleolus

nucleoplasm

Π Which subunits are common to all these polymerases?

The subunits of 27, 23 and 14.5 kDa are identical in all three polymerases. The two large subunits in each enzyme are also similar and bear some resemblance to the β and β' subunits of *E. coli* RNA polymerase.

RNA polymerase I (kDa)	RNA polymerase II (kDa)	RNA polymerase III (kDa)
190	220	160
135	150	128
27	27	27
23	23	23
14.5	14.5	14.5
40	44.5	40
19	12.6	19
		82

Table 5.2 Subunits of yeast RNA polymerases. Subunits are described in terms of their molecular masses (kDa).

α-amanitin

actinomycin D

Apart from their different structures and functions, the three RNA polymerase activities can be distinguished by their different sensitivities to certain antibiotics. For example, RNA polymerase II is specifically inhibited by the fungal toxin α-amanitin, whereas transcription by RNA polymerase I is much more sensitive to actinomycin D than is transcription by the other two enzymes. Table 5.3 summarises these details of eukaryotic RNA polymerase properties.

	RNA polymerase I	RNA polymerase II	RNA polymerase III
Product	28S + 18S rRNA	mRNA some snRNAs	tRNA 5S rRNA some sRNAs
Location	Nucleolus	Nucleoplasm	Nucleoplasm
α-amanitin inhibition of enzyme activity	Not inhibited	Inhibited by low concentrations	Variable effect in different organisms
actinomycin D inhibition of transcription	Inhibited by low concentration	Inhibited by high concentration	Inhibited by high concentration

Table 5.3 Major properties of eukaryotic RNA polymerases.

[^{3}H]uridine

RNA synthesis in eukaryotic cells can be monitored by measuring the incorporation of radioactive uridine into RNA. (NB. Why would uridine be a better choice than one of the other three nucleosides found in RNA? Think about the other macromolecules into which radioactive nucleosides might be incorporated.) After a short treatment (about 5 minutes) of mammalian cells in culture with [^{3}H]uridine, incorporated radioactivity can be detected in RNA isolated from the cells. It is possible to analyse this labelled RNA in a number of different ways. The size of the RNA can be determined by, for example, sucrose gradient centrifugation as shown in Figure 5.1. The location of the RNA can be determined by isolating RNA from fractionated cells. Furthermore, the kinetics of RNA synthesis and accumulation can be examined by using different labelling times and employing various 'pulse-chase' techniques which will be described later. Figure 5.2 shows the sedimentation of cytoplasmic RNA through a sucrose gradient after labelling of cells for 2 hours with [^{3}H]uridine. The radioactivity (ie newly synthesised RNA) and the UV absorbance (total RNA) are shown.

Π What would you expect the gradient profiles to look like if the cells were treated with a low amount (0.04μg ml^{-1}) of actinomycin D?

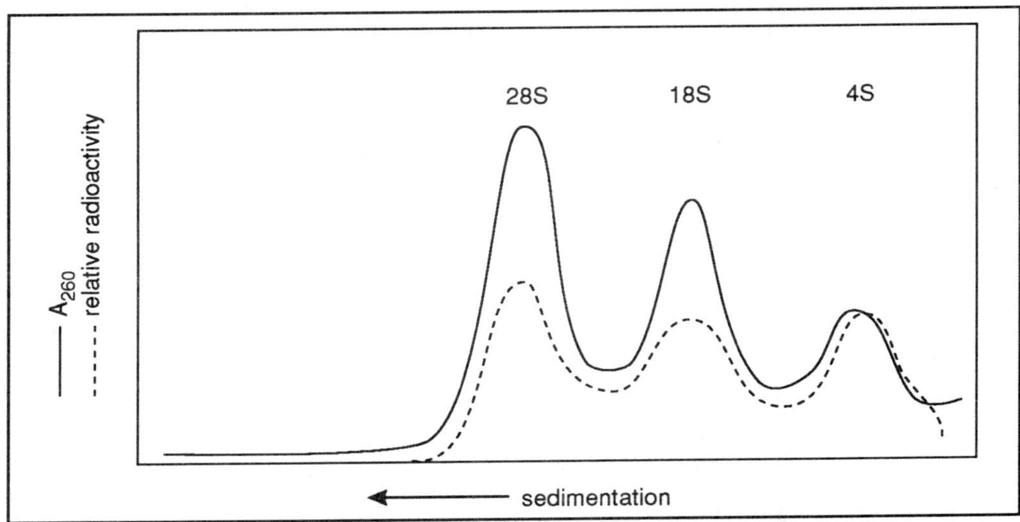

Figure 5.2 Sedimentation of cytoplasmic RNA radioactivity labelled with [^{3}H]uridine for 2 hours *in vivo.*

Actinomycin D will inhibit RNA polymerase I transcription and therefore radioactivity will not be incorporated into 28S and 18S rRNA. 5S rRNA and tRNA, and also mRNA, will still synthesised. The gradient profile will look like Figure 5.3, with the 28S and 18S rRNA peaks missing from the profile of radioactivity, though not from the total RNA profile.

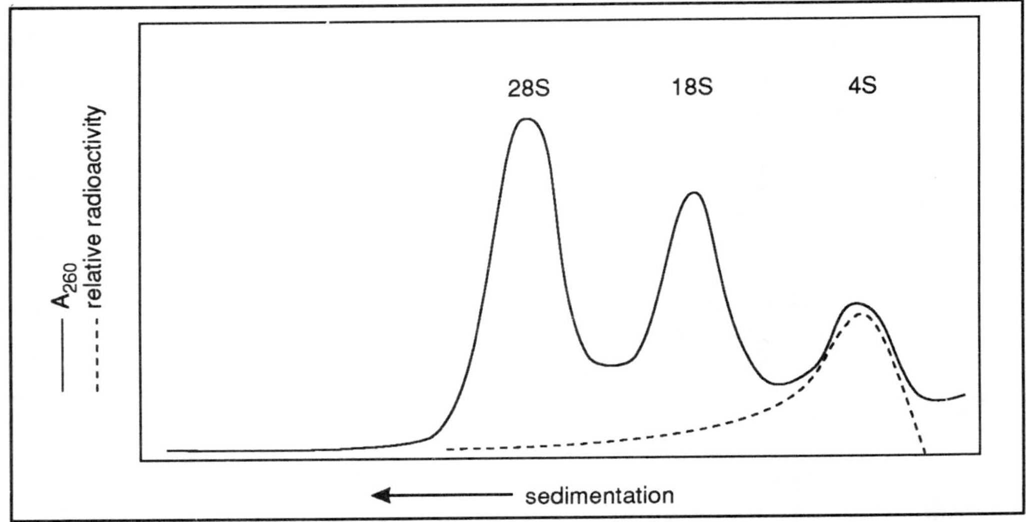

Figure 5.3 Sedimentation of radioactivity labelled cytoplasmic RNA isolated from cells treated with 0.04μg ml^{-1} actinomycin D.

5.4 Transcription units

Since we have compared the three RNA polymerases principally in terms of the transcription units which they copy, it is logical to move on to describe the three major types of transcription units in eukaryotes.

5.4.1 Ribosomal RNA transcription units

Having explored one of the basic techniques in some of the early work on transcription units this is a useful place to examine in more detail the types of results which lead to our understanding of the ribosomal RNA transcription unit. When RNA is labelled for only 5 minutes with $[^3H]$uridine, very little radioactivity is found in cytoplasmic RNA other than 5S rRNA and tRNA (Figure 5.4a).

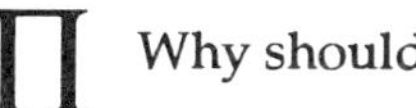

Why should this be the case?

Consider the relative sizes of the RNA classes - it takes much longer to synthesise, process and transport the larger RNA classes. After 60 minutes labelling, the labelled RNA profile resembles the total RNA much more. However, the labelled RNA in the nucleus is also shown in Figure 5.4a. If the nucleus is divided into nucleolar and nucleoplasmic fractions, it can be seen that a specific peak of 45S is found in the nucleolar RNA (Figure 5.4b), whereas the nucleoplasmic RNA is extremely heterogenous, with sizes from much greater than 45S down to about 10S (Figure 5.4b). This rapidly labelled nucleoplasmic RNA is called heterogeneous nuclear RNA (hnRNA).

The more immediate question is why does the nucleolus contain RNA of 45S size, but not 28S or 18S RNA? When cells are labelled for longer periods, the radioactivity in this peak increases and a second peak of 32S RNA also appears. One explanation for this is that the 45S RNA is a precursor of 32S RNA which in turn is a precursor of 28S and 18S rRNA.

How can this be proved?

pulse, chase experiments

One way is to employ a 'pulse-chase' techniques by labelling cells for a short time (say 15 minutes) and then adding actinomycin D to inhibit further rRNA synthesis. The 'pulse' of radioactivity is then 'chased' into its final destination. Figure 5.4c shows the nucleolar RNA labelled for 15 minutes with prominent peaks at 45S, 32S and 20S. After a 60 minute 'chase', much of this radioactive RNA has disappeared from the nucleolus (Figure 5.4d). However, it is now possible to detect radioactively labelled 28S and 18S rRNA in the cytoplasm which was not present at the start of the 'chase' period.

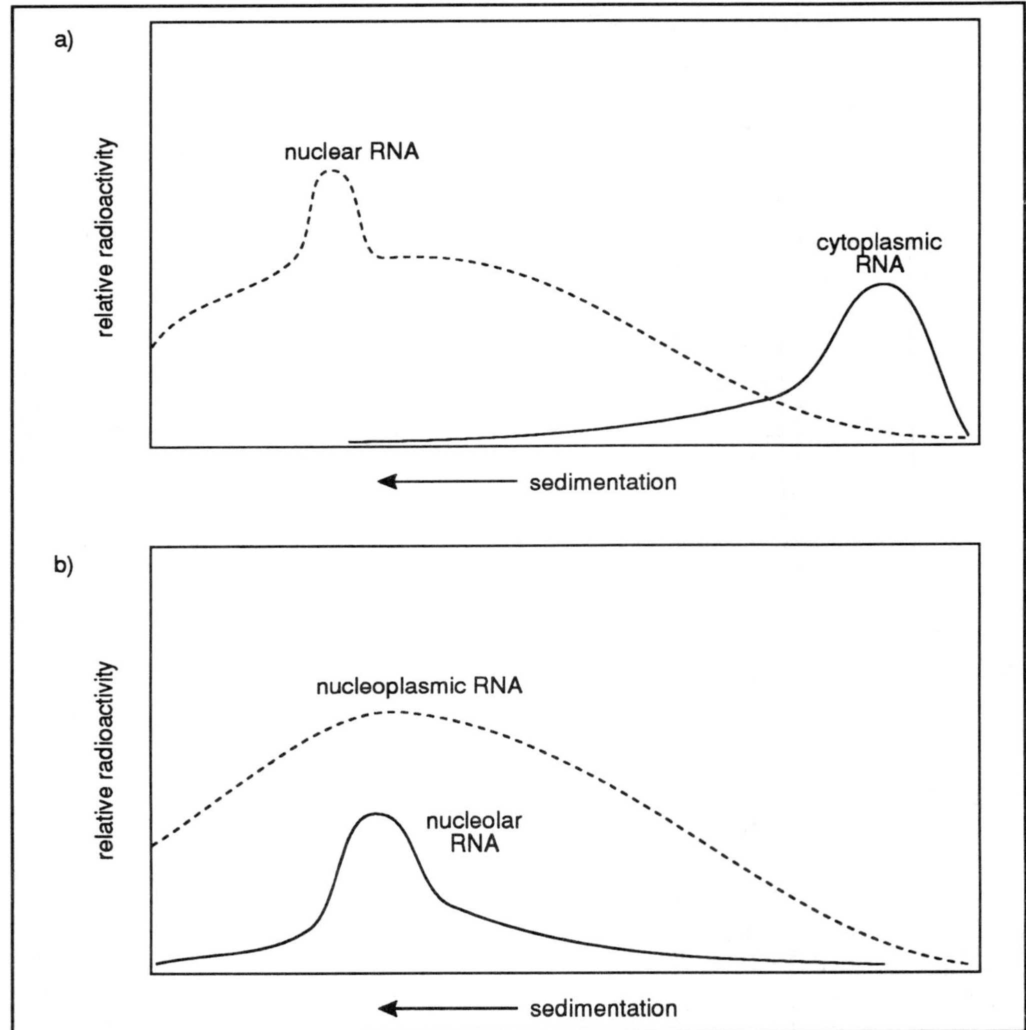

Figure 5.4 Sedimentation of cellular RNA radioactivity labelled *in vivo* with [^{3}H]uridine for 5 minutes. a) Nuclear and cytoplasmic RNA. b) Nucleolar and nucleoplasmic RNA. (continued)

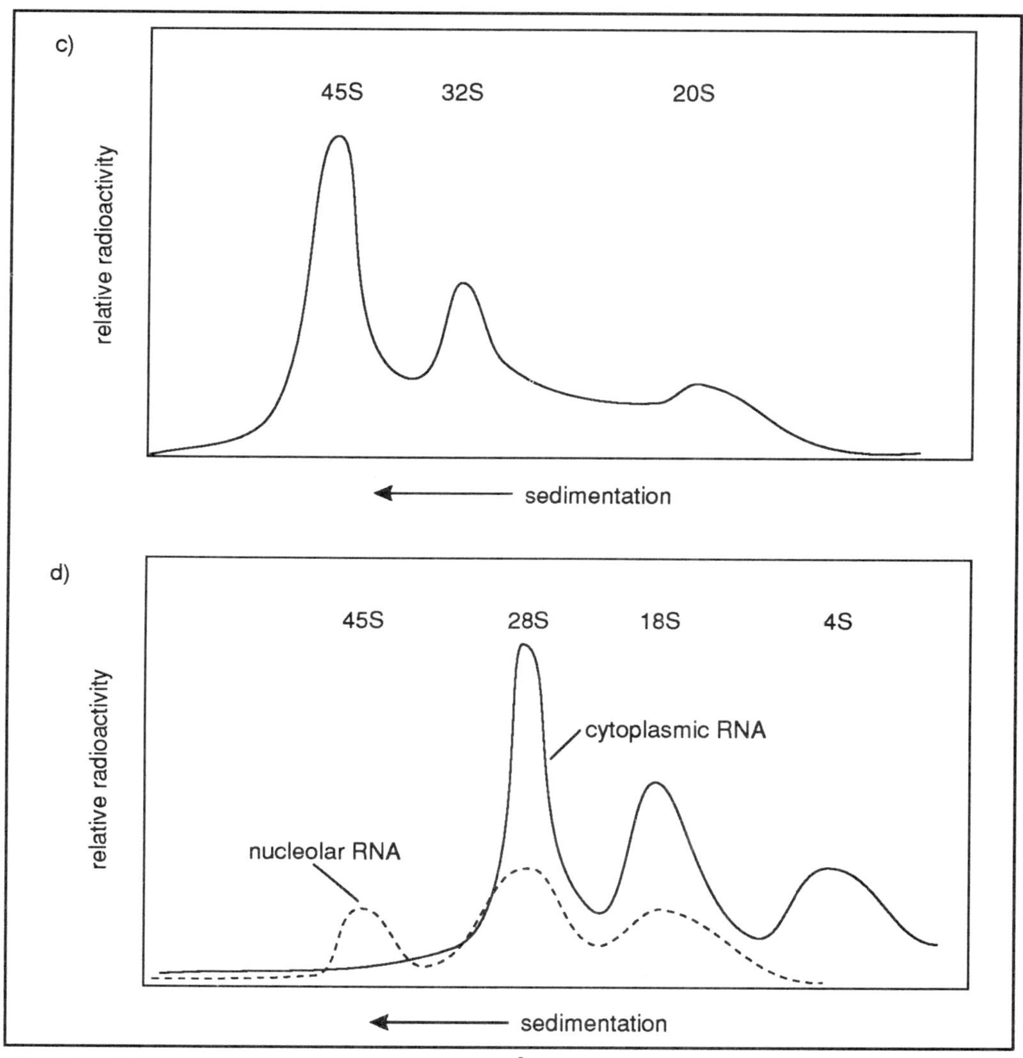

Figure 5.4 Sedimentation of RNA labelled *in vivo* with [^{3}H]uridine for 15 minutes. c) Nucleolar RNA isolated immediately after labelling. d) Nucleolar and cytoplasmic RNA isolated after a 60 minute 'pulse-chase' with 0.04 μg/ml actinomycin D.

Π What does this experiment show?

primary transcript

The implication is that the 45S RNA molecule is a precursor to both 28S and 18S rRNA, and may be the primary transcript of the rRNA transcription unit.

Much more recent work on the structure of ribosomal RNA genes from many different eukaryotes has confirmed this conclusion. The structure of ribosomal transcription units is as shown in Figure 5.5. Both the 18S and 28S rRNA sequences (and also 5.8S rRNA which is hydrogen bonded to 28S rRNA in ribosomes) form part of a longer transcription unit. The rRNA sequences are separated by regions called transcribed spacers (T.S.) which contribute about half of the 45S primary transcript. These regions have no known function and are discarded during the processing of the primary transcript, as shown in Figure 5.5 for human cells. The processing involves a sequence of precise nucleolytic cleavages generating intermediates such as the 32S RNA previously seen in Figure 5.4. The synthesis and processing take place in the nucleolus, as does the packaging of rRNA with ribosomal proteins prior to transport into the cytoplasm. 5S rRNA is transcribed on a separate transcription unit (by which RNA polymerase ?) and is incorporated into the 60S ribosomal subunit in the nucleolus.

transcribed spacerss

nucleolytic cleavages

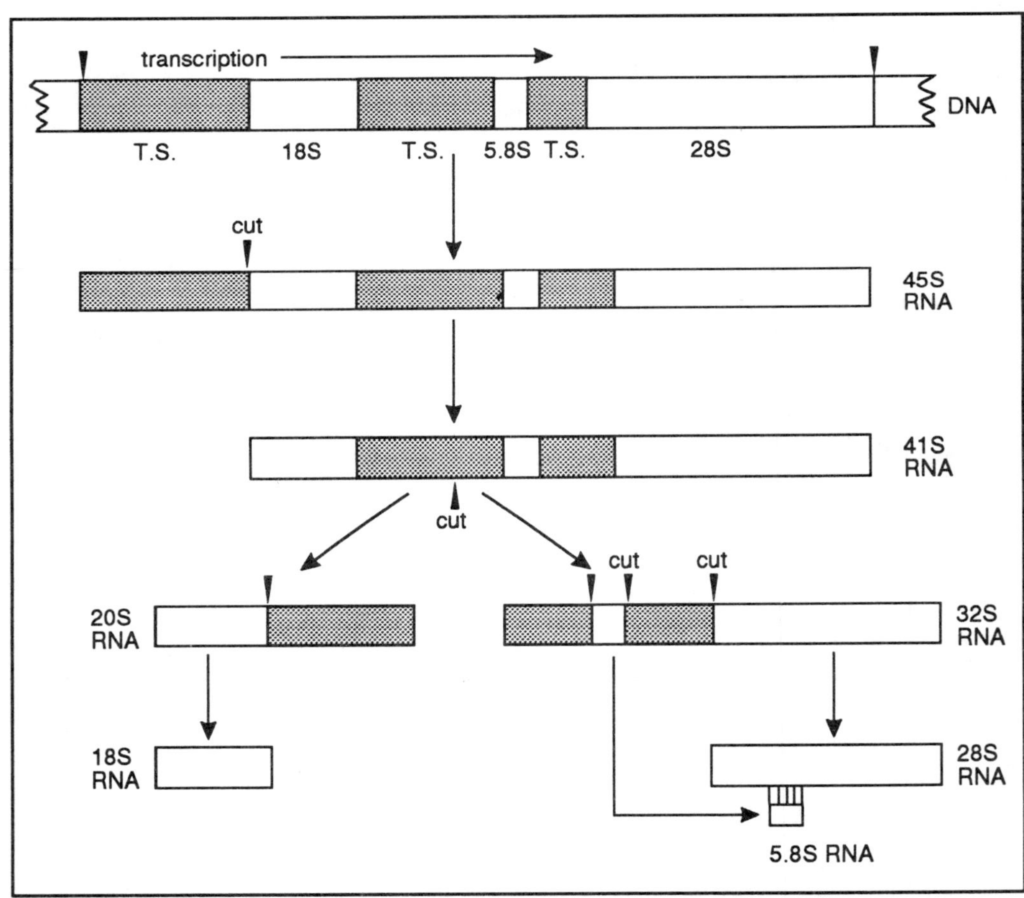

Figure 5.5 The structure of a human rRNA transcription unit. T.S. = transcribed spacers.

tandem array

non-transcribed spacers

The nucleolus, then, is the site of activity of rRNA transcription and processing and is formed at the site of a large number of rRNA genes. All eukaryotes contain multiple copies of rRNA genes and these are organised in a repeating fashion called a tandem array (Figure 5.6). The transcription units are separated by a non-transcribed spacer region. This region contains the promoter sequence and other sequences involved in directing high levels of transcription from these transcription units. One of the important questions concerning this type of organisation is how do we know that these spacer regions are not transcribed? How can we distinguish between a model in which transcription at each transcription unit is initiated and terminated independently, and one in which the RNA polymerase I starts at one end of the tandem array and transcribes the entire region, with the individual 45S transcripts generated by rapid cleavage of the primary transcript?

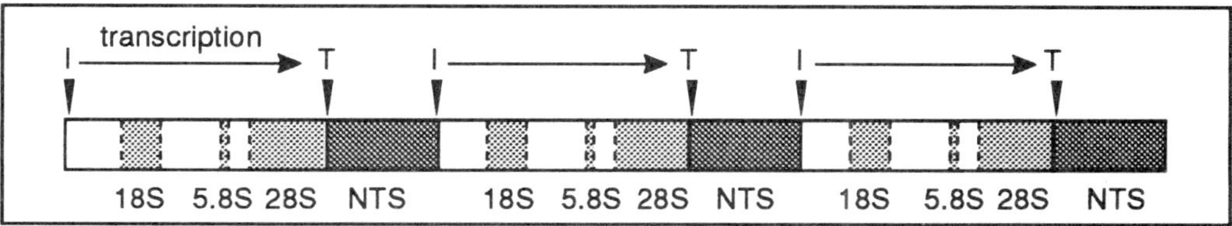

Figure 5.6 A tandem array of rRNA transcription units. The non-transcribed spacer region (NTS) separates individual transcription units. I = initiation site, T = terminator site.

Some of the most visually compelling analyses of transcription units comes from electron microscopical examination of ribosomal genes from cells in which there are greatly increased amounts of ribosomal DNA and nucleoli, such as *Xenopus* oocytes (see Chapter 4).

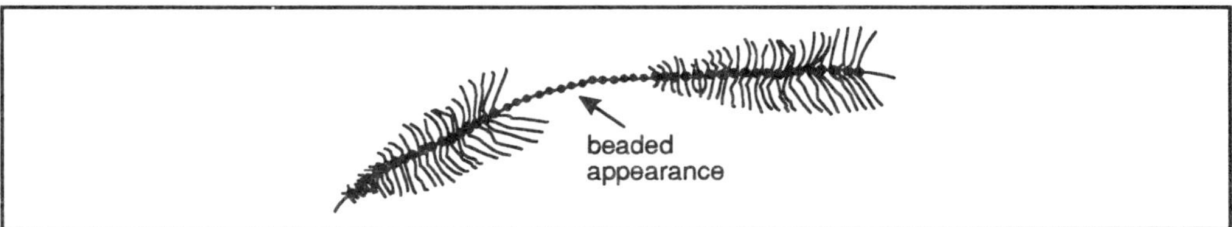

Figure 5.7 Drawing of a typical electron micrograph of ribosomal RNA transcription units from *Xenopus* oocytes.

Π Figure 5.7 shows a drawing of a typical electron micrograph of a tandem array of ribosomal RNA transcription units 'caught in the act' of RNA synthesis. The 'feather' or 'Xmas tree' like patterns result from the nascent RNA (RNA in the process of being transcribed) molecules each extending from an RNA polymerase I molecule. In what direction is transcription proceeding? On the diagram indicate where you think the initiation and termination points and the non-transcribed spacer regions are.

The direction of transcription is from the smallest to the largest RNA molecule (obvious, if you think about it!). These transcription units are densely packed with polymerase molecules due to the massive demand for rRNA in these cells. (Not all transcription units are transcribed so frequently). It is therefore relatively simple to extrapolate back to the putative initiation point, and to assume that the longest RNA molecule is very close to the termination site. The RNA polymerase I molecules can be just detected as dark spots on the DNA fibril on the original electron micrographs. The non-transcribed region also has a beaded appearance, but with smaller beads. This is due to the chromatin structure of the DNA fibril, which you have already met in a previous chapter.

Π What else do you notice about the length of the RNA fibrils compared to the length of DNA fibril from which they have been transcribed?

They appear to be shorter. What does this suggest about the packaging of RNA during transcription? The RNA appears to be associated with proteins and is therefore packaged, rather than fully stretched out. It is likely that processing of the rRNA and assembly of ribosomal subunits occurs while the RNA is packaged as a ribonucleoprotein complex.

5.4.2 RNA polymerase III transcription units

RNA polymerase III transcribes small RNAs

You saw earlier that one of the RNA components of ribosomes, 5S rRNA, is transcribed by RNA polymerase III in the nucleoplasm. In general, RNA polymerase III is responsible for transcription of a variety of small RNAs including tRNA and some of the snRNAs and scRNAs.

In some ways, the 5S rRNA transcription unit is simpler than the other transcription units which we shall cover in this chapter, since the primary 5S RNA transcript undergoes little or no processing.

Π The structure of the 5′ end of 5S rRNA indicates that transcription begins at this point. What structure would you expect to find at the 5′ end of a transcript?

A 5′ triphosphate group from the first nucleotide triphosphate of the RNA chain. In some species the 5S rRNA primary transcript is not processed, whereas in others a small number of bases are removed from the 3′ end. The 5S transcription unit is particularly interesting, however, in that the regulation of transcription has been extensively characterised, particularly with reference to promoter structure and the role of transcription factors, which will be dealt with in the next chapter.

splicing

Eukaryotes contain multiple copies of 5S rRNA genes, as they also do of tRNA genes. The tRNA primary transcripts are longer than the final tRNA molecules and may then be processed by cleavage and also splicing, a process which will be examined in much more detail when we look at mRNA production. The bases in tRNAs are also extensively modified after transcription.

Π Before we finish this section on RNA polymerase III transcription units, visualise the scale of the machinery involved. Compare the size of the 5S transcription unit with the size of RNA polymerase III as judged from its structure in Table 5.2.

The RNA polymerase molecule is comparable in size to the entire transcription unit!

5.5 RNA polymerase II transcription units

5.5.1 hnRNA as mRNA precursor

base composition

As shown in Figure 5.4, the initial nuclear RNA product to be rapidly labelled with [^{3}H]uridine, apart from 45S rRNA and the small RNA polymerase III products, is not mRNA itself but hnRNA. Using the types of kinetic analyses described for rRNA, it was possible to demonstrate that hnRNA is produced by RNA polymerase II activity and that it is a precursor to mRNA. Other evidence included the similarity in base composition of hnRNA and mRNA, although it remained something of a puzzle that 95% of labelled hnRNA sequences are apparently degraded in the nucleus. Early work also showed that hnRNA and mRNA molecules both have poly A tails, a stretch of upto 200 A residues at the 3′ end of the RNA chain.

poly A tails

Π Given that both mRNA and hnRNA have a 3′ poly A tail, but hnRNA molecules are in general much larger than mRNA, how would you expect hnRNA to be processed?

If your answer is that the hnRNA is cut down from the 5′ end, this is logical given the information you have been supplied with. Unfortunately, you may need to reconsider in the light of the next piece of evidence. Both hnRNA and mRNA are also found to have an unusual structure at the 5′ end, called a cap (Figure 5.8). This structure is formed by addition of a G residue to the first nucleotide of the RNA chain by a novel 5′-5′ triphosphate link. The G residue of the cap is methylated in the 7-position and the ribose groups of the first two nucleotides may also be methylated. The cap structure is now known to play a role in eukaryote translation (Chapter 8).

5′-5′ triphosphate link

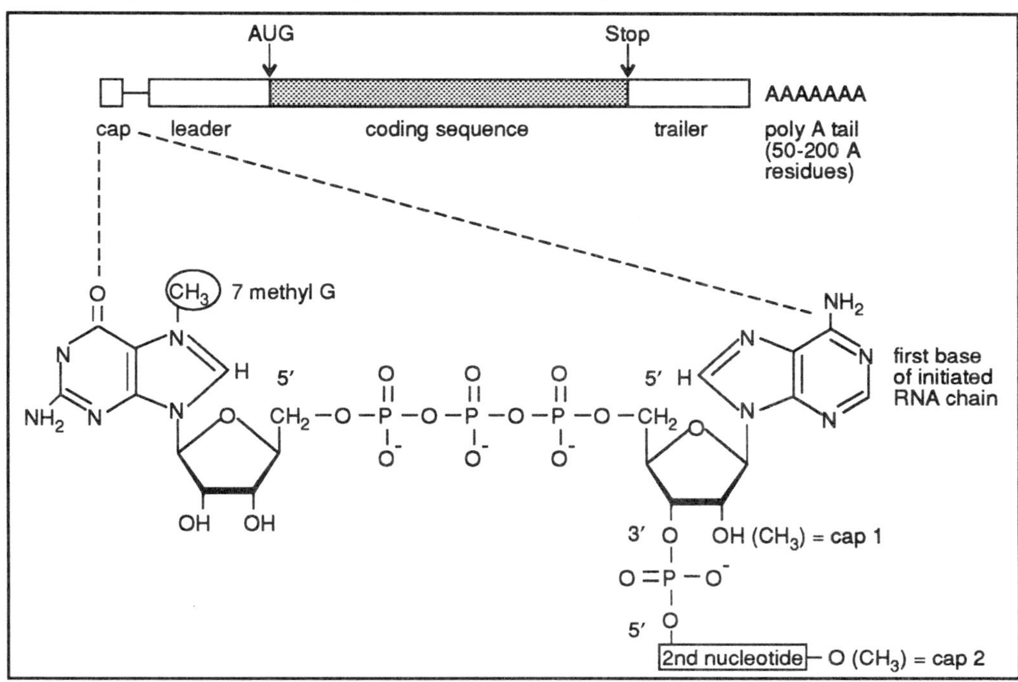

Figure 5.8 The structure of the 5′ cap on hnRNA and mRNA. Three types of caps exist. Cap 0 is the basic structure shown in the figure. Cap 1 has an -OCH3 residue on the ribose of the first base, cap 2 also has an -OCH3 residue on the ribose of the second base.

Π The poly A tail has proved to be extremely useful in eukaryote molecular biology, even though its role is still not well understood (see Chapter 6). The poly A tail was initially used to separate the relatively small populations of mRNA and hnRNA (see Table 5.1) from the bulk of cellular RNA. How might this be done? The following list contains some affinity agents which have been successfully used to isolate mRNA: poly A-Sepharose; protein A-Sepharose; oligo dT-cellulose; poly U-Sepharose. Which ones could be used to isolate mRNA?

oligo dT cellulose

poly U-Sepharose

The poly A tail will hybridise to a complementary DNA or RNA sequence, ie poly U or poly dT. In practice oligo dT cellulose (short DNA chains containing only T residues bound to a cellulose matrix) has been most widely used, although poly U-Sepharose may also be used (Figure 5.9).

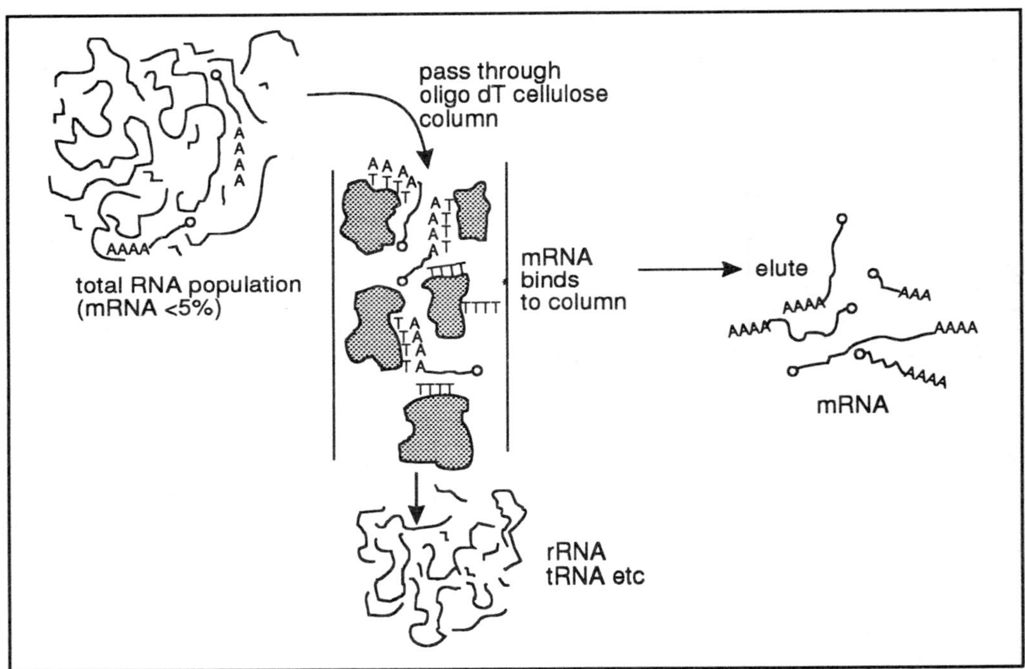

Figure 5.9 Isolation of mRNA by affinity chromatography on oligo (dT) cellulose.

Π How would you elute the mRNA?

Think about conditions which will melt double stranded nucleic acid chains eg low salt, high temperature, formamide etc.

cDNA synthesis, reverse transcriptase, DNA polymerases, primer

Another way in which the poly A tail has been exploited is in cDNA synthesis. The enzyme used to copy the RNA into a cDNA (copy DNA) strand is called reverse transcriptase. This enzyme is involved in the replication of a group of eukaryotic RNA viruses called retroviruses and, like other DNA polymerases, it requires a primer. Suggest a possible primer sequence that could be used for cDNA synthesis from eukaryotic mRNA.

olgio dT

Again, oligo dT would be suitable, as shown in Figure 5.10.

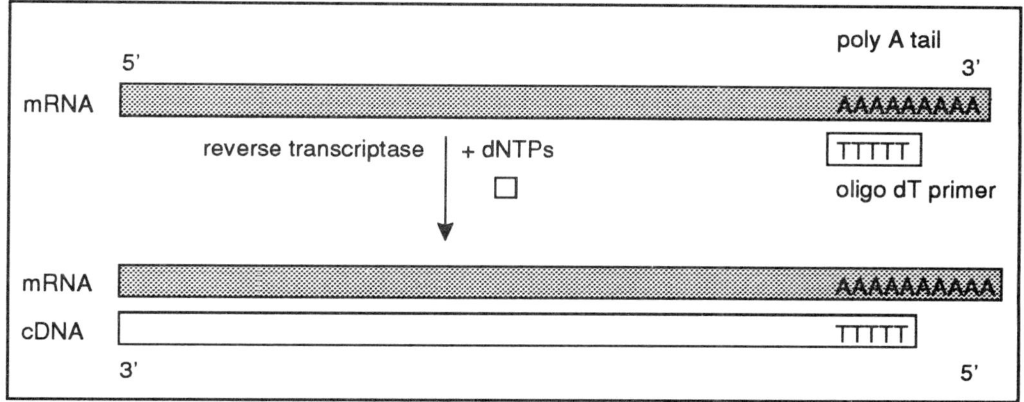

Figure 5.10 cDNA synthesis by reverse transcriptase using an oligo (dT) primer.

5.5.2 RNA polymerase II transcription units contain introns

The problem of how hnRNA could be the precursor of mRNA when both the 5′ and 3′ end were effectively blocked by the cap and tail structures had to await the development of gene cloning techniques for its resolution. These techniques made it possible to study individual transcription units and the first results came as something of a bombshell. One type of result came from the comparison of cDNA (ie mRNA) and genomic sequences.

Π One way of doing this is by Southern blotting, a procedure in which DNA fragments separated by agarose gel electrophoresis are blotted onto a membrane. The blot can then be hybridised to a radioactively labelled nucleic acid sequence called a probe. Figure 5.11 shows the result of such a blot in which fragments of a particular cDNA and total genomic DNA cut with the restriction enzyme *Eco*RI have been probed with an internal *Eco*RI fragment of the cDNA. How do you interpret these results?

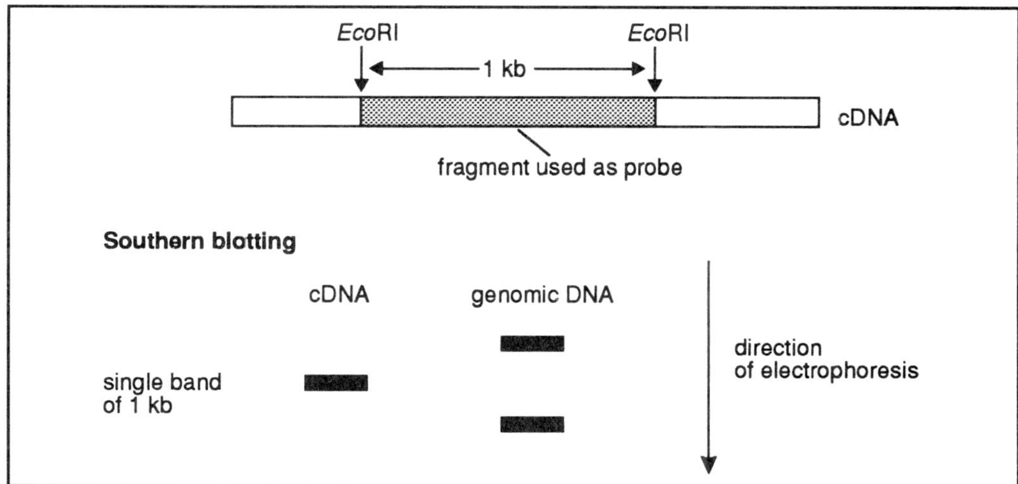

Figure 5.11 Southern blotting analysis of a cDNA and genomic cloned sequence cut with *Eco*RI and probed with an internal *Eco*RI fragment of the cDNA sequence.

The genomic sequences encoding the mRNA are longer than the mRNA and contain internal restriction sites which are not present in the cDNA ie the gene must contain intervening sequences which are not present in the mRNA. Figure 5.12 shows the structure of a well characterised transcription unit, the gene coding for the β globin chain of haemoglobin, compared to the structure of the globin mRNA. The most obvious difference is the presence of two intervening sequences, or introns, which interrupt the globin coding sequence between codons 31 and 32, and 105 and 106. The interrupted coding sequences are called exons. Notice also the typical structure of the globin mRNA. Like virtually all eukaryotic mRNAs it is monocistronic, with a single coding sequence sandwiched between 5′ and 3′ untranslated regions.

intervening sequences

globin

haemoglobin

introns

exons

monocistronic

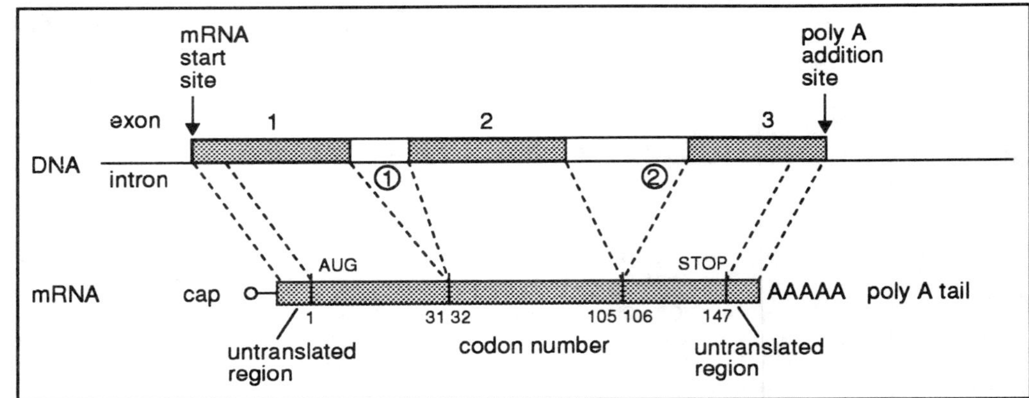

Figure 5.12 Comparison of the organisation of sequences coding for β globin in genomic DNA and mRNA. Note that introns ① and ② are removed allow the joining together of exons 1, 2 and 3.

Π When they were first discovered, there was much speculation as to how these intron sequences could be removed between transcription and translation. Sketch a possible mechanism for removal of introns at the level of:

- transcription;
- post-transcriptional processing.

It is certainly possible to envisage a mechanism acting at the level of transcription (eg the introns might be looped out of the DNA and the RNA polymerase could jump from one exon to the next). We have also already seen post-transcriptional processing in operation in rRNA production, though admittedly not involving the removal of introns, by the action of specific RNA cutting enzymes. In other words, either process is mechanically feasible.

Π Which mechanism would appear to be more likely, given the accumulated evidence for the role of hnRNA?

Northern blotting

One technique which can be used to distinguish between these two levels at which intron removal might operate is similar in principle to Southern blotting. Since it involves separation of RNA rather than DNA it is called Northern blotting, and allows the detection of specific RNA sequences by hybridisation.

Π Figure 5.13 shows two hypothetical Northern blots in which cytoplasmic and nuclear RNA samples have been probed with two different cDNA clones, X and Y. Do these results confirm the post-transcriptional processing model which previous information about hnRNA strongly predicts?

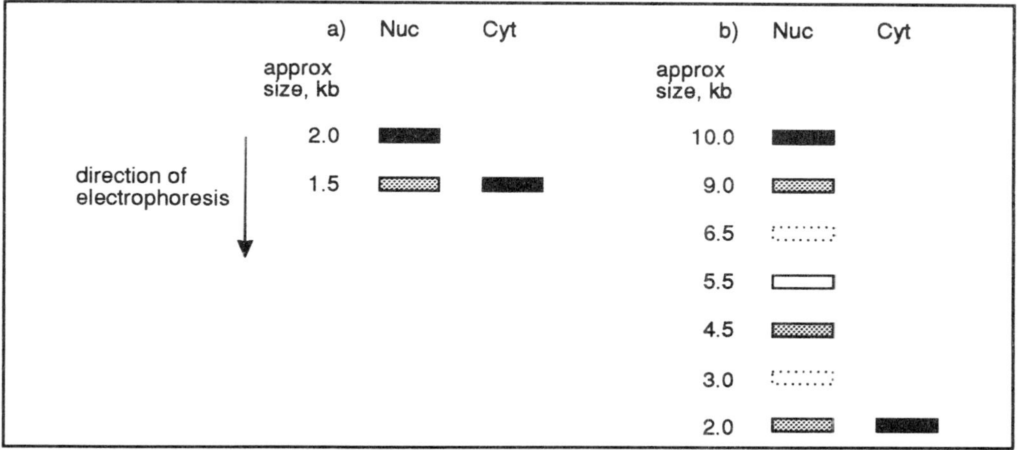

Figure 5.13 Northern blotting analysis of nuclear and cytoplasmic RNA probed with a) cDNA clone X and b) cDNA clone Y.

Yes, since the nuclear RNA contains hybridising molecules which are longer than the mRNA.

Π How many introns do you think gene X has?

Gene X seems to have only one intron, which is removed from the large band of 2000 bases in the nuclear RNA to form the mature mRNA of 1500 bases. A scheme of how this might occur by a splicing reaction is shown below.

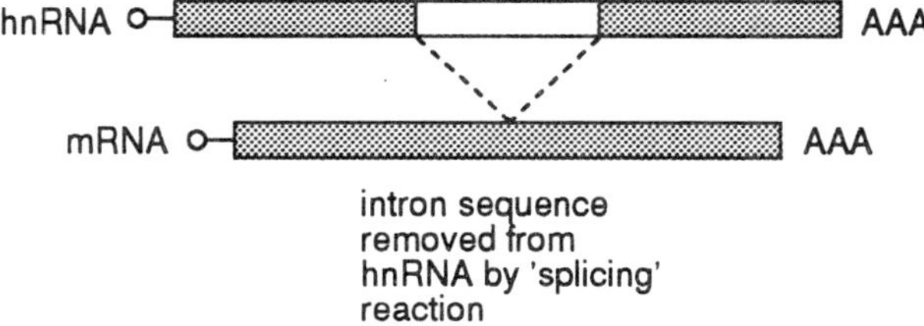

Π Why is the intron band of 500 bases not visible on the blot?

Because a cDNA probe containing only the mRNA sequence was used.

Π Where in the cell does the splicing reaction take place?

It is apparently nuclear since some mature mRNA is also found in the nucleus. This is presumably then transported out of the nucleus into the cytoplasm.

Π How many introns does gene Y have?

The primary transcript is probably the 10 kilobase (10 000 bases) band. The pattern fits a splicing reaction in which 3 introns (A, B, C) of sizes 4.5kb (kilobases), 2.5kb and 1.0kb are removed. There are 6 possible splicing intermediates (-A, -B, -C, -(A+B), -(A+C) and -(B+C)) between the primary transcript and the mRNA. The gel shows two things about the splicing reaction. It is sequential (ie each intron is removed independently) and there is a preferred, though not obligatory, order of splicing (1.0kb, then 4.5kb and then 2.5kb).

Π What else strikes you about the size of gene Y?

chick ovalbumin

It is 5 times larger than the mature mRNA! This is by no means an improbable example. One of the other genes to be first characterised, chick ovalbumin, was shown to have 7 introns in its 7700 base pairs of length, less than a quarter of which codes for the ovalbumin protein.

Π Why do you think globin and ovalbumin were amongst the first genes to be characterised?

collagen

Think about the likely abundance of the mRNAs coding for these proteins in their cells of origin. These proteins are produced in large amounts and thus the mRNAs are abundant. This enabled these RNAs to be isolated and cDNA copies to be made. Thus we were able to probe these genes. More recently even this example of many introns interrupting a gene has been shown to be relatively restrained. For example, the gene encoding collagen is over 38 kbp (kilobase pairs) long and contains more than 50 introns which separate very small exons of 45-108 (usually 54) base pairs long. Collagen is the major structural protein of connective tissue and has an unusual structure based on a repeating structure of 3 amino acids (Pro- Hypro- Gly).

Π What is the significance of the exon size being a multiple of 9 base pairs?

Each exon codes for an exact number of repeat units. In other words, the positions of introns are apparently not random, but reflect the structure of the protein product of the gene. We will return to this point later.

Duchenne muscular dystrophy, dystrophin, mammalian interferon

Most recently, work being done to isolate the gene responsible for Duchenne muscular dystrophy on the X chromosome revealed a transcription unit of more than 2000 kbp with more than 50 huge introns. (The mRNA coding for the protein product of the gene, dystrophin, is only (!) 18 kbp). At the other end of the scale, it should be noted that some genes do not contain any introns, eg mammalian interferon genes and many genes in lower eukaryotes.

Π 1) How much larger than the globin gene is the dystrophin gene?

Π 2) If RNA polymerase II transcribes at most 50 nucleotides/sec, how long does it take to copy the entire dystrophin gene?

1) 2000 000/2000 = 1000 times longer; 2) 2000 000/50 = 40 000 seconds = 11 hours!

SAQ 5.2

Samples of nuclear and cytoplasmic RNA were run on an agarose gel and blotted onto a membrane. Three replicas of this membrane were probed with radioactive: 1) cloned 18S rRNA sequences, 2) cloned 28S rRNA sequences and 3) cloned transcribed spacer RNA sequences from between the 18S and 5.8S rRNA regions. With reference to Figure 5.5, draw the pattern of bands for each Northern blot.

SAQ 5.3

Match the words and phrases in list A with list B:

List A	List B
rRNA precursor	intron
mRNA precursor	5′ triphosphate
5′ end structure of mRNA	5′ phosphate
5′ end structure of 5S rRNA	exon
5′ end structure of 28S rRNA	45S RNA
3′ end structure of mRNA	hnRNA
intervening sequence	cap
interrupted coding sequence	poly A tail

5.6 Processing of RNA polymerase II transcripts

Given that there are at least 3 processing events which we know must occur to produce the mature mRNA molecule from the primary transcript (capping, polyadenylation and splicing) we should spend a little time looking at when and how these events occur.

5.6.1 Capping

capping factors associated with RNA polymerase II

From the early information that hnRNA is both capped and polyadenylated, it is no surprise that these modifications occur early after transcription. Indeed capping occurs almost immediately after the initiation of transcription by the sequential mechanism shown in Figure 5.14. Although it is possible to isolate the enzymes involved in capping and perform the capping reaction *in vitro*, it appears that *in vivo* the capping factors are associated with RNA polymerase II.

Π Why does it make apparent good sense for the capping reaction to be apparently obligatorily co-transcriptional?

It may be one way of ensuring that all mRNAs are capped, but that the other major classes of RNA molecules are not.

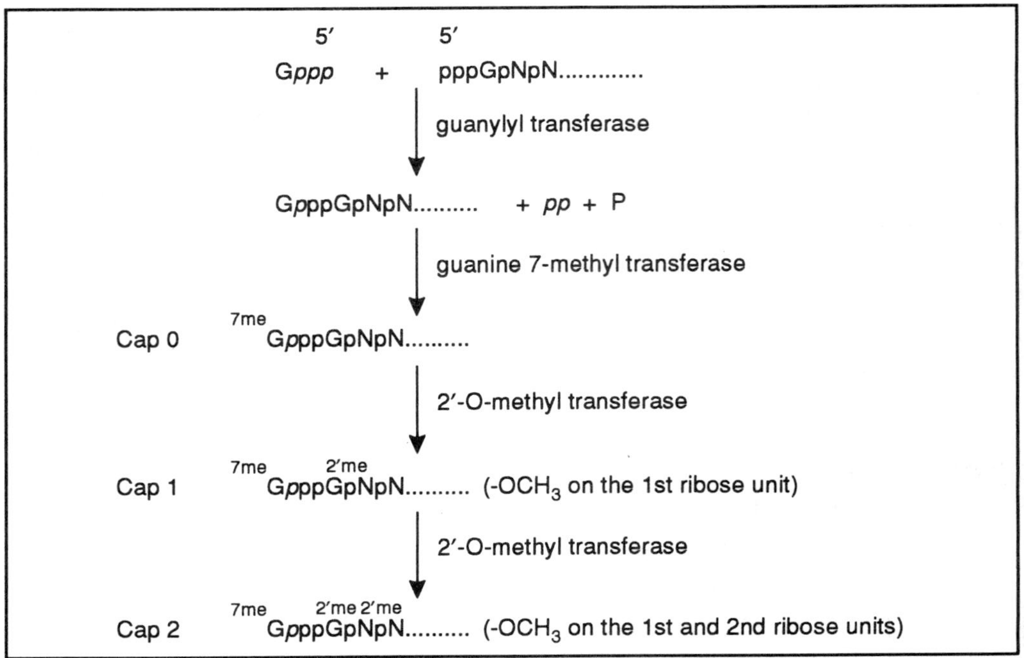

Figure 5.14 Sequential capping of RNA polymerase II transcripts.

5.6.2 Polyadenylation

histones

poly A polymerase

signal sequence

consensus

Most mRNAs are polyadenylated at the 3′ end. (Not all of them are. One important group of mRNAs which are not polyadenylated code for the histones). It would be easy to assume that the poly A tail was added on to the last base of the transcription unit ie at the termination site. In fact, this is not the case and polyadenylation involves first creating an internal site by nucleolytic cleavage, followed by addition of the poly A tail base by base by the poly A polymerase enzyme. Again, this is actually done during transcription, and some of the factors involved are also associated with the RNA polymerase II molecule. How does the polyadenylation mechanism 'know' where to cut?

snRNP

U1 snRNA

There is a signal sequence in the transcription unit, with the consensus: AAUAAA, which serves as a marker for the polyadenylation event. The RNA chain is cut some 15-20 bases downstream from this site and the poly A tail is then added (Figure 5.15). One of the factors involved in recognition of the AAUAAA sequence is a snRNP particle (Small Nuclear RiboNucleoProtein particle), a particle containing a number of specific proteins and one of the snRNAs (in this case U1 snRNA). We will cover the role of snRNPs in more detail in the next section.

The term 'consensus sequence' will be returned to later. Using the polyadenylation signal as an example, a sequence closely resembling AAUAAA is found in virtually all eukaryotic mRNAs about 15-20 bases upstream of the poly A tail. However, the sequence need not be identical to AAUAAA. It is possible to quantify the degree of consensus by comparing many different mRNAs and calculating the frequency at which each base appears at each position in the sequence.

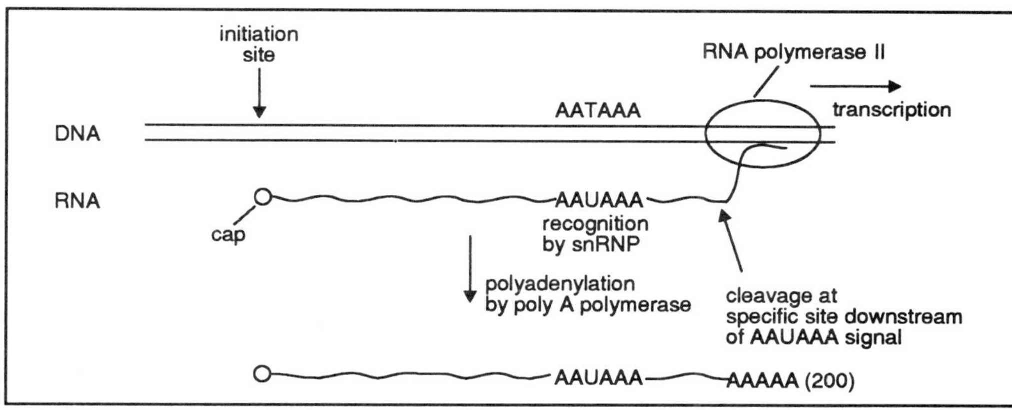

Figure 5.15 Co-transcriptional polyadenylation of pre-mRNA.

The polyadenylation particle appears to drop off the RNA polymerase II at this point, since subsequent AAUAAA signals between this point and termination are ignored. In effect the polyadenylated hnRNA molecule which results is not truly the primary transcript of the transcription unit, since the RNA polymerase II carries on transcription after this point.

Π How could you determine where the actual termination site of a transcription unit is?

This still remains a largely unsolved problem, not least because the transcript beyond the poly A site is degraded rapidly!

5.6.3 Splicing

Whilst capping and polyadenylation are co-transcriptional events, splicing may occur post-transcriptionally. The mechanism of splicing has not been easy to ascertain but steady progress towards an understanding of this complex process has been made by amassing evidence from several different sources.

Consensus splice site sequences

sequence motifs

consensus sequences

Now that increasing amounts of gene sequence data are being accumulated, it is possible to look for the presence of sequence motifs common to a number of different splice sites which might be involved in the splicing mechanism. One surprising fact about the consensus sequences around the splice sites is how short the highly conserved regions are (Figure 5.16). The only absolute requirement is for the 2 bases at each end of the intron to be GU--- and ---AG respectively. Notice that the two splice junctions are different, ie the intron has direction with a defined left and right hand junction. Figure 5.16 also shows the % frequency of each position eg in 77% of exon-intron junctions, G appears as the 3′ terminal base of the exon.

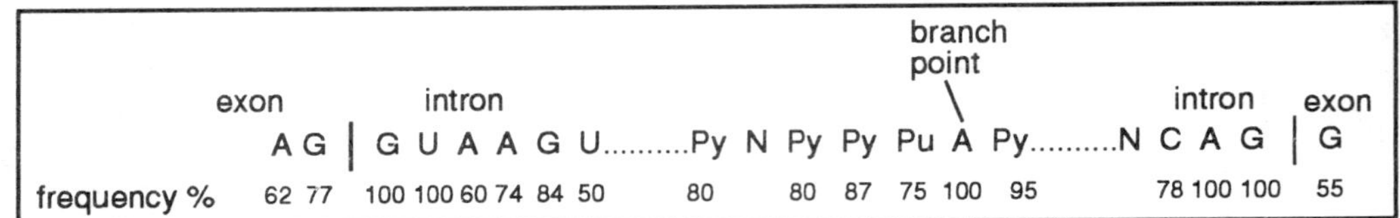

Figure 5.16 Consensus splice site sequences.

pre-mRNA

lariat

5'-2' linkage

It is possible to perform splicing reactions *in vitro* using cell or nuclear extracts and a specific pre-mRNA (an hnRNA molecule which is known to be the precursor of a particular mRNA). Using *in vitro* splicing reactions it is possible to determine the reaction pathway of splicing and also find out about the specific factors involved. It can be shown, for example, that the pre-mRNA requires a cap in order to be spliced. It is also possible to show that the intron is not removed as a linear molecule, but rather as a branched or lariat configuration. Figure 5.17 shows the pathway of splicing, starting with cutting the 5′ exon-intron boundary, subsequent circularisation to form an unusual 5′-2′ linkage to a specific A residue within the intron, and then cleavage and ligation of the two exons. The removed lariat-form intron sequence is linearised and degraded. The branch point is recognised by virtue of a specific pyrimidine-rich branch point consensus sequence, also shown in Figure 5.16.

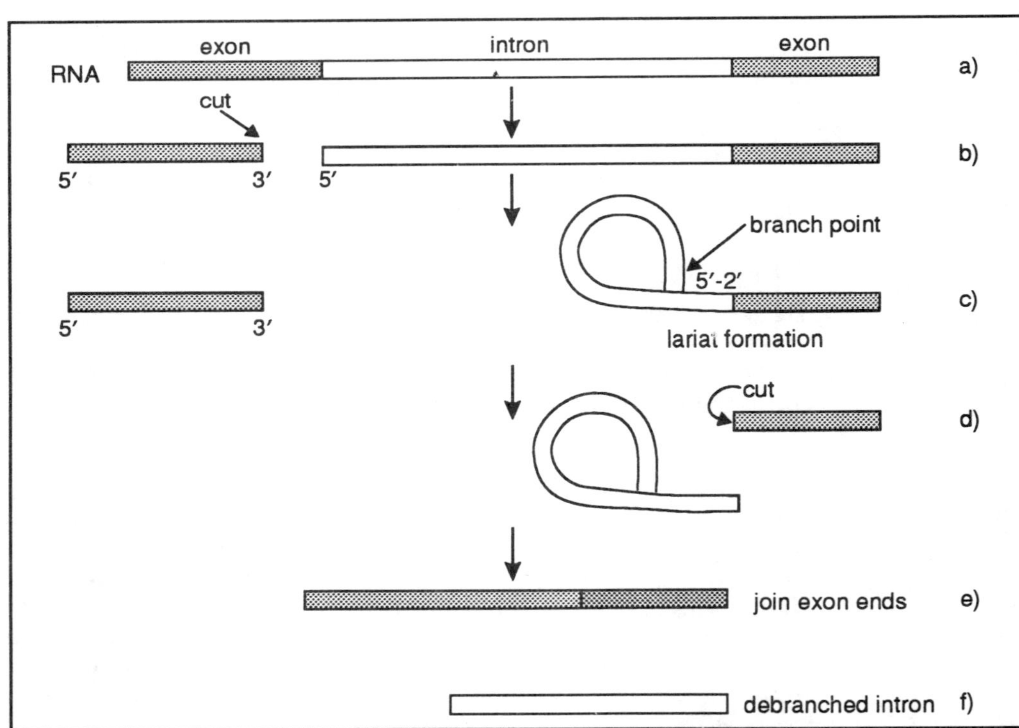

Figure 5.17 The pathway of pre-mRNA splicing, showing lariat formation (see text for details).

Π Would it be possible to form a lariat with a single stranded DNA molecule?

U-rich snRNAs

No! Check the structure of DNA if you are not sure why. (Does DNA have a functional OH group at position 2′?). The absence of this functional group means that a 5′-2′ branch point cannot be produced. If the information for directing the position of splice sites and the branch point resides in the primary transcript, what is it in the splicing machinery which recognises these? A number of pieces of evidence have implicated some of the U-rich snRNAs in this process.

Π 1) The sequence at the 5′ end of U1 snRNA is shown in Figure 5.18. Compare this sequence with the splice site consensus sequence shown in Figure 5.16. Can you spot anything unusual about these sequences?

5′ ^{7me}Gppp A^{me} U^{me} A C ψ ψ A C C U G ....
ψ = pseudo uridine, which can basepair with A

Figure 5.18 The 5′ end sequence of U1 snRNA.

The sequence at the 5′ end of U1 snRNA is complementary to the 5′ exon-intron boundary consensus sequence, AGGUAAGU. (Remember that basepairing is anti-parallel).

Π 2) Patients with the autoimmune disease, lupus erythematosus, produce antibodies to proteins of snRNP particles, the particles in which the snRNAs are found. Some antibodies react with particular particles, eg U1 snRNP, whilst others react with a number of different particles, eg any of the particles containing U1-U6 snRNA. Presumably in the latter case they are reacting with a protein common to all six snRNP particles. When antibodies specific for U1 snRNP are added to an *in vitro* splicing reaction, the splicing reaction is inhibited. If the pre-mRNA substrate is then digested with RNase, a fragment 15-20 bases long at the 5′ exon-intron junction remains. How do you interpret these results?

The implication is that the U1 snRNA is bound to this part of the pre-mRNA and therefore protects it from RNase attack. When this experiment is repeated with antibodies raised against U2 snRNP, a region around the branch point is protected, whereas U5 snRNP appears to bind near to the 3′ intron-exon boundary. Figure 5.19 summarises these results and others which suggest that splicing reaction requires the co-ordinate activity of several snRNPs within a large aggregate called a spliceosome.

spliceosomes

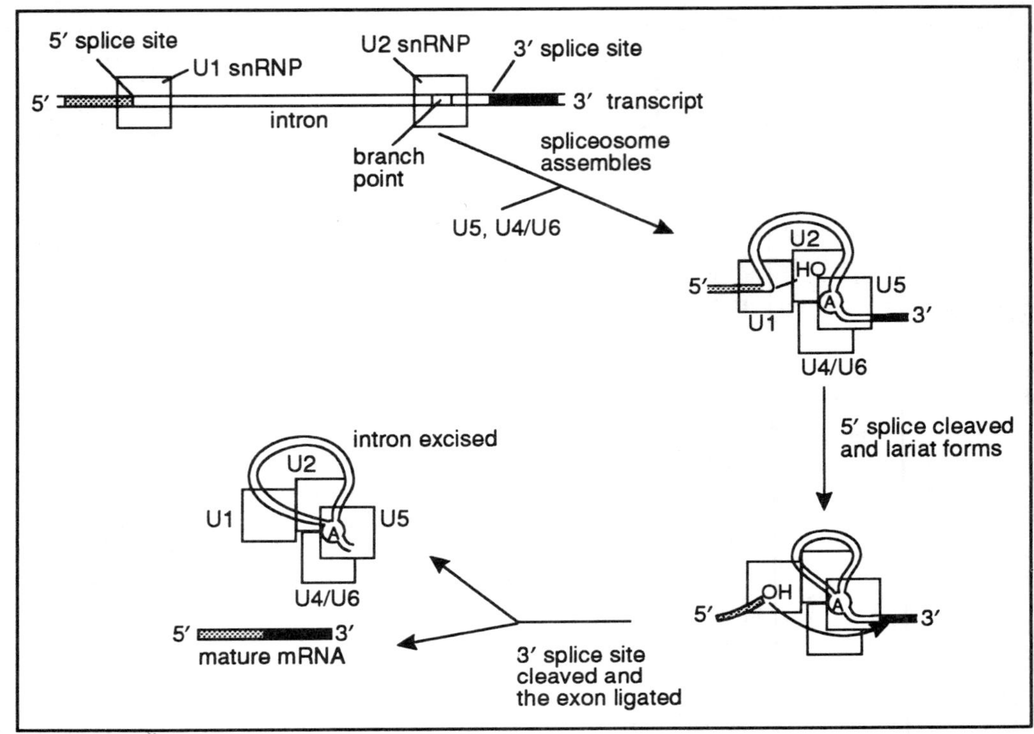

Figure 5.19 The role of snRNPs in the splicing of pre-mRNA.

SAQ 5.4

From the diagram, summarise the roles of U1, U2 and U5 snRNP particles in the splicing reaction.

5.7 Complex transcription units

calcitonin, thyroid gland, inactive precursor protein, pre-sequences

Within the description of the processing events which you have just covered, there is an implication of a mechanical process in which the presence of the correct signals in the pre-mRNA determines that polyadenylation or splicing will inevitably follow. However, for a number of transcription units there is nothing automatic about these processes. For example, the processing scheme for the transcription unit which encodes calcitonin is shown in Figure 5.20. Calcitonin is a hormone involved in the regulation of calcium levels which is produced by the thyroid gland. The hormone is synthesised as a larger inactive precursor protein, the 'pre-sequence', which is removed prior to release from the thyroid cells.

Π Which exons of the transcription unit correspond to: a) the untranslated region of the calcitonin mRNA; b) the pre-sequence of the precursor protein; c) the active calcitonin coding region?

a) exon 1; b) exons 2 and 3; c) exon 4.

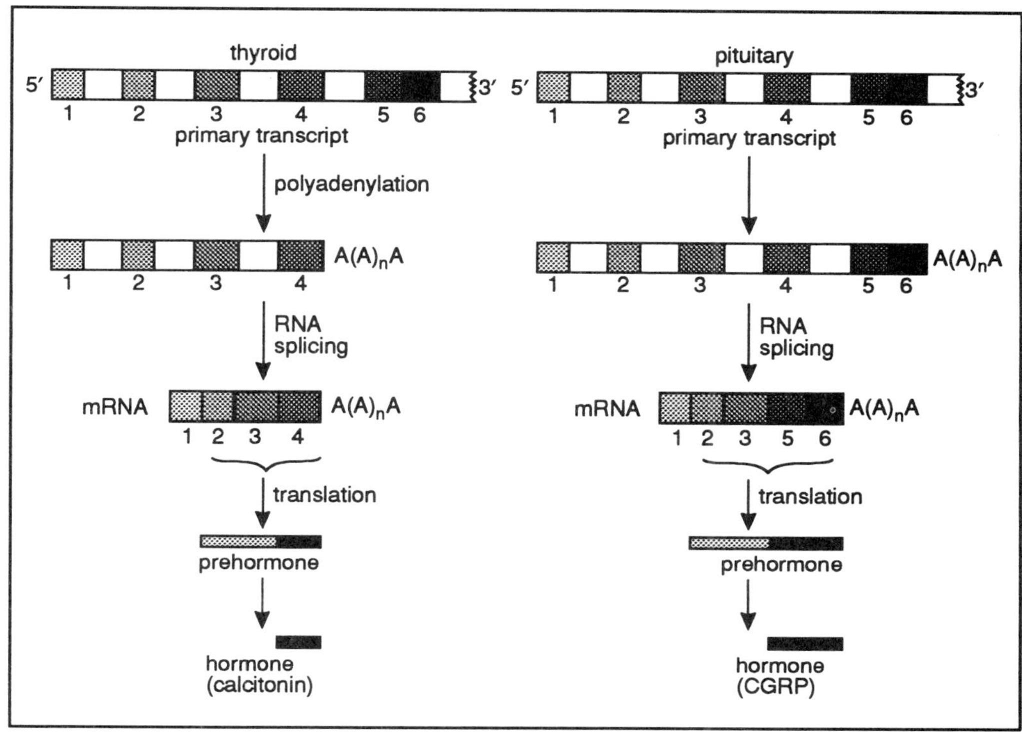

Figure 5.20 Production of two different mRNAs from the calcitonin gene.

Π What is significant about the answer to the last question?

The fact that there is an exact correspondence between the exons and the different functional regions of the mRNA or protein. This has been observed in many genes (although it does not apply to every gene). Even where there is not a clear separation of functions within a protein there are often distinct structural regions, called domains, which are encoded by separate exons. The calcitonin gene is not transcribed in other tissues apart from the pituitary gland. However, the pituitary does not produce calcitonin but a different protein called CGRP (Calcitonin Gene Related Protein). Figure 5.20 also shows how this occurs.

domains

pituitary gland

CGRP

Π A specific processing signal has apparently been overridden to result in the formation of this different protein. Which signal?

It would seem that the poly A signal which defines the 3′ end of the calcitonin mRNA in the thyroid is ignored in the pituitary and transcription carries on until a second signal is reached. The larger transcript is subsequently spliced such that the calcitonin coding region is treated as part of an intron and removed. Since the one transcription unit can give rise to more than one protein, this has been described as a complex transcription unit, as opposed to the simple ones met so far.

complex transcription unit

Π Given the structure of the CGRP mRNA and protein which result from this scheme, what would you predict the fate of the coding regions of exon 2 and 3 would be?

Since the initial translation product will also contain the pre-sequences from exons 2 and 3 it would be a safe prediction that these will also be removed from the CGRP coding region prior to secretion of this protein from the pituitary. (ie they will perform the same role for both proteins).

Π Figure 5.21 shows a different processing scheme indicating how the α and β forms of the muscle protein troponin T are produced. Which processing signals are being read differently in this case?

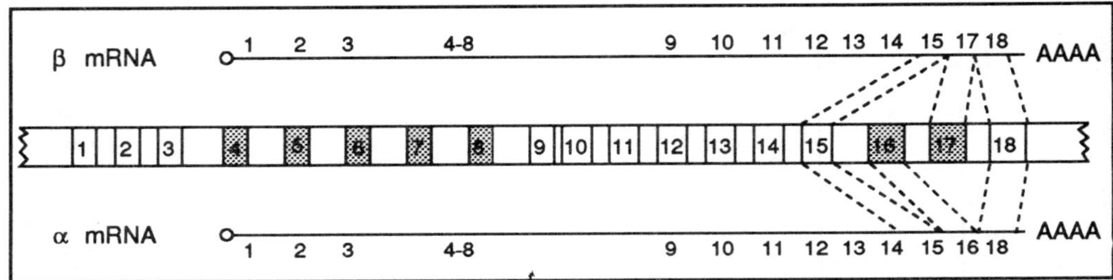

Figure 5.21 Alternative processing of rat troponin T gene products. Different troponin T molecules are found in different muscle types. The diagram highlights the alternative splicing of exons 16 (α) and 17 (β). Exons 4-8 are also spliced in various combinations, giving rise to 32 possible arrangements of both the α and β type.

In this instance it is the alternative splicing out of either the α exon or the β exon which produces the two proteins, which are identical except for the amino acids between codons 229 and 242. The study of complex transcription units is a good place to finish this chapter since it brings us full circle to a point raised at the beginning of the chapter - how does the transcription unit relate to the classical concept of the gene?

SAQ 5.5

In the complex transcription unit shown below, at what stages are alternative choices being made?

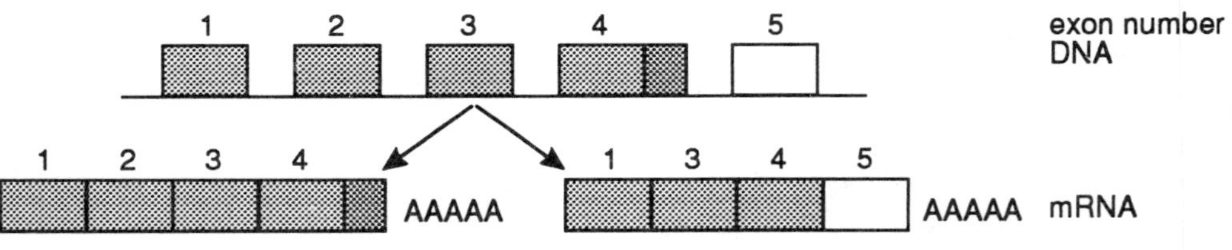

SAQ 5.6

Fill in the blank spaces in the summary table of RNA polymerase II transcription units shown below.

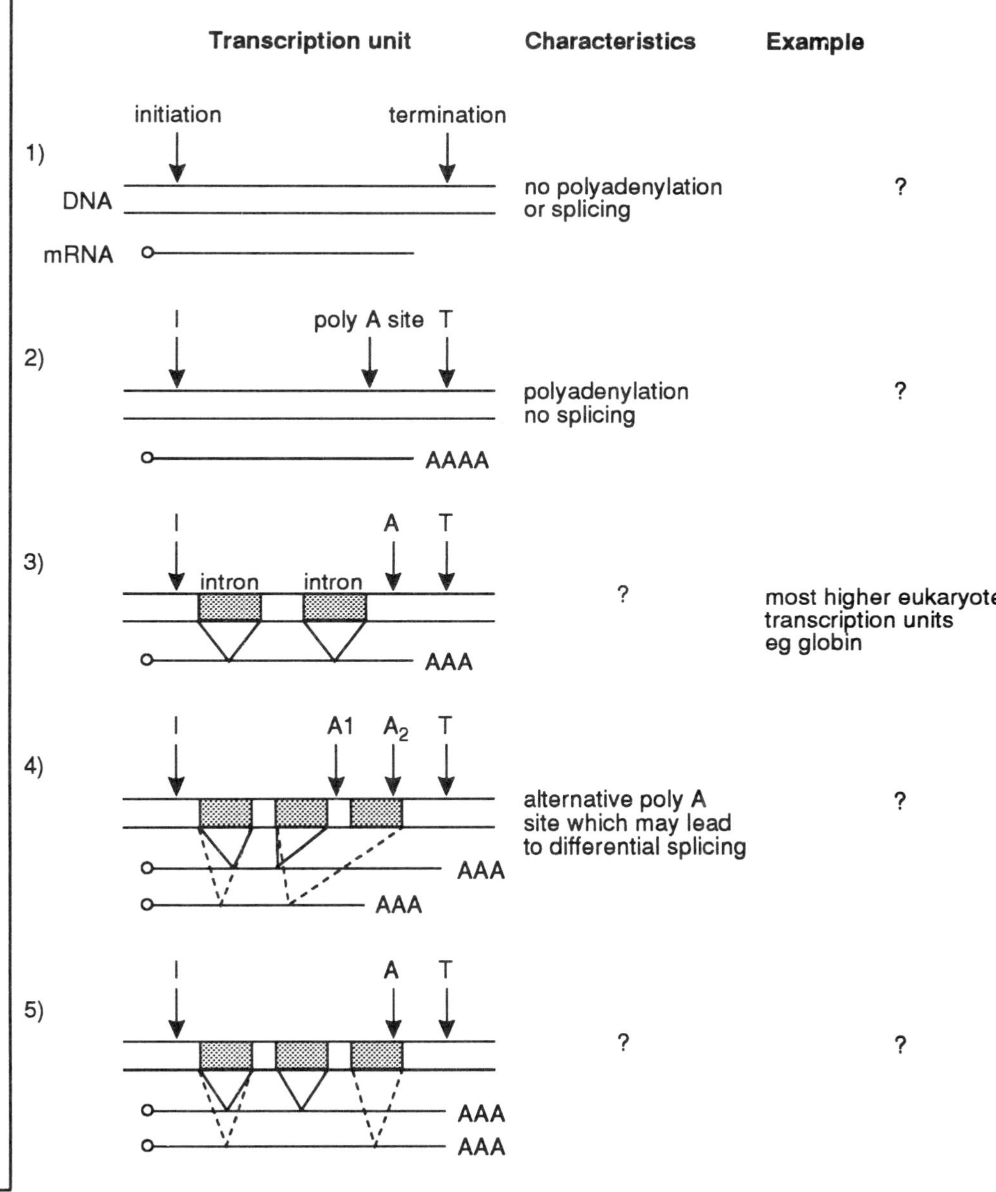

Summary and objectives

In this chapter we have examined the processes of transcription in eukaryotic cells. We began by briefly reviewing the major classes of RNA molecules and examined the 3 types of RNA polymerases produced by eukaryotes and introduced the concept of transcriptional units. We then examined the transcription of ribosomal RNA genes by RNA polymerase I and the subsequent processing of the primary transcript. Then we explained the function of RNA polymerase III in the transcription of small RNA molecules. The major part of the chapter was devoted to a description of RNA polymerase II and the production of hnRNA and the processes involved in the maturation of hnRNA to produce functional mRNAs. Here we particularly emphasised the importance of introns. We also described how nucleotide sequences within pre-mRNA are implicated in the polyadenylation and splicing of these RNA molecules.

- distinguish between the concept of a gene and a transcription unit;
- compare the organisation of the transcription units transcribed by RNA polymerase I, II and III;
- differentiate between the functions of RNA polymerase II and III;
- recognise the significant features of an RNA polymerase II transcription unit;
- relate the structure of mRNA to the structure of a protein coding gene;
- describe the processing steps involved in the production of mRNA;
- explain the mechanisms involved in mRNA splicing;
- demonstrate the significance of complex transcription units in eukaryote gene regulation.

The regulation of transcription

The regulation of transcription

6.1 Introduction

In this chapter we continue our examination of the process of transcription by looking at the ways in which transcription is regulated in eukaryotes. In order to do this, it is necessary first to consider the physical structure of the DNA template. It should always be borne in mind that diagrams of transcription units such as those shown in the previous chapter are stylised representations of chunks of information, rather than of the DNA molecules themselves. What you should remember from Chapter 4 is that the large amount of DNA in the eukaryotic nucleus is extensively packaged as chromatin. It will therefore be useful to start by refreshing your memory of the material which you met earlier in this book by completing the following statements.

SAQ 6.1

1) DNA in the nucleus is associated with [ ] to form chromatin.

2) The regular coiling of [ ] turns of the DNA molecule around a particle comprising [ ] molecules each of histones H [ ], H [ ], H [ ] and H [ ] forms a structure called a [ ].

3) A fifth histone molecule, histone H [ ], appears to play a role in chromatin condensation.

4) Partial digestion of chromatin with micrococcal nuclease results in a ladder of DNA fragments with a repeat length of [ ] base pairs.

5) More extensive digestion of chromatin with micrococcal nuclease results in fragments of DNA approximately [ ] base pairs in length.

6.1.1 Transcriptionally active chromatin is packaged in chromatin

If you look back to the previous chapter, you will see that one of the more spectacular ways of visualising active transcription units is by electron microscopical examination of dispersed chromatin. In particular, ribosomal transcription units are heavily transcribed, such that the RNA polymerase I molecules are effectively packed 'nose to tail' along the transcription unit, making it difficult to examine the state of the DNA template.

Π In contrast, look at Figure 6.1, an E.M. picture of an RNA polymerase II transcription unit from *Oncopeltus fasciatus*. What do you notice about the structure of the DNA template between RNA transcripts?

It should be apparent that the template retains a bead-like appearance reminiscent of chromatin structure. The next question is whether it is actually the case that DNA which is being transcribed is still packaged in chromatin.

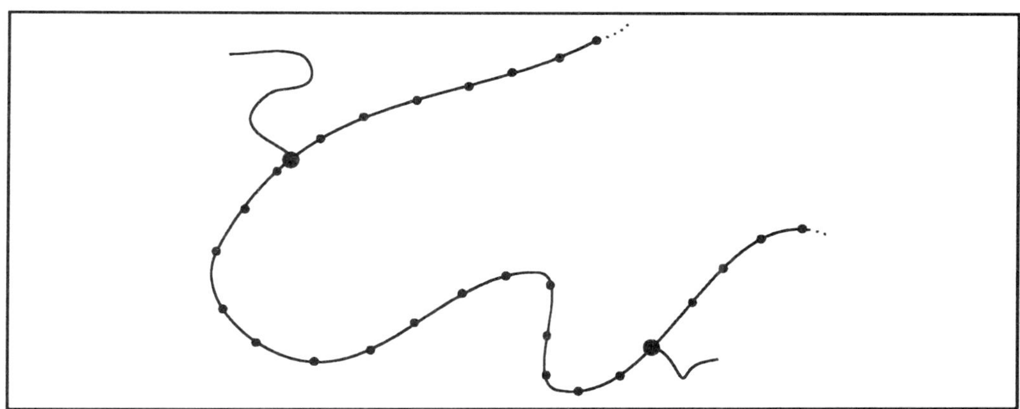

Figure 6.1 Tracing of an electron micrograph of an RNA polymerase II transcription unit from *Oncopeltus fasciatus* (compare this with Figure 5.7)

6.1.2 Nuclease digestion studies of transcriptionally active chromatin

digestion by micrococcal nuclease

Some of the most effective experiments which helped to elucidate the nucleosome structure of chromatin were relatively simple enzyme digestion studies which exploited the different properties of various nucleases. Figure 6.2a shows the type of result expected from a partial digestion of chromatin with micrococcal nuclease. The most obvious feature is the 200 base pair unit length ladder of DNA fragments which result from random nicks between nucleosome beads.

Π How do you think this type of experiment could be extended to look at the distribution of a specific gene sequence within this population of DNA fragments?

Think about the technique which was described in the previous chapter as a way to examine the organisation of mRNA sequences within genomic DNA. If you are not certain, refresh your memory by looking back at the section on Southern blotting.

Figure 6.2b shows the result of a Southern blot of the gel shown in Figure 6.2a probed with an mRNA sequence from the same tissue as the chromatin.

Π What can you deduce from this experiment about the nature of transcriptionally active chromatin? Does it show that transcriptionally active chromatin is or is not packaged by nucleosomes?

The result is clearly consistent with the packaging of transcribed chromatin in a nucleosome-like structure, since the mRNA coding sequences are still cut into 200 base pair unit length fragments.

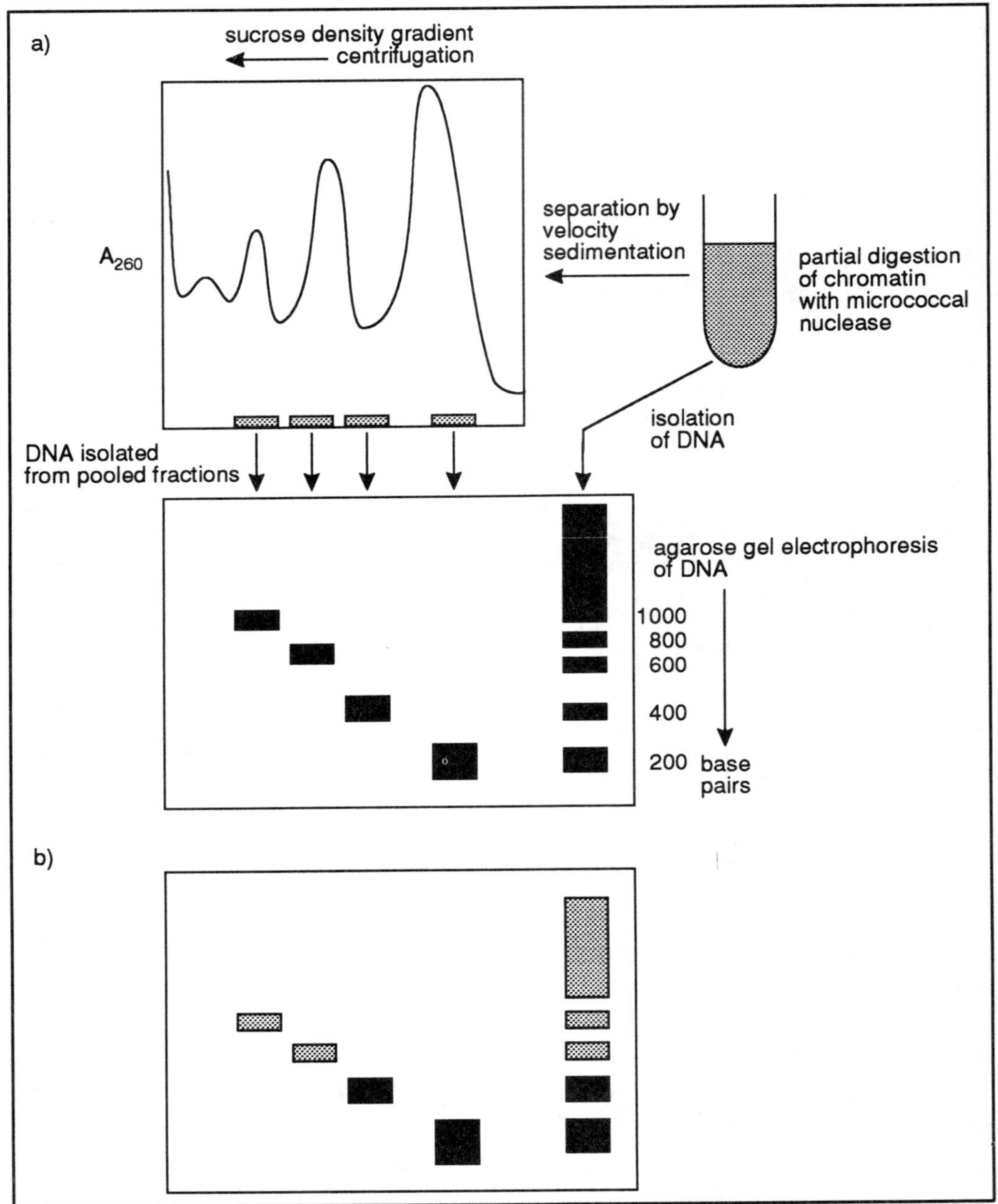

Figure 6.2 a) Separation of nucleosome beads after partial digestion of chromatin with micrococcal nuclease. Multimeric nucleosome particles contain mutliples of 200 bases of DNA when examined by agarose gel electrophoresis. b) Southern blot analysis of the agarose gel in a) probed with a specific mRNA sequence.

Π In order to pause and reflect on why the result is not necessarily the expected one, compare the size of RNA polymerase II with the size of a nucleosome, using the information which you should have met earlier in this book.

RNA polymerase II is large compared to nucleosomes

The histones are small proteins, so even a particle with 8 histones like a nucleosome will be much smaller than RNA polymerase II which contains many large protein subunits. It is therefore difficult to visualise the transcription of a DNA template wound tightly around a nucleosome bead by an enzyme as large as RNA polymerase II, particularly considering the fact that about 50 bases of the template strand are in contact with the enzyme during transcription. How can we account for this, given that the results in Figure 6.2b tend to rule out one possibility - that transcribed DNA is completely 'unpackaged' prior to transcription?

Π What other models can you envisage to explain this apparently paradoxical result?

1) It is possible that just the region being transcribed is 'naked' ie either the RNA polymerase 'nudges' the nucleosomes along ahead of it, or they are removed one by one, and then replaced after the polymerase has passed on.

2) Alternatively, the chromatin could retain the same basic nucleosome repeat (as reflected by the micrococcal digestion studies) but the structure is altered sufficiently to allow the free passage of RNA polymerases along the template. We shall see in the next section that the balance of evidence is in favour of the second proposition.

6.2 Transcriptionally active chromatin differs from inactive chromatin

6.2.1 Micrococcal nuclease digestion of active chromatin

If you look at Figure 6.3, you will see that the experiment shown in Figure 6.2 has been extended to show a range of chromatin samples digested for different times with micrococcal nuclease. Figure 6.3a again shows the total population of DNA fragments separated by electrophoresis and stained with ethidium bromide whilst Figure 6.3b shows a Southern blot of the same gel probed with a specific radioactive mRNA sequence from the same tissue.

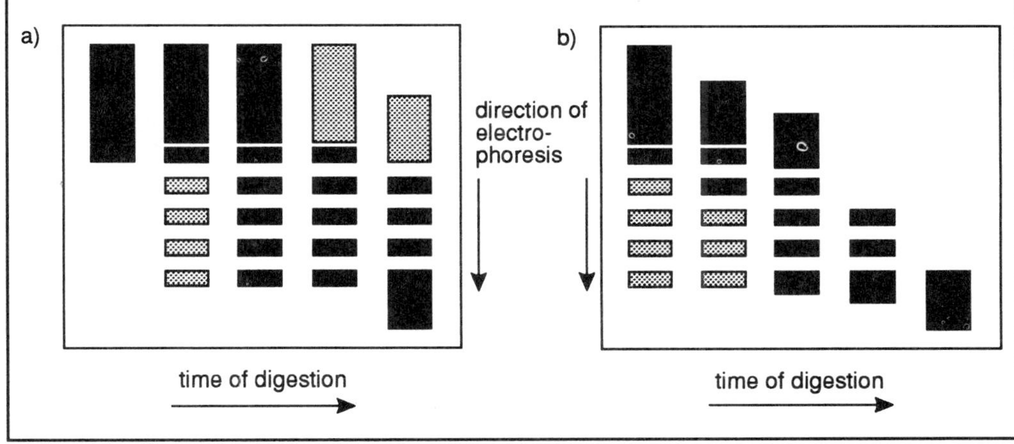

Figure 6.3 Time course of digestion of chromatin with micrococcal nuclease. a) DNA from the digest analysed by agarose gel electrophoresis. b) Southern blot of 6.3a probed with a specific mRNA from the same tissue. NB. These diagrams are highly stylised. In practice the tracks of the gel would contain a continuous background smear of fragments.

Π What do you notice about the size distribution of fragments in the total population compared to the specific fragments which code from the mRNA? What is your interpretation of this observation?

The gel shows that transcribed DNA is more susceptible to nuclease digestion than the average for the entire nuclear chromatin population, ie the structure of active chromatin differs in some way from the rest of the chromatin.

6.2.2 DNase I digestion of active chromatin

The same type of result can be obtained using a different nuclease enzyme, DNase I When chromatin is digested with DNase I, the region of DNA between the nucleosomes is not specifically cut, as it is by micrococcal nuclease. Therefore, instead of a 200 base pair ladder being produced, the DNA is gradually digested down to very small fragments. However, when chromatin is only partially digested by the enzyme, it appears that some regions of chromatin are more susceptible to degradation than are others and are digested more rapidly. We saw in the previous chapter that globin gene structure has been extensively characterised and the packaging of globin genes was one of the first to be analysed. Table 6.1 shows the sensitivity to degradation of globin gene sequences during DNase I treatment of chick erythrocyte chromatin, compared to ovalbumin gene sequences. (Ovalbumin, the major egg white protein is expressed only in the chick oviduct).

Gene sequence	Concentration of DNase (μg ml^{-1})						
	0	0.01	0.05	0.1	0.5	1.0	1.5
β globin	+++	+++	+	+	+	+/-	-
Ovalbumin	+++	+++	+++	+++	+++	+	+

Table 6.1 Digestion of chromatin from chick 14 day erythroid cells with varying concentration of DNase I. The disappearance of β globin and ovalbumin sequences was analysed by Southern blotting. The + and - symbols indicate the level of sequences detectable by hybridisation to appropriate probes.

Π Are the transcriptionally active globin genes more or less susceptible to DNase I digestion than the inactive ovalbumin genes?

transcriptionally active chromatin is DNase I sensitive

It should be obvious that the globin genes are more susceptible to DNase I activity than the ovalbumin genes or indeed the bulk of the nuclear DNA, ie transcriptionally active genes are more accessible to DNase I and are said to be packaged in DNase I sensitive chromatin.

SAQ 6.2

Since this experiment refers only to globin genes at the time of transcriptional activity, how would you modify this basic experiment to answer the following questions?

1) Is DNase I sensitivity a property of the chromatin structure only of highly expressed genes such as the globin genes, perhaps related to the packing of the transcription unit with RNA polymerase II molecules?

2) Is DNase I sensitivity only a property of genes which are in the process of being transcribed?

3) Is DNase I sensitivity a property only of the precise region of DNA which corresponds to the transcription unit?

In order to indicate the sort of experiment which might provide a relevant answer, decide which of the experimental results stated below best answers each of the questions asked above.

a) A region of around 15 kilobasepairs around the β globin gene locus is found to be DNase sensitive is chick erythrocytes.

b) This region remains sensitive in mature erythrocytes, when transcription of the β globin transcription unit has ceased.

c) So-called 'housekeeping' genes which require only a few protein molecules expressed per cell, are also DNase I sensitive.

You should be starting to form a view of the sort of relationship which exists between the structure of chromatin, as revealed by DNase I sensitivity, and transcriptional activity. The following pieces of information should enable you to crystallise this view.

As mentioned above, the egg white protein, ovalbumin, is expressed only in the chick oviduct. The ovalbumin gene sequence is therefore not DNase I sensitive in any other chick tissue except oviduct. Expression of the ovalbumin gene in the oviduct is induced by the steroid hormone oestradiol (see next chapter).

Π Table 6.2 compares the DNase I digestion of ovalbumin gene sequences in oviduct and erythrocyte chromatin with or without oestradiol stimulation of the tissue. Does oestradiol stimulation of transcription affect the chromatin structure?

	Concentration of DNase µg ml^{-1}					
Source of chromatin	0	0.01	0.05	0.1	0.5	1.0	1.5
Oestrogen stimulated oviduct	+++	+++	+	+	+	+	-
Non-stimulated oviduct	+++	+++	+	+	+	+	-
Erythrocyte	+++	+++	+++	+++	+++	+	+
Oestrogen treated erythrocyte	+++	+++	+++	+++	+++	+	+

Table 6.2 Digestion of chromatin from chick oviduct and erythrocyte cells with DNase I. The effect of oestrogen treatment on ovalbumin gene sequences analysed by Southern blotting. Symbols indicate the level of detectable sequences.

It should be apparent that the ovalbumin gene is sensitive to DNase I digestion in the oviduct even before oestradiol stimulates transcription of the gene. However, oestradiol treatment of erythrocytes neither alters the chromatin structure, nor induces ovalbumin synthesis.

SAQ 6.3

Taking this information together with the other pieces of evidence covered upto this point, which of the statements given below best describes the relationship between DNase I sensitivity of chromatin and transcription.

1) DNase I sensitivity of chromatin is caused by the act of transcription.

2) Transcription inevitably follows the change in chromatin structure which is reflected by DNase I sensitivity.

3) DNase I sensitivity is a necessary but not sufficient condition for transcriptional activity.

4) There is a precise correspondence between regions of DNase I sensitive chromatin and transcription units, but no clear cause and effect relationship between them.

The conclusion then is that the 'opening up' or relaxation of a large region of chromatin (making it more accessible to DNase I) is a primary stage of gene activation in eukaryotes which makes transcription units available for transcription. However at least one further stage is necessary before transcription actually occurs within a DNase I sensitive region (eg oestradiol induction of ovalbumin gene transcription). Figure 6.4 summarises this model.

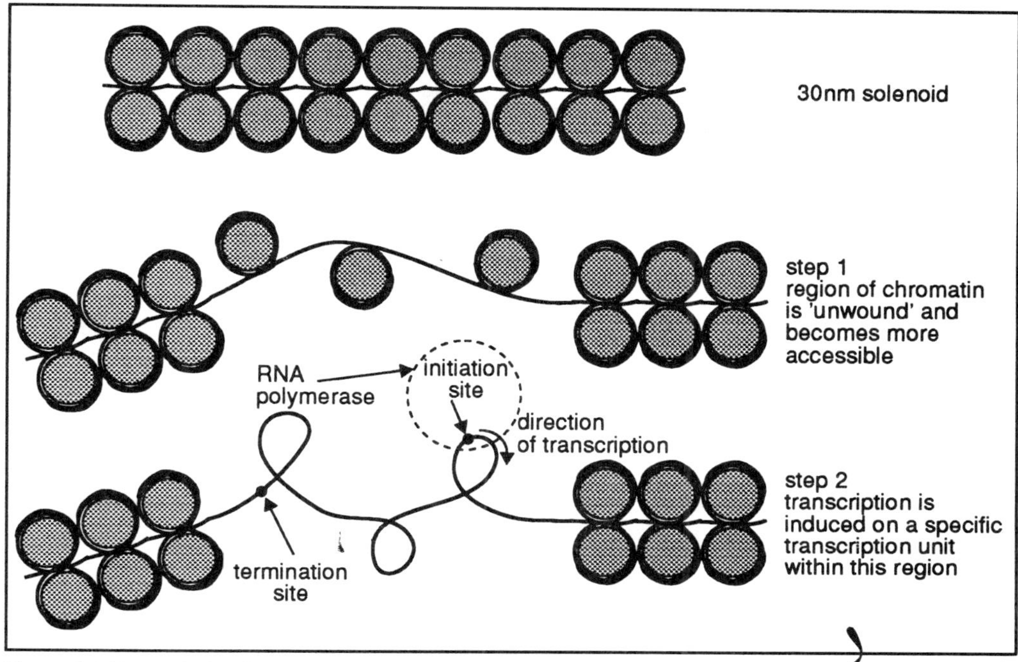

Figure 6.4 Transcriptional regulation in eukaryotes can be described as a two stage process.

6.2.3 Chromatin structure and long term commitment

long term commitment may occur early in development

Another related conclusion is that the pattern of chromatin structure is maintained over the long term within particular cells or tissues. This pattern appears to be determined early in the development of a particular cell type before transcriptional activity is required and may be retained after transcriptional activity has ceased. In other words, changes in chromatin structure may be involved in the long term commitment of cells during differentiation as described early in this book. A decision early in development to make certain genes available for transcription may be executed via chromatin structure. The question then is how is this chromatin state maintained throughout the subsequent cell divisions which occur.

6.2.4 DNA methylation

One clue to this problem comes from analysis of the pattern of DNA methylation in different tissues. Eukaryotic DNA is methylated in a number of positions, but the particular modification of interest here is the addition of a methyl group to the 5 positions of the cytosine ring to form 5 meC (Figure 6.5). This normally occurs at some of the sites where C is preceded by G.

Π What might the significance of specifically methylating the sequence 5′-GC-3′ be in terms of the methylation state of the opposite strand of DNA?

GC doublets may be methylated

The sequence 5′-GC-3′ will also occur in the opposite strand of DNA and may also be methylated. Not all GC doublets in DNA are methylated. Specific sites are methylated, which vary between different cell types in the same organisms.

5-azacytosine

The drug 5-azacytosine has the structure shown in Figure 6.5. The base can be incorporated into DNA in place of cytisine and can base pair quite normally with guanosine.

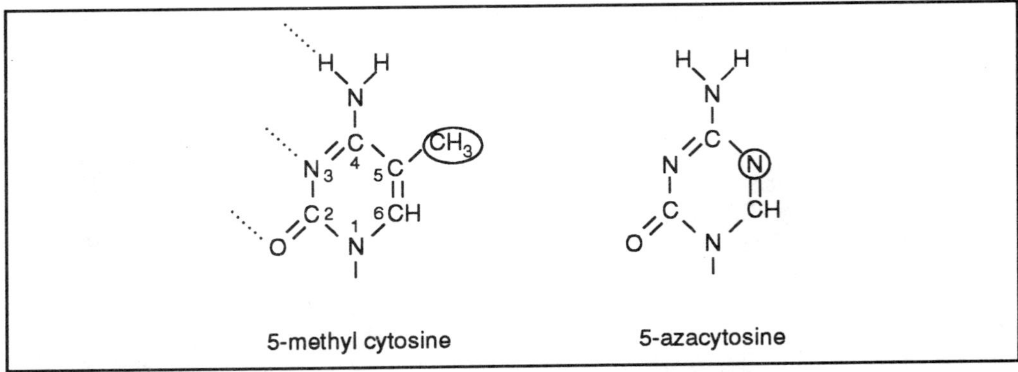

Figure 6.5 The structures of 5-methyl cytosine and 5-aza cytosine. Chemical groups which differ between the two structures are circled.

Π What effect would you expect incorporation of 5-azacytosine into DNA to have on the methylation pattern?

5-azacytosine causes under-methylation

Since the methyl group is also added to the 5-position, sites where 5-azacytisine is incorporated will not be capable of being methylated. In fact, the effect of 5-azacytisine treatment results in more extensive under-methylation than can be explained by this fact alone, presumably because it acts as an inhibitor of the DNA methylase.

Π When mouse epithelial cells in culture were treated with 5 azacytisine, some of the cells differentiated to form muscle-like cells in culture, capable of contraction! Propose a hypothesis to explain these results.

methylation determines commitment

In very broad terms it would appear that DNA methylation is involved in determining the long term commitment of cells. There is evidence to show that the methylation state of regions of DNA is related to transcriptional inactivation of the level of chromatin structure.

We have therefore seen evidence that both DNA methylation and chromatin structure are related to the long term commitment of cells to a particular path of development and differentiation.

SAQ 6.4

Draw a scheme to show how the symmetrical nature of the methylation site might allow a specific pattern of DNA methylation to be maintained through a series of cell divisions.

6.3 Eukaryote gene regulation as a two stage process

Given the conclusions which we reached about the role of chromatin in gene regulation - effectively a means of limiting the possibilities for gene activation in the future life of a differentiated cell line - we now need to look at the second stage of the process, ie how are individual transcription units regulated?

You should already be familiar with the types of mechanism involved in the regulation of transcription in prokaryotes. We will now discuss the extent to which eukaryote genes are regulated in the same fashion.

6.3.1 DNase I hypersensitive sites

Before we leave the role of chromatin packaging in gene regulation it is worth mentioning a further piece of evidence from the type of experiment already described. When chromatin is digested with very low levels of DNase I it is possible to detect the cutting of DNA at specific sites in the genome before the more random digestion of the bulk of the DNA. These specific sites are called DNase I hypersensitive sites. Their location differs between different cell types or tissues within the same organism. For example, the DNase I hypersensitive sites upstream of the adult β globin gene region of chick erythrocyte chromatin are shown in Figure 6.6. These sites are only found when the adult β globin gene is being transcribed. In the chick embryo, these sites are not detectable in erythrocyte chromatin, but other sites upstream from the embryonic β globin gene can be seen.

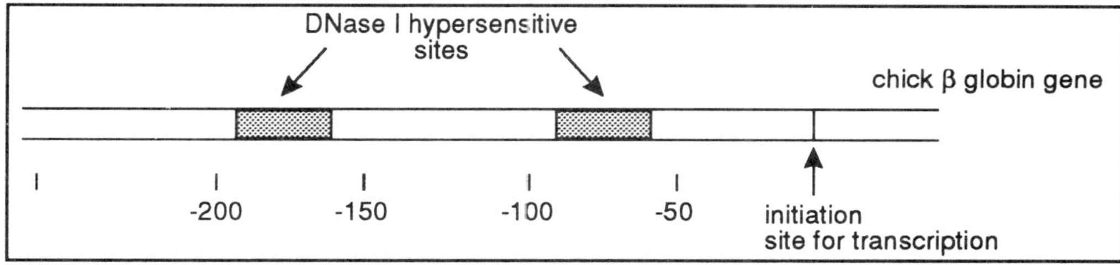

Figure 6.6 DNase I hypersensitive sites upstream of the chick β globin gene in chick erythrocytes.

Π What relationship is there between the position of the DNase I hypersensitive sites and transcriptional activity?

location of DNase I hypersensitive sites

You may need to look back at the previous chapter which describes the structure and developmental regulation of the globin gene family. Although the pattern is complex, one obvious feature is that DNase I hypersensitive sites are found precisely at the 5' end of the transcription units.

We will see in this chapter and the next that these DNase I hypersensitive sites probably represent specific regions of DNA which are not packaged with histones, although there may be other proteins bound to the DNA at these sites.

6.4 Promoter structure

6.4.1 Promoter consensus sequences

The location of some of the DNase I hypersensitive sites corresponds to the position expected for the promoters of these transcription units, (if we assume that eukaryote promoters are analogous to prokaryote promoters).

Whilst the promoter regions of prokaryotic transcription units can often be mapped by genetic analysis this is not possible in eukaryotes. The first type of information available about potential eukaryote promoters has come from the steady accumulation of DNA sequence information from a large number of different genes. It has therefore been possible to compare the 5′ upstream regions of several genes to try to identify similarities between them which might be features of the promoter region.

bases upstream from start site		-34 to -26									-18 to -26 →		start site
consensus sequence					T	A	T	A	A/T	A			
base frequency %	A	17	22	13	7	97	7	85	63	88	50	33	18
	T	17	27	10	82	2	93	10	37	10	33	12	15
	C	50	38	53	2	2	0	0	0	0	13	38	48
	G	15	13	23	10	0	0	5	0	2	3	17	18

Figure 6.7 Similarities (consensus sequences) found upstream of eukaryotic genes. These are the so called TATA box (see text).

consensus sequences

TATA box

Consensus sequences are sequences which occur recognisably, though not necessarily identically in a number of genes. For example, one of the promoter consensus sequences which appears in virtually all animal promoters is called the TATA box with a consensus sequence of TATAAA. Figure 6.7 gives the TATA consensus derived from a large number of genes. The numbers represent the percentage of all the genes studied which contained the base shown in that position (ie 82% of all genes contained T as the first base of the TATA sequence). The important point to note is that this is a consensus, not a precise identity, and therefore many promoter sequences differ in some positions from the TATA consensus, but a sequence which resembles the TATA consensus will still be identifiable.

The TATA box is relatively easy to spot when comparing promoters because it is always in the same position and orientation relative to the initiation site (ie it always reads 'TATA' towards the initiation site). This is not true of other common promoter consensus sequences.

particular consensus sequences found with particular types of genes

Table 6.3 shows a number of other recognised promoter consensus sequences. Apart from the TATA box, these may be found at a variable distance from the initiation point and in either orientation. They are also not universal consensus sequences but are typically associated with particular types of genes. For example, the GC box is not found in all genes but appears to be particularly related to the promoter region of 'housekeeping genes' (see Chapter 7). These are the sort of genes which encode enzymes or proteins essential for the normal maintenance of cells and are typically expressed constitutively at relatively low levels.

TATA box	TATAAA
CAAT box	GGCCAATCT
GC box	GGGCGG
Octamer	ATTTGCAT
κB	GGGACTTTCC
ATF	GTGACGT
AP-1	TGANTCA
AP-2	CCCCAGGC

Table 6.3 Promoter consensus sequences in mammalian genes.

6.4.2 The role of promoter consensus sequences

'surrogate' genetics

Having reviewed the evidence that there are several short consensus sequences common to all or some potential promoter regions it is necessary to ask whether these play a role in promoter function. There are a number of ways in which the relationship between consensus sequences and promoter function has been elucidated. It has already been mentioned that mapping transcription unit promoters by classical genetic techniques is virtually impossible in eukaryotes. However, the powerful techniques of gene cloning and manipulation have allowed a form of 'surrogate genetics' to be developed. Basically, it is possible to isolate a gene along with a large upstream region in which the promoter is presumed to reside. This can be checked by transferring the gene back into a eukaryote host cell or an *in vitro* transcription system to determine whether the gene is still transcriptionally functional. By systematically 'whittling down' the promoter region from the 5′ end, chopping out defined fragments of it and analysing individual base mutations throughout the promoter, it is then possible to identify those parts of the upstream region which are required for transcription to occur.

Π Figure 6.8 shows the regions upstream of the β globin gene which can be shown to be essential for promoter activity. Compare this with Table 6.3 and determine whether there is any correspondence between the essential promoter regions and previously identified consensus sequences.

Two of the regions identified upstream of the globin gene as essential for promoter functions contain the TATA box and CAAT box. Most of the rest of this region is apparently not essential for promoter function.

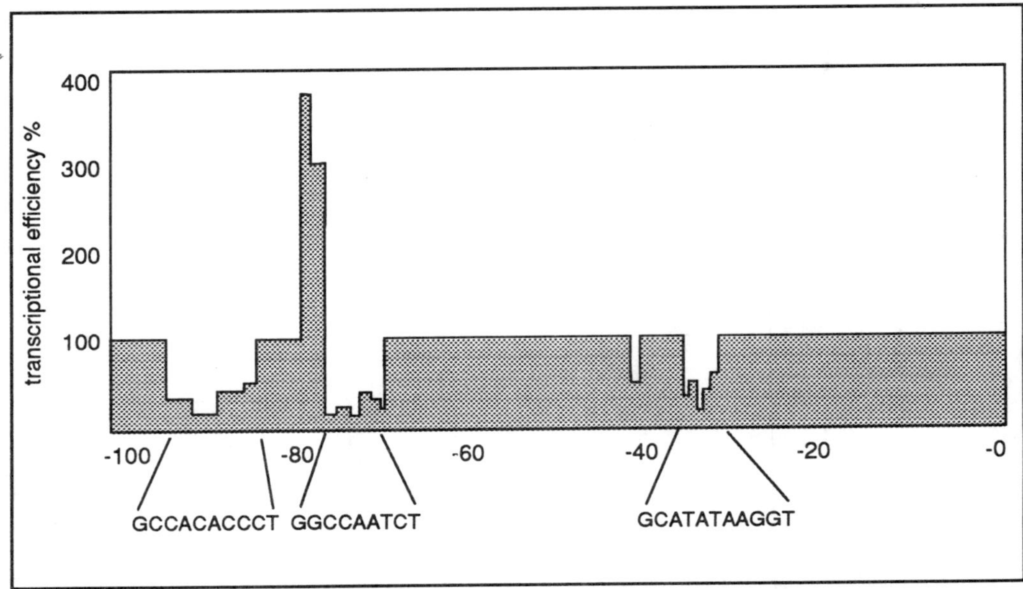

Figure 6.8 The effect of single base mutation on the promoter activity of the β globin promoter. Three main regions are identified where alteration of a single base has a marked inhibiting effect on transcription.

RNA polymerase II promoter related to consensus sequences

The experimental evidence from surrogate genetic analysis strongly indicates that the function of an RNA polymerase II promoter is related to the presence of relatively short sequence elements, many previously identified as consensus sequences. It is also possible to surmise that the TATA box, given its fixed position and (apparent) universal occurrence is rather different in function to the other sequence elements. It may not have escaped your notice that the TATA box in many ways resembles the -10 or Pribnow box found in prokaryote promoters. The role of the Pribnow box is binding and hence positioning and orientation of RNA polymerase in prokaryotes. Is the TATA box, then, also involved in binding RNA polymerase II in eukaryotes? As we shall see in the next section, although the TATA box may have a similar function, it does not operate in the same way as the Pribnow box. Indeed, many of the mechanisms of transcriptional activation in eukaryotes differ from prokaryotes.

6.4.3 Consensus sequences and specific DNA binding proteins

Having described the sort of evidence which indicates a role for a variety of different short sequences within a promoter, we will move on to examine what sort of role these sequences might have. One simple hypothesis is that these sequences are involved in binding specific proteins, either associated with RNA polymerase itself or some other type of transcriptional factors. It is therefore worthwhile spending a short time looking at some of the techniques available for studying the interaction between DNA binding proteins and specific DNA sequences.

footprinting

One technique for determining the precise region of a DNA molecules which is actually bound to a protein is the 'DNA footprinting' technique. This is basically a nuclease protection assay which is similar in concept to the nuclease digestion of chromatin. If a fragment of DNA is radioactivity labelled at one end and partially digested by DNase I, random cutting by the enzyme will produce a radioactive population of fragments of every possible length from 1 - n bases (Figure 6.9). It is possible to separate this population of labelled fragments by polyacrylamide gel electrophoresis in a long thin slab of polyacrylamide capable of resolving DNA fragments which differ in length by only one base. If the DNA molecule is actually protected from attack in one particular region due to the binding of a protein, this region will appear blank on the gel since the enzyme will not be able to cut in this region.

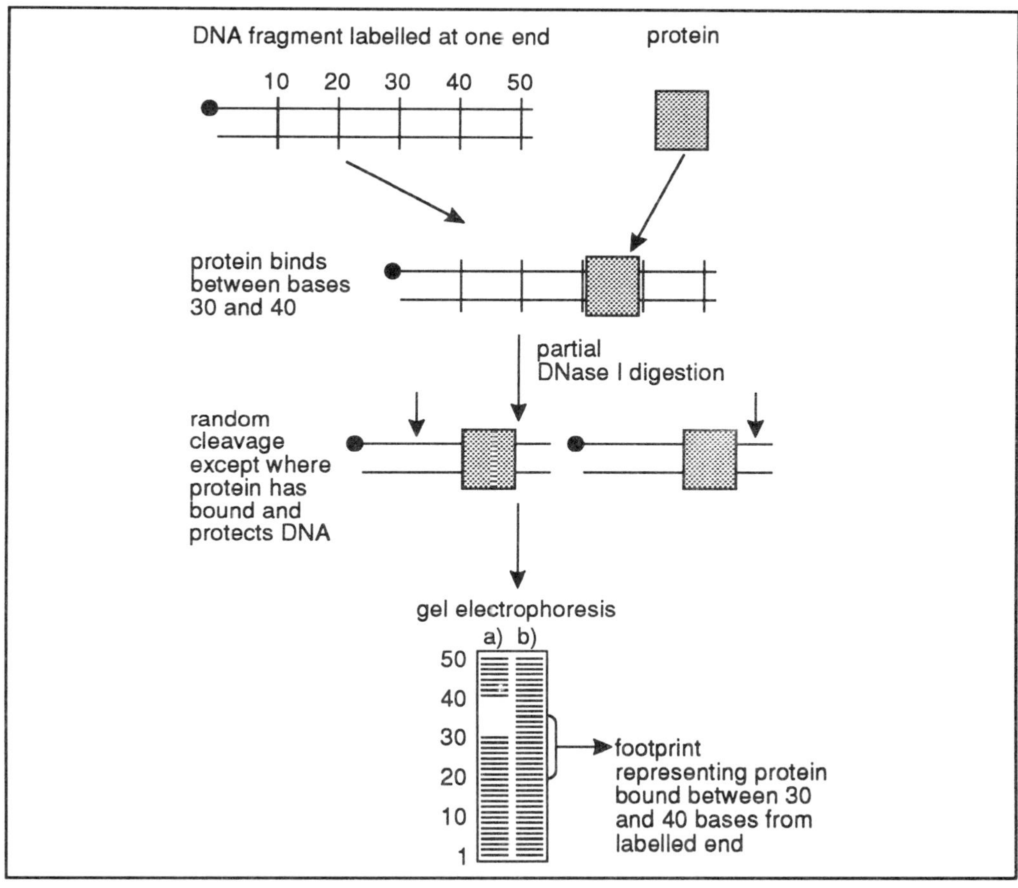

Figure 6.9 DNase I footprinting assay. Specific binding of a protein to a site in the labelled DNA fragment protects the DNA from DNase I digestion. The specific binding site will appear as a footprint in the ladder of bands produced by the DNA being cut at all other points by DNase I. a) DNA fragment incubated with binding protein. b) Control DNA fragment without protein.

While DNA footprinting allows the DNA sequence involved in protein binding to be analysed, another relatively simple technique can be used to assay for the specific DNA binding activity of proteins. The gel retardation assay makes use of the fact that a fragment of DNA will migrate more slowly in an agarose gel if it has a protein bound to it. The ability of protein samples to bind to particular DNA sequences can therefore be assayed using radioactive DNA fragments of known sequence. Figure 6.10 illustrates a gel retardation assay.

gel retardation assay

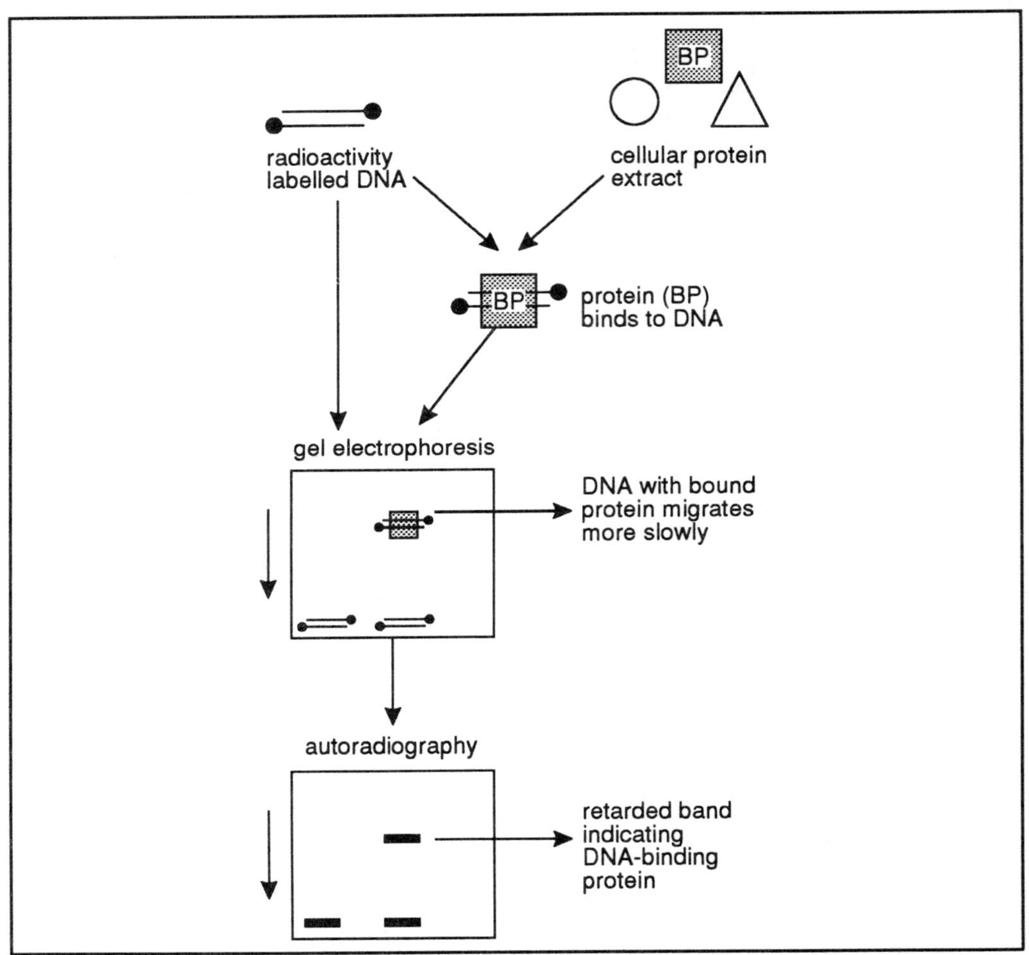

Figure 6.10 Gel retardation assay. Binding of a specific DNA binding protein (BP) to the radioactively labelled DNA causes it to move more slowly during gel electrophoresis. A retarded band running slower than the naked DNA fragment can be detected by autoradiography.

SAQ 6.5

One problem with the gel retardation assay is that non-specific DNA binding proteins will also retard the mobility of the labelled DNA fragment. One way of determining the specificity of binding is to include a large excess of an unlabelled second DNA sequence in one of the binding reactions. Draw the autoradiographic result if: a) the protein in the assay is a non-specific binding protein and b) the protein specifically binds to the radioactive fragment of DNA.

6.5 The role of the TATA box

role of specific transcription factors

The invariant position and orientation of the TATA box suggest a role in the correct positioning and choice of start point for RNA polymerase II. This has been confirmed by examining the effects of various manipulation of the TATA box region on transcription *in vivo* or *in vitro*. For example, deletion of the TATA box causes the site of initiation to become erratic. It is now known that the TATA box does not act by directly binding RNA polymerase II. Rather, a number of specific transcription factors bind to this region in a sequential assembly of the transcription complex, as shown in a simplified fashion in Figure 6.11.

The first step in the formation of the complex is the binding of the transcription factor TFIID. DNA footprinting shows this initial binding to protect a region that extends from the TATA box to about 20 bases upstream of the box. A second factor TFIIA is added further upstream, followed by addition of TFIIB, which gives additional protection to the region around the start site. RNA polymerase II can then join the complex at this stage, followed by downstream binding of TFIIE. The complex is then able to initiate transcription upon addition of ribonucleotide triphosphates. The transcription factors are presumed to be released after initiation of transcription.

SAQ 6.6

Compare this sequence of events with the initiation of transcription in prokaryotes.

'core' promoter

The binding of the TFII factors to the TATA box region appears to be part of a general transcription apparatus necessary for assembling the transcription complex in which RNA polymerase II is capable of initiation. The sequences around the TATA box and transcription start site thus comprise a 'core' promoter at which this general transcription apparatus is assembled. It has been shown that the core promoter is sufficient to direct the correct initiation of transcription *in vitro*.

However, transfer of genes containing only the core promoter back into cells shows the core promoter is not normally sufficient to produce detectable levels of transcription *in vivo*. In general, upstream sequence elements are also required for transcription to take place *in vivo*.

Π What might the role of these upstream elements be, given that many of them are further upstream than the region around the core promoter protected by RNA polymerase II? Can you propose a model to explain the possible role of the upstream elements?

One possibility is that RNA polymerase II initially binds to the site upstream and then moves along the DNA to the initiation site. Another possibility is that although the upstream sites appear to some distance from the core promoter when drawn as positions on a linear molecule, the sites may be in close proximity depending upon the conformation of the DNA protein complex which comprises the chromatin. Much of the evidence points to this second alternative, as does the information to be discussed in the next section.

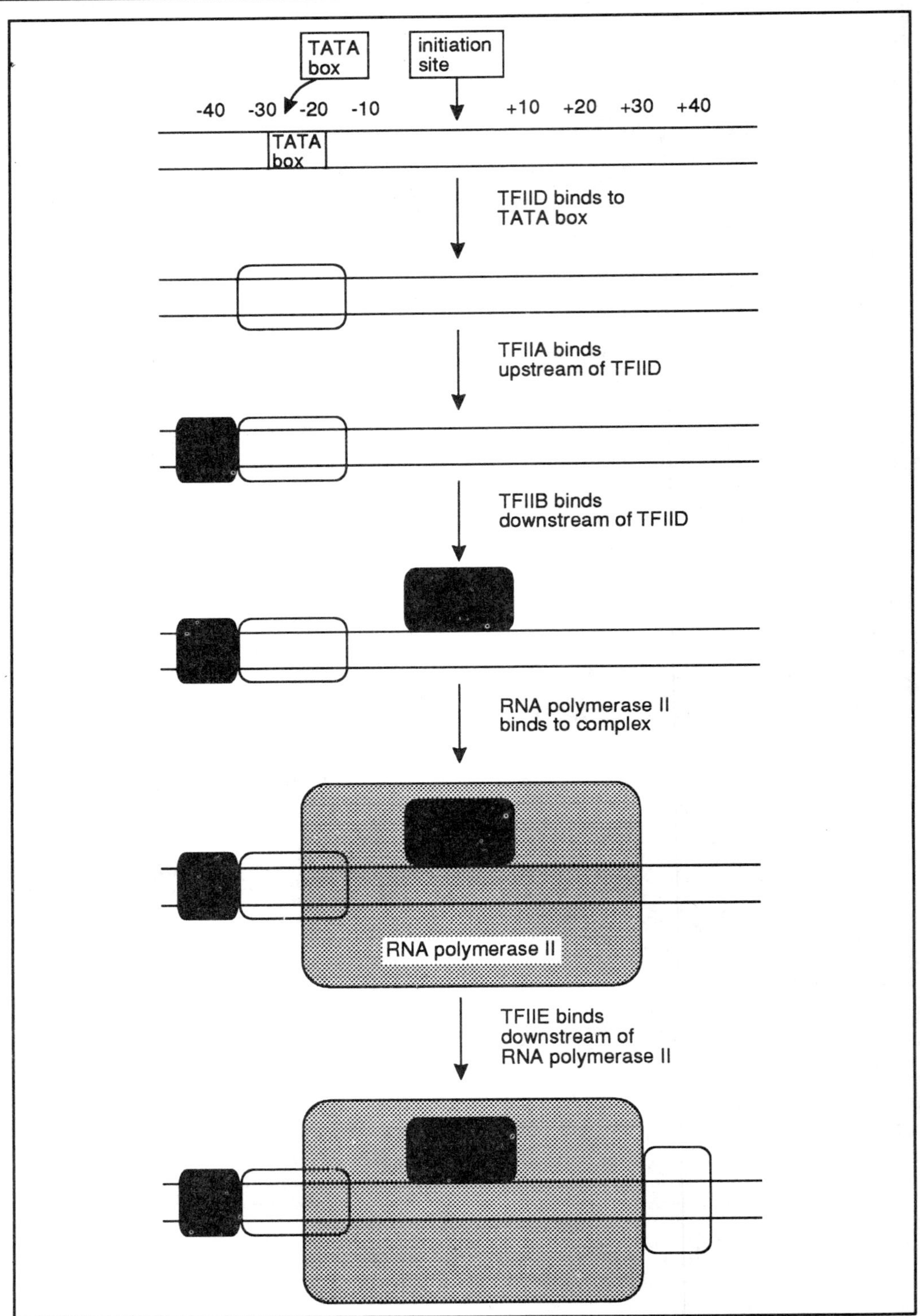

Figure 6.11 Role of transcription factors in initiation of transcription (see text for further details). TFIID, TFIIA, TFIIB and TFIIE are transcription factors. These will be described in more detail in Chapter 7.

6.6 Enhancer elements

enhancers may be remote from the start site

It should already be apparent that the promoter itself is a complex structure comprising the core promoter region with several additional sequence elements upstream. It is also known that the activity of a promoter can be greatly increased by other groups of regulatory sequence elements at a much greater distance from the start site. These regions are called enhancers. Manipulations of DNA show that enhancers can stimulate any promoter in their vicinity (where the vicinity can mean several thousand base pairs!).

The distance from the promoter can vary, as can the orientation of the enhancer element. Indeed some enhancers are found downstream of the gene or actually within the transcription unit which is affected. In general enhancers also contain a number of short sequence elements which have also been shown to bind specific DNA binding proteins. Some of the same sequence elements occur in both promoters and enhancers. The next chapter looks in more detail at the possible mode of action of transcription factors and short regulatory sequence elements in promoters and enhancers.

Summary and objectives

In this chapter we have described the molecular evidence that supports the idea that chromatin becomes relatively unpacked when it is transcribed although the basic nucleosomal structure of chromatin is retained. We also described a role for DNA methylation as a device for controlling gene expression in eukaryotes.

We then described the evidence that segments of DNA upstream from the start site contain consensus sequences. We described the so called TATA box and a variety of other consensus sequences associated with eukaryotic genes. We discussed the role of these consensus sequences as promoters and enhancers. This led us to describe techniques used to investigate DNA-binding proteins and the role of transcription factors in the initiation of transcription.

Now that you have completed this chapter you should be able to:

- assess the evidence that transcriptionally active DNA is packaged in chromatin;
- evaluate the evidence that transcriptionally active chromatin differs structurally from inactive chromatin;
- draw conclusions from the lack of a direct relationship between DNase sensitivity of chromatin and the act of transcription;
- relate chromatin structure and DNA methylation to long term commitment is differentiated eukaryote cells;
- evaluate the significance of promoter consensus sequences;
- demonstrate that promoter consensus sequences do play a role in the regulation of transcription;
- relate regulatory sequence elements to specific DNA binding proteins;
- explain the difference between promoter and enhancer elements;
- compare current models for promoter and enhancer action.

Transcription factors and the co-ordination of gene expression

Transcription factors and the co-ordination of gene expression

7.1 Introduction

constitutive and modulated gene expression

Transcription factors were introduced to you in the previous chapter. In this chapter they are discussed in the context of the co-ordination of gene expression. Two forms of regulation will be considered. First, many genes in eukaryotes are apparently constitutive (ie they are expressed all the time in all tissues) but the regulation of the level of their expression is dependent upon transcription factors that are separate from RNA polymerase II. In the second form, gene expression is regulated either for individual genes or for groups of genes in response to various stimuli, and during growth and development. Current evidence indicates that transcription factors play the most important role in regulating these genes and that the biology of the transcription factors can display quite complicated forms of interaction with DNA, with RNA polymerase, with intermediate messengers, and with other transcription factors.

In this chapter we will therefore describe transcription factors and how are they able to perform many important functions?

First let us consider the number of genes in the eukaryote genome and the number of transcription factors which are required to regulate their expression.

Π How many genes are encoded by a higher eukaryote?

A reasonable estimate would be in terms of 10s of thousands, perhaps 30-40 000 in a mammalian genome.

Π Is it likely that each gene is activated by its own specific transcription factor?

Almost certainly not - this would require a further 40 000 genes coding for transcription factors! Each of the transcription factor genes would then by regulated by its own transcription factor, and so on *ad infinitum*!

Π How many transcription factors are there likely to be?

The answer to this question is of course unknown. However the number is not likely to reflect a direct one-to-one relationship between transcription factors and regulated genes. Instead, it is becoming clear that single transcription factors can regulate a number of genes. In this chapter we will investigate how this is possible.

7.2 Types of transcription factors and gene structure

Prior to looking at the complexities of the subject we will first look at a few simple examples to describe how TFs (as transcription factors will be known from now on) actually work and how they differ in their structure and binding properties.

This chapter will concentrate exclusively on TFs which regulate protein-coding genes.

Π Which RNA polymerase is involved in the transcription of protein-coding genes in eukaryotes?

RNA polymerase II is the enzyme involved. If you had forgotten this, check back to Chapter 5 to revise the roles of RNA polymerase I, II and III. RNA polymerase II initiates transcription after forming on initiation complex with TFs in a region known as the TATA box (see previous chapter).

A eukaryote gene, chick β-globin, that is transcribed by RNA polymerase II is shown in Figure 7.1. In this figure we have shown a variety of TFs which interact with the promoter and enhancer regions. Note that the transcription factors are often given initials which are taken from the system from which they were demonstrated or the activity they display. Here we will not burden you with all of their derivations. It is more important that you understand their main properties. Perhaps when we know more about these factors, we will able to use more systematic names.

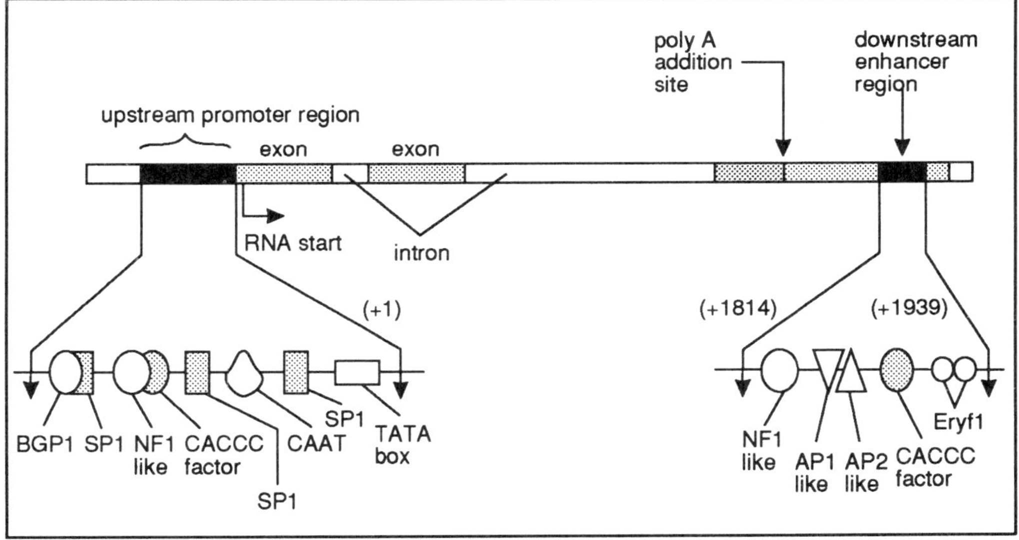

Figure 7.1 Diagrammatic representation of chick β-globin gene showing the transcription factors involved with initiation of transcription. Some of these transcription factors are discussed in further detail in the text.

Π What parts of this transcription unit will appear in chick β-globin mRNA?

The three exons between the RNA start site and the poly A addition site will be spliced together to form the mRNA.

The physical relationship between the TATA box and start site for RNA synthesis is clear from this diagram. What is also clear is the complex nature of the promoter region upstream from the TATA box, which contains many separate regulatory elements. There is also a downstream enhancer region which contains similar types of regulatory elements.

Π Which elements are common to both the upstream promoter and downstream enhancer region?

some TFs common to both promoter and enhancer regions

The NF1-like and CACCC elements appear in both regions. The complexity of transcriptional regulation should be apparent from this diagram and you should realise it is still not clear what the effects of binding of each factor is on the regulation of this gene. What is clear, however, is that transcriptional activation requires the concerted binding of several different factors to both the promoter and enhancer regions and the gene may be inactive if any single key TF is absent. This requirement for binding of several different TFs to the promoter region appears to be typical of many eukaryote transcription units. What is not yet clear is how the binding of proteins to upstream promoter regions and enhancer elements actually activates transcription.

We will briefly consider as an introduction to transcription factors, two factors at this point; Sp1, a constitutive factor and the AP1 complex, an inducible element, (the biology of which we will consider later in much more detail).

7.2.1 Sp1 transcription factor

Sp1 binds to constitutive gene promoters

First we will deal with the so called constitutive factor Sp1 as this will introduce us to many of the basic properties of TFs. This protein interacts with DNA upstream from the start site within the promoter region (see, for example, Figure 7.1). In the example shown there are three possible specific binding sites to which Sp1 can attach. The Sp1 binding site has the consensus sequence GGGCGG.

Π What is the name given to this promoter consensus sequence?

This is the GC box which was described in the previous chapter. It is found in regulatory elements in a number of genes that are expressed in all cell types and as one would expect, the transcription factor is also found in all cell types. The binding of the transcription factor is required to stimulate gene expression. Without the transcription factor the genes are only expressed at a basal level.

zinc finger DNA binding motif

Binding is mediated through three copies of a particular structure within the transcription factor called a zinc finger DNA binding motif. The structure of this and other DNA binding motifs will be described in the next section. Activation of transcription is presumed to occur through an interaction of the bound TF and the RNA polymerase initiation complex. This requires the presence of two glutamine rich regions or 'domains' in Sp1. This interaction is made possible because Sp1 bound to upstream promoter elements is apparently closely associated with the initiation complex. So we visualise the following situation.

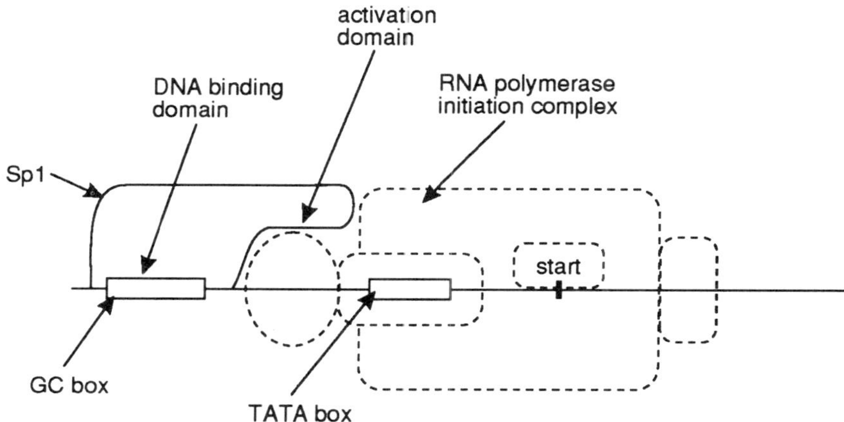

Note that the RNA polymerase initiation complex has been drawn with dashed lines. The initiation complex is composed of RNA polymerase and the transcription factors TFIIA, TFIIB, TFIID and TFIIE (see Figure 6.11).

7.2.2 Transcription factors and enhancer elements

Π The binding site of the AP1-like transcription factors are a long way downstream from the promoter, so how could binding of this TF to the enhancer stimulate transcription?

See if you can draw a model to explain this.

One possible mechanism is shown in Figure 7.2. It has been proposed that the intervening DNA is looped out allowing an intimate contact to be made between the transcription factor and the initiation complex. (There are other possible models of enhancer action which do not require this intimate contact but instead propose long distance effects on DNA or chromatin structure. However, it is not easy to accommodate the fact that the same TFs may bind to promoter and enhancer regions into these models).

SAQ 7.1

Which of the following statements are true?

1) In a typical eukaryotic gene, transcription factors only need to bind to specific nucleotide sequences (regulatory elements) in the promoter region in order for the initiation of transcription to occur.

2) The promoter region of a gene binds a completely different set of transcription factors from the enhancer region.

3) Genes that contain a GC box in their promoter region bind the transcription factor Sp1.

4) The molecules of the transcription factor Sp1 have two distinctive components. One binds with a specific nucleotide sequence in the DNA within the promoter region. The other interacts with the complex involved in the initiation of transcription.

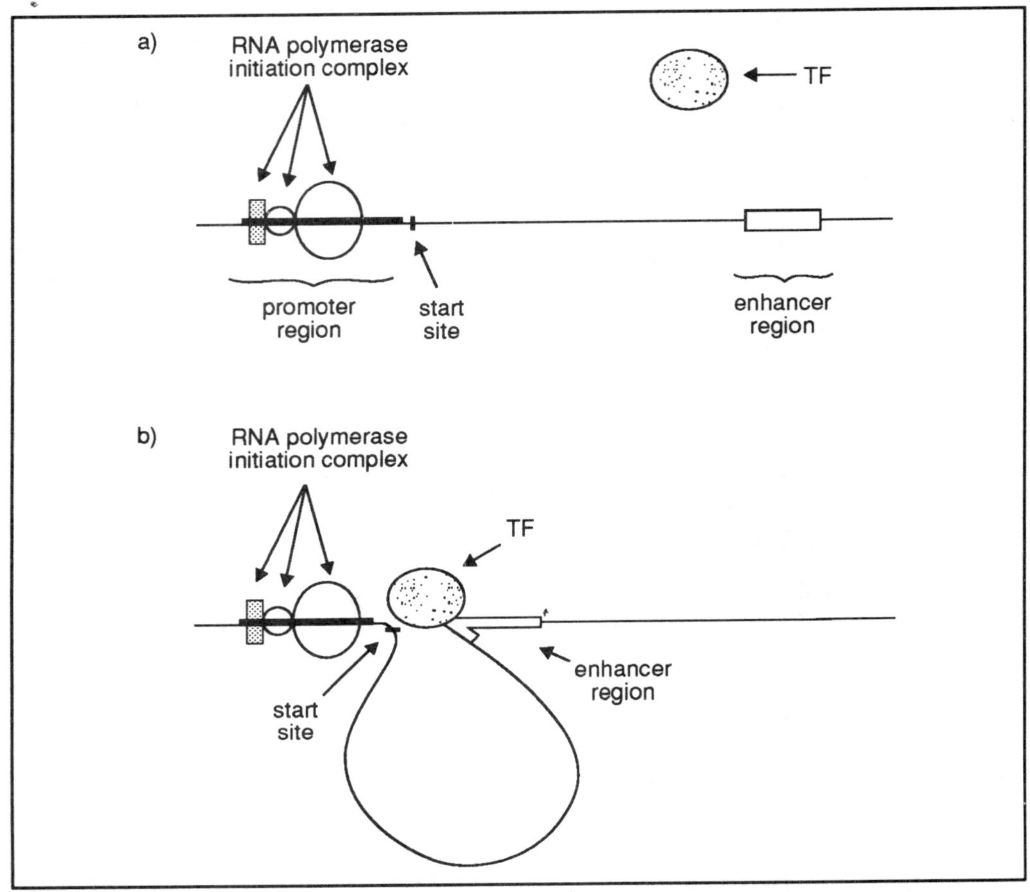

Figure 7.2 Model of how a TF binding to an enhancer region at a distance from the promoter region may influence transcription. a) No TF bound to the enhancer region. b) TF bound to the enhancer with the intervening DNA looped out. The enhancer bound TF is brought into close contact with the RNA polymerase initiation complex thereby influencing its activity.

7.3 General properties of transcription factors

At this point it may be prudent to consider some of the specific terms and properties of TF in a more general way. So far we have introduced three very important properties of TF Sp1:

- DNA binding motif;
- activation domain;
- the site at which the TF binds to the DNA.

Let us now briefly consider these for other TFs: are they the same or are they different?

Clearly there must be some features TFs have in common but others that must be distinct.

7.3.1 DNA binding motifs

Read Table 7.1 carefully, it contains a lot of information about so called DNA binding motifs.

DNA binding motif	Structure of motif	Transcription factors containing motif	Comments
Homeobox	Helix-turn-helix	*Drosophila* homeotic genes, mouse *Hox* genes	Structurally related to similar motif in bacteriophage proteins
POU	Helix-turn-helix and adjacent helical region	Mammalian Oct-1, Oct-2, Pit-1, nematode *unc*86	Related to homeodomain
Cysteine-histidine Zinc finger	Multiple fingers, each co-ordinating a zinc atom	TFIIIA, Sp1 etc	May form β-sheet and adjacent α-helical structure
Cysteine-cysteine Zinc finger	Single pair of fingers, each co-ordinating a zinc atom	Steroid-thyroid hormone receptor family	Related motifs in EIA, GAL4 etc
Basic domain	α-helical	*c-fos*, *c-jun*, c-Myc, MyoD etc	Linked with leucine zipper (BZIP) and/or helix-loop-helix dimerisation motifs

Table 7.1 Some important DNA binding motifs (see text for explanation).

Π You saw in Chapter 4 proteins like histones binds tightly to DNA. What sort of interaction occurs between the histones and DNA which results in the strong binding?

The interaction is largely ionic bonding between the positively charged lysine and arginine sidechains of the histones and the negatively charged phosphate groups of the DNA backbone.

Π Is this likely to be the basis of the DNA binding properties of the transcription factor motifs?

specific binding involves nucleotide bases

No. The interaction between histones and DNA is non-specific. In other words they do not require a specific DNA sequence for binding. Any interaction between a specific DNA binding protein and its binding sequence must involve the base pairs themselves, not just the phosphate sugar backbone. However, it is possible that non-specific interaction between a binding protein and DNA will occur to stabilise the DNA-protein complex.

Π What sort of interactions between a protein and DNA bases are likely to provide specific recognition and binding?

We will see the answer to this when we consider the structure of DNA binding motifs, ie the region of DNA binding proteins that actually interact with the DNA.

A number of many different types of DNA binding motif have been identified. In Table 7.1 we listed several different DNA binding motifs that have been identified in a number of different transcription factors. Perhaps the most important to discuss here in any depth, as we will refer to them again, are the homeobox motif, zinc fingers (which we have already mentioned) and the basic domain. We will briefly consider the structure of these domains.

helix-turn-helix motif

The homeobox domain is a form of helix-turn-helix motif. It is 60 amino acid residues in length and forms a stable folded structure that can bind DNA. The motif is shown in Figure 7.3.

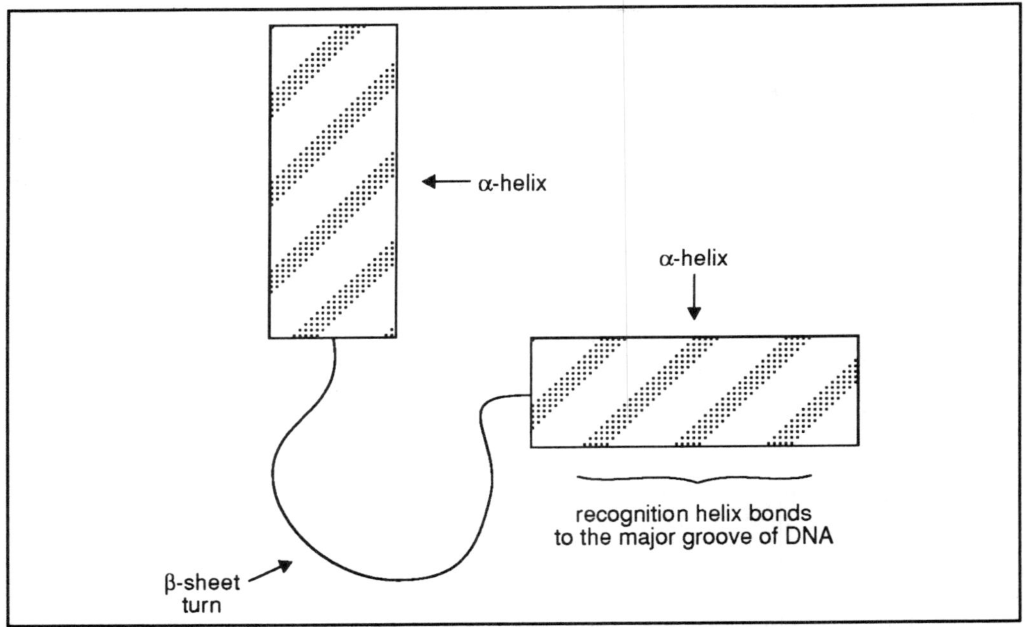

Figure 7.3 Stylised representation of the helix-turn-helix motif found in the homeobox types of TFs.

recognition helix

For the homeobox protein one helix (the recognition helix) is involved in specific DNA binding within the major groove of the DNA (B-form). This seems to be the case for many of the various types of DNA binding motifs and some general rules apply to all of them as regard to sequence specific recognition. Within the major groove, hydrogen bonding seems critical for recognition and this is usually between amino acid side chains in the recognition helix of the TF and the bases within the major groove. However, there does not appear to be a single relationship between the side chains and the bases. Rather it appears that folding and docking of the entire protein helps to control site specific recognition.

Π Several well characterised prokaryote repressor proteins contain helix-turn-helix motifs, including the λ repressor (cI) and Cro proteins. It is possible to create hybrid proteins in which the recognition helix of one DNA binding proteins has been replaced by the short stretch of recognition helix from a different protein with different sequence specificity. Predict the binding properties of such a hybrid protein.

When such experiments have been performed, the hybrid protein has been shown, in at least some cases, to bind to the sequence specified by the recognition helix.

Many DNA binding motifs cannot function as single units but function as multiple forms. This is the case with zinc fingers and the basic domain.

zinc fingers have loop structure

Zinc fingers are domains that have a loop structure (approximately 12 amino acids) stabilised by interactions between a zinc atom and four amino acids.

At least two different forms of this structure exist where the amino acids are either 2 cysteines and 2 histidines or 4 cysteines (Figure 7.4). The two structures are quite distinct. In the case of the former, multiple tandem copies of the zinc finger are formed in the transcription factors.

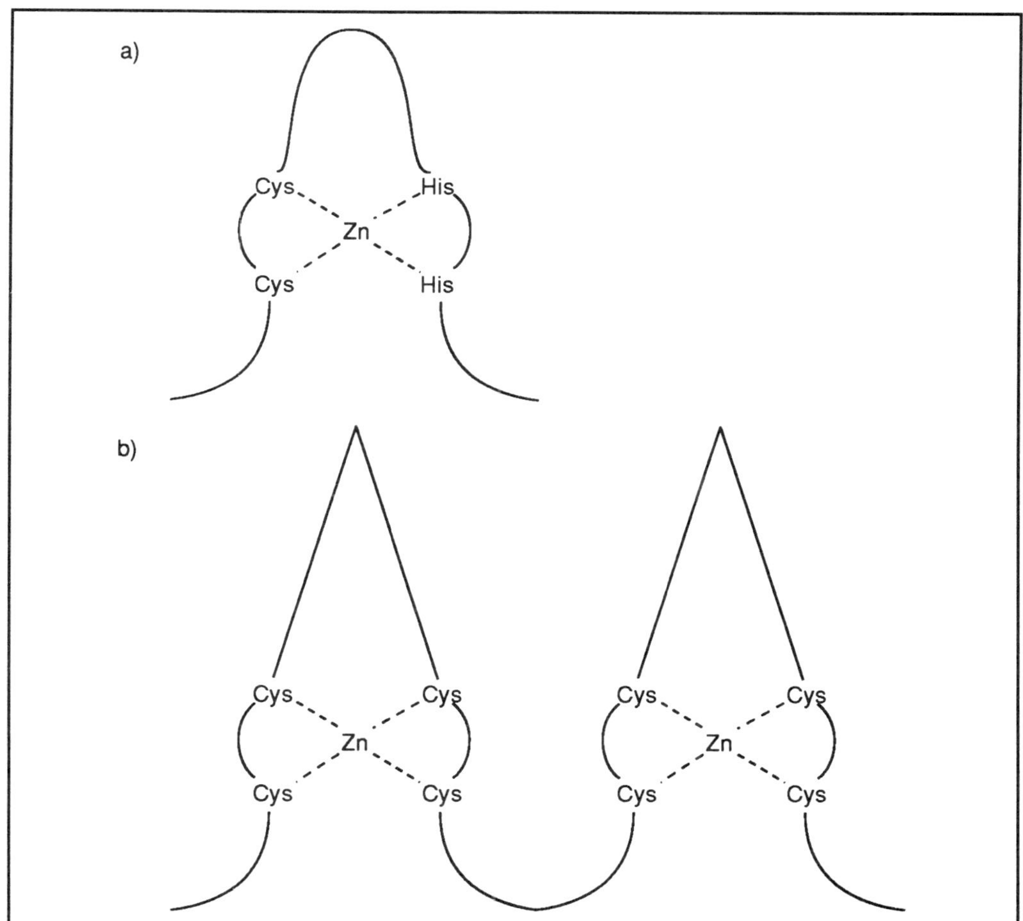

Figure 7.4 a) Four cysteine zinc fingers. b) Two cysteine two histidine zinc finger motif.

Π Which TF have you already met which contains zinc fingers?

multi-cysteine zinc fingers typical of steroid hormone receptors

Sp1 contains three copies of Cys_2/His_2 fingers in tandem. Interaction between the zinc fingers seems to be important for DNA binding, because isolated single zinc fingers cannot bind to DNA. In the case of the multi-cysteine zinc fingers it seems that two zinc fingers form a functional DNA binding unit and the presence of adjacent structures is important to the maintenance of the domain. Examples of TFs with this sort of motif are the steroid hormone receptor proteins, for example the glucocorticoid and oestrogen receptors.

Π When the Cys_2/Cys_2 fingers of the oestrogen receptor are deleted and replaced with those of the glucocorticoid receptor, the hybrid protein recognised the GRE sequence (glucocorticoid response element) but not the ERE (oestrogen response element). What does this indicate about the role of the Cys_2/Cys_2 fingers?

It shows that the region containing the zinc fingers is involved in binding DNA and establishing the specificity of DNA recognition.

Π When the steroid receptor genes are mutated so that the second two Cys residues of the zinc finger are mutated to His residue (ie the protein now has a Cys_2/His_2 structure) the protein is unable to activate target genes. What does this indicate about the two types of zinc finger motif?

They are not interchangeable and presumably interact with DNA rather differently.

basic domains are rich in arg and lys

Finally we will briefly discuss the basic domain. This is a region of about 30 amino acids that is rich in arginine and lysine and binds DNA. However DNA binding only occurs once the transcription factor containing the domain has dimerised. Formation of dimers is often driven by other motifs or dimerisation regions either of the leucine zipper type (see Figure 7.5) or amphipathic helix-loop-helix motifs.

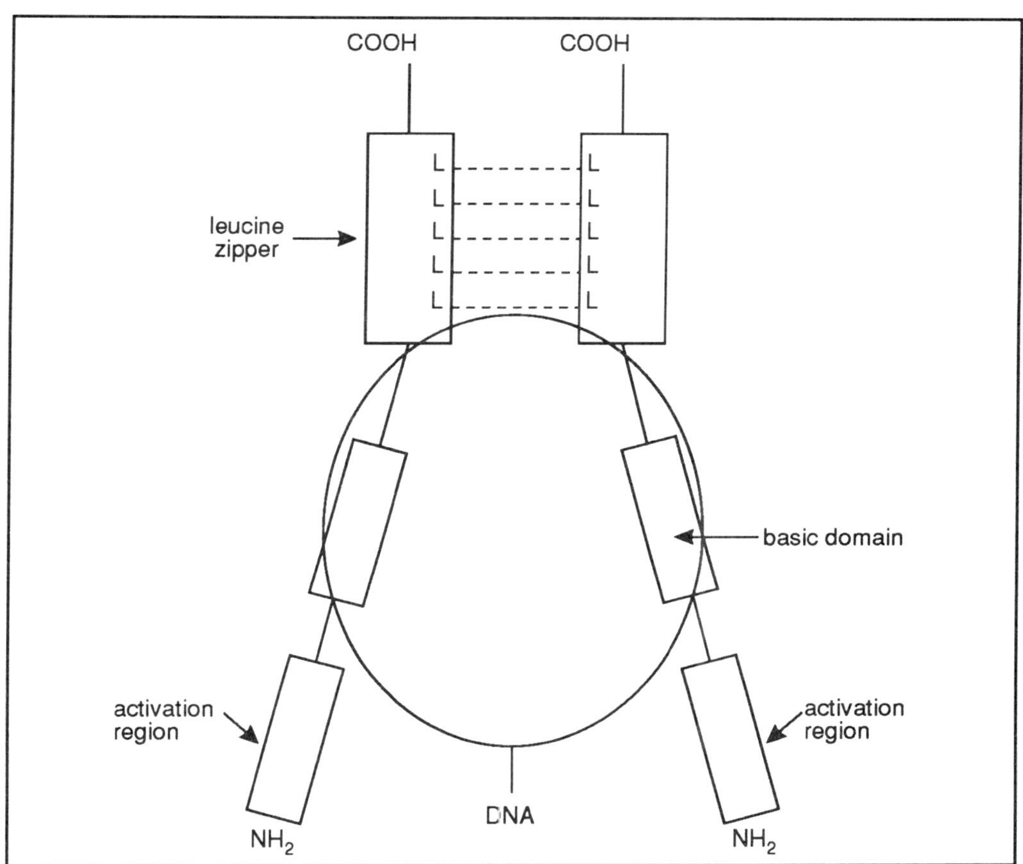

Figure 7.5 A BZIP protein. This figure shows a stylised relationship between the basic domains and a 'leucine zipper'. In reality the two helices in the leucine zipper regions of the molecules wrap around each other and are held together predominantly by hydrophobic interactions.

leucine zippers have leucines at every 7th residue

The leucine zipper is composed of a heptad repeat of leucines (ie a leucine residue occurs every 7th residue) over a region of 30-40 residues with a conserved repeat of hydrophobic residues occurring three residues on the N-terminal side of the leucines. The dimer is formed by two leucine zipper region helical coils wrapping around each other (Figure 7.5).

The amphipathic helix-loop-helix dimerisation motifs serve the same function as the leucine zipper and encourages dimerisation. In amphipathic helix-loop-helix dimerisation, basic (positively charged) residues and acidic (negatively charged) residue on the helices interact. There are also hydrophobic interactions. The important thing about these dimerisation regions is that they encourage the formation of homodimers (dimer with 2 identical subunits) and allow the transcription factors to bind to the DNA. They also enable the formation of heterodimers between different transcription factors and this, as we shall discuss later, can have profound effects on the regulation of transcription factor function.

7.3.2 DNA binding sequences

CCAAT box found in many promoters

A number of consensus sequences for TF binding sites are listed in Table 7.2. It is obvious that not a lot can be learned from studying these sequences in isolation. In many cases the sequences are specific for the binding of particular transcription factors. However the CCAAT box (Chapter 6) which has been found in the promoters of a lot of genes including thymidine kinase and *hsp*70, binds a number of different proteins, some of which are tissue specific and some of which are expressed in all tissues.

Consensus sequence	Gene containing sequence	Response to	TF
CTNGAATNTT CTAG	*hsp*27, *hsp*70, *hsp*83	Heat	Heat shock transcription factor
T/GT/ACGTCA	α-gonadotrophin, *hsp*70, fibronectin, somatostatin	cAMP	CREB
TGAGTCAG	α-antitrypsin, collagenase, metallothionein IIA	Esters (mainly of phorbol)	AP1 (Fos/Jun)
GATGTCCATATT AGGACATC	*c-fos*, γ-actin	Growth factors	Serum response factor (SRF)
AGGTCANNN TGACCT	Ovalbumin, conalbumin	Oestrogen	oestrogen receptor (ER)
TCAGGTCAT	Growth hormone, myosin heavy chain	Thyroid hormone, retinoic acid	Thyroid hormone (THR) and retinoic acid receptors (RAR)
GGTACANNN	Tryptophan oxygenase, uteroglobin, lysozyme	Glucocorticoid hormone, progesterone	Glucocorticoid (GR) and progesterone receptors (PR)

Table 7.2 Examples of consensus consequences for TF binding sites.

7.3.3 Activation domains

acid blobs and negative noodles

The model of TF action shows that the part of the protein involved in activating transcription is quite separate from the DNA binding motif. The most common activation domains found in transcription factors are acidic domains with interesting names such as 'acid blobs' and 'negative noodles'!

Figure 7.6 shows that they consist of a short α-helix that has high negative charge along one side and hydrophobic residues along the other. This amphipathic helix allows specific contact with other proteins to bring about activation. It is not yet clear whether the interaction is directly with RNA polymerase II or with TFIID or some other protein that subsequently interacts with the initiation factors. These conclusions have been reached in part because over expression of factors such as the well characterised yeast transcription factor, GAL4, can interfere with the initiation of transcription from genes not normally regulated by GAL4.

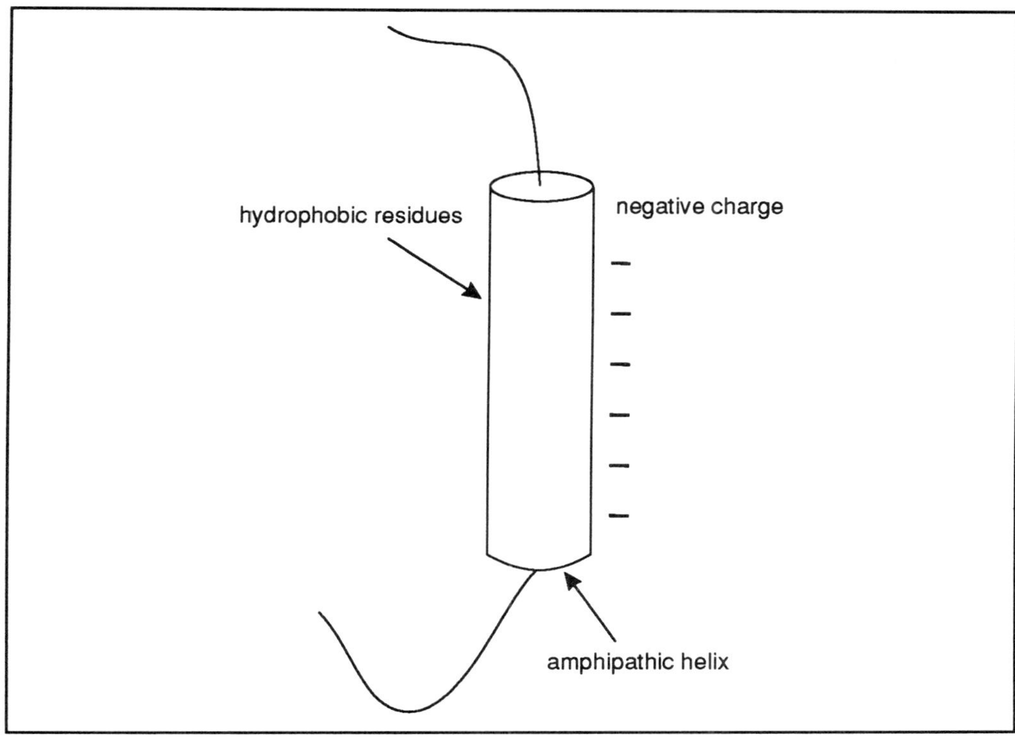

Figure 7.6 The amphipathic nature of an acidic activation domain.

glutamine rich domains

Two other activation domains can be considered here. First, the glutamine rich domain mentioned earlier. These types of domains are found in Sp1 and a number of other proteins including homeobox transcription factors, Oct1 and Oct2 and CREB.

proline rich domain

Finally, proline rich domains are found at the C-terminus of CTF/NF1 factor, Jun and Oct2. It consists of a region containing approximately 25% proline residues.

More information will emerge on the properties of TFs as we progress through the chapter. In particular, we will discuss in more detail how TFs regulate gene expression, activation and repression, and we will consider what regulates the activity of the TFs themselves.

SAQ 7.2

Transcription factors bind DNA in a specific way using a variety of DNA binding motifs. These binding motifs may be found on a single molecule or be provided by dimers. Identify, from this list below, those binding motifs which enable DNA binding when they are present on a single molecule.

1) Cys_2/His_2 zinc fingers.

2) Cys_4 zinc fingers.

3) Homeobox motifs.

4) Basic domains, with leucine zippers.

5) Basic domains with amphipathic helix-loop-helix motifs.

SAQ 7.3

Use the text and Table 7.2 to help you to respond to the following.

1) The consensus sequence GGTACANNN has been introduced into the promoter region of gene X. Is it true that the expression of this gene will now be under the control of glucocorticoid hormone?

2) The promoter of the gene encoding growth hormone contains a nucleotide sequence which binds what type of hormone receptors?

3) In the absence of oestrogen, would we expect a TF to be bound to the nucleotide sequence AGGTCANNNTGACCT in the promoter regions of oestrogen regulated genes?

7.4 Transcription factors and the co-ordination of gene expression

In the example of Sp1 described above, we have considered each gene being activated by a single transcription factor. This may happen with some genes in all the cells of a complex eukaryote organism, or all of the cells in culture in response to some stimulation or it may be more specific and only occur in a single cell type.

multiple TF binding by genes

The co-ordination of gene expression in eukaryote organisms is usually much more complex than this. As one can see in the example illustrated in Figure 7.1, there may be a number of TF binding sites in the gene displayed indicating a complex system of regulation. The growth and development of a particular cell may be largely governed by a pre-determined pattern of gene expression but it can also be influenced by external factors such as hormones, growth factors and responses to environmental stress. In the next part of this chapter, we will consider the complexities of the biology of transcription factors.

7.4.1 Transcription factors and the response to external stimuli

In this section, we will begin by examining the biology of transcription factors that respond to a particular stimulus. We have identified several in Table 7.2. You will notice some have been mentioned already and some will be discussed later in a different context but as an example we will discuss the heat shock transcription factor.

Heat shock is a good example to start with because there is specific and identifiable causative agent.

heat shock factor and heat shock element

The heat shock response appears to be universal and shows many common features across the range of eukaryotes. In response to a rapid rise in the ambient temperature, a small number of heat shock proteins (hsp) are synthesised. This is often accompanied by a temporary reduction in normal protein synthesis. Some of the heat shock proteins are very highly conserved, being found in yeasts, plants and animals. The transient rise in temperature (heat shock) turns on a cascade of genes. The phenomenon works through a protein transcription factor - heat shock factor (HSF), and a regulatory heat shock element (HSE) found in the promoter region of heat shock inducible genes.

Figure 7.7 displays a situation in which a single protein, from a regulatory gene is able to activate and/or repress the transcription of a number of genes to which the protein binds. In this example, the regulatory gene codes for HSF which activates the transcription of a number of heat shock genes.

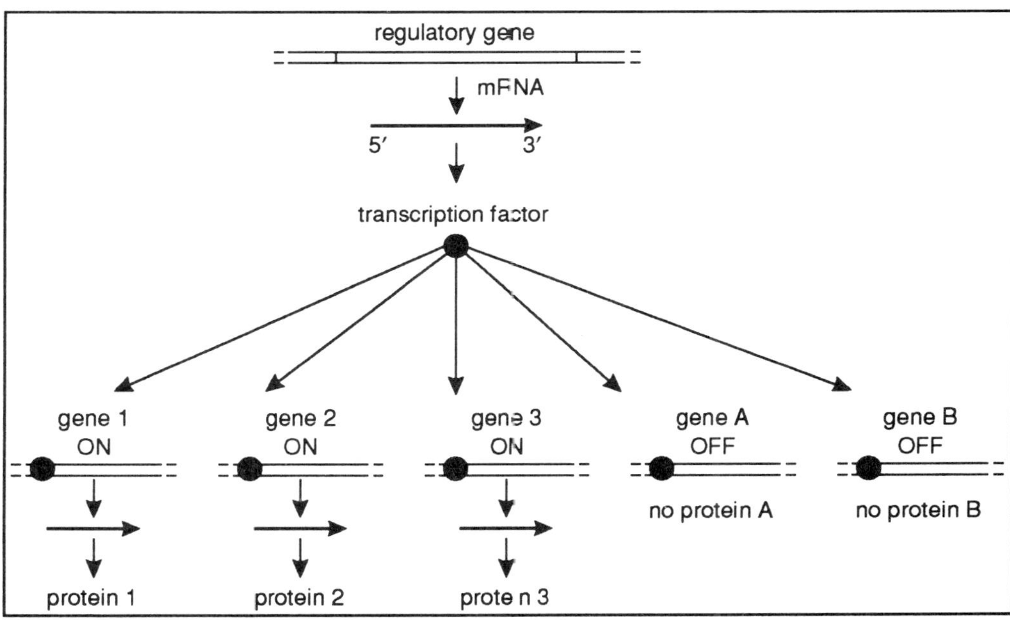

Figure 7.7 A single TF is able to activate or repress a number of genes. This number can be quite large.

The consensus heat shock element (HSE) sequence is CTNGAATNTTCTAGA. Such elements have been shown to be involved in the response and to be the binding site for transcription factors by some elegant experiments. For example, the HSE sequence from a *Drosophila* heat shock gene has been fused to an indicator gene (HSV thymidine kinase) and expressed in mammalian cells. Expression of the indicator gene only occurred when the temperature of the mammalian cells was raised from 37°C to 42°C (Figure 7.8).

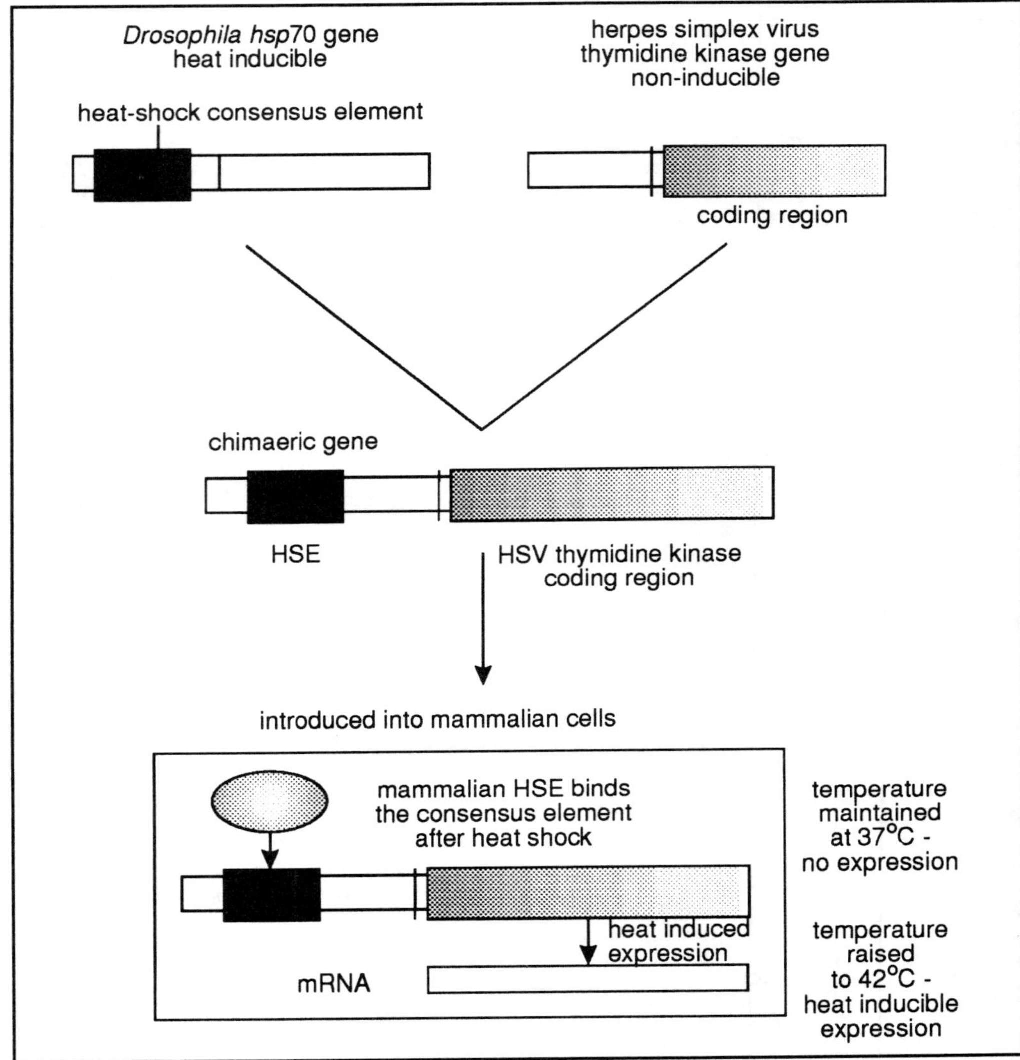

Figure 7.8 The introduction of a chimaeric heat shock gene into mammalian cells. The outcome of this type of genetic manipulation is that the mammalian cell will now express HSV thymidine kinase as a response to heat shock.

Π What does this indicate about the *Drosophila* heat shock element?

This indicates that:

- the element is controlling transcription;

- the element is responding to a signal/message in the cell and not itself acting as a thermostat (since 37°C would represent a considerable heat shock in *Drosophila*);
- the HSE and HSF must be very highly conserved since the mammalian HSF must be binding to the *Drosophila* HSE.

In yeast, the HSF is an 833 amino acid protein that contains a domain of 118 amino acids that mediates the protein's ability to bind DNA. The protein acts by binding to DNA as a single protein, not as a dimer. The *de novo* synthesise of HSF protein is not induced by heat or any other stress. Rather, activation is regulated post-translationally. The mode of regulation is similar in humans, *Drosophila* and fission yeast but slightly different in the budding yeast, *Saccharomyces cerevisiae*. Heat shock genes contain regions which are hypersensitive to DNAse I digestion.

Π What does this indicate?

hs gene transcription requires HSF binding and phosphorylation

The chromatin in the region of the promoter is accessible and ready for transcription (see Chapter 6). It has been shown, in part by DNAse protection studies of *hsp70*, that in non-induced cells, the TATA box is initially protected from digestion by binding of the protein factor TFIID to this region. However, the presence of this factor alone is not enough to allow transcription of *hs* genes to occur. Following heat stress, the presynthesised HSF binds to the heat shock element which, in *hsp70*, is further upstream of the RNA start site (Figure 7.9). This is only the first stage and transcription still does not occur until a second stage is complete. In the second stage, the HSF is phosphorylated. This creates a heavily phosphorylated activation region that interacts with the TFIID complex, activating transcription. In *S. cerevisiae* the HSE is already bound to the promoter and only the second stage is required.

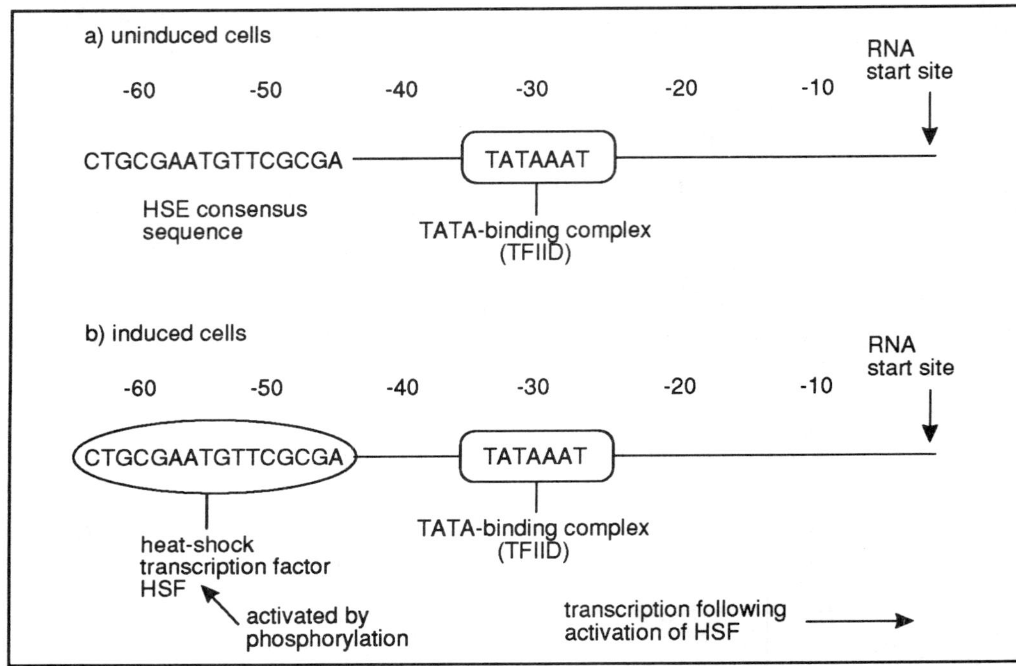

Figure 7.9 The stages in expression of *hs* genes.

SAQ 7.4

A fission yeast, which is normally cultivated at 25°C is unable to use lactose as a sole source of carbon and energy. A chimaeric gene containing the HSE consensus sequence and TATA box from *Drosphila* and the coding sequence of the enzyme β-galactosidase from *Escherichia coli* has been introduced into this yeast. Explain the likely physiological consequence of this gene transfer.

7.4.2 Transcription factors in cellular differentiation

The heat shock system shows activation of transcription by DNA binding and post-translational modification of a TF in response to environmental stress. MyoD is a good example of a transcription factor involved in the co-ordination of gene expression during cellular differentiation. This TF is an example of a master regulatory gene that is involved in turning on specific genes in only one cell type.

MyoD is involved in myoblast and myotubules development

It is also an interesting system because it takes us to another level of complexity regarding the regulation of TF activity. Here, activation is at the level of transcriptional regulation and a form of inhibition is found at the level of protein interactions. MyoD is a TF found only in skeletal muscle precursor cells (myoblasts) and in myotubules. The protein itself is interesting because it does not function as a single monomer but as a dimer. This, as we shall see later, is an important feature of the regulation of activation. The protein contains a helix-loop-helix motif that is involved in the dimerisation of the subunits, and a basic region that is directly involved in DNA binding. The target genes of this TF include creatine kinase (an upstream enhancer element of creatine kinase has been shown to bind the MyoD) and the *MyoD* gene itself.

muscle lineage and muscle terminal marker genes

Transcription of *MyoD* is turned on early in development of muscle cell lines and, once expressed, the TF apparently maintains its own synthesis by a feedback loop mechanism and switches on so called muscle lineage marker genes (MLG). This initial step of differentiation occurs with an increase in the concentration of MyoD. The subsequent differentiation of myoblasts into myotubules does not display any further increase in the level of MyoD, yet a further set of MyoD dependent genes are expressed, the MTDMs (muscle terminal differentiation markers). This paradox has been explained by suggesting that it is the concentration of available protein that is important. It has been shown that a further protein factor, Id, occurs in the myoblast cell that has a similar helix-loop-helix structure to MyoD and so has the potential to form dimers with MyoD. However, it does not have a basic region and therefore is unable to bind DNA. It has been suggested that Id either binds to MyoD, inhibiting DNA binding, and preventing the expression of terminal markers, or that Id binds to another protein factor that is required to interact with MyoD and, only on the removal of Id, is this factor able to interact with MyoD, allowing expression of later genes.

MyoD dependent genes influenced by available MyoD protein

These observations are summarised in Figure 7.10.

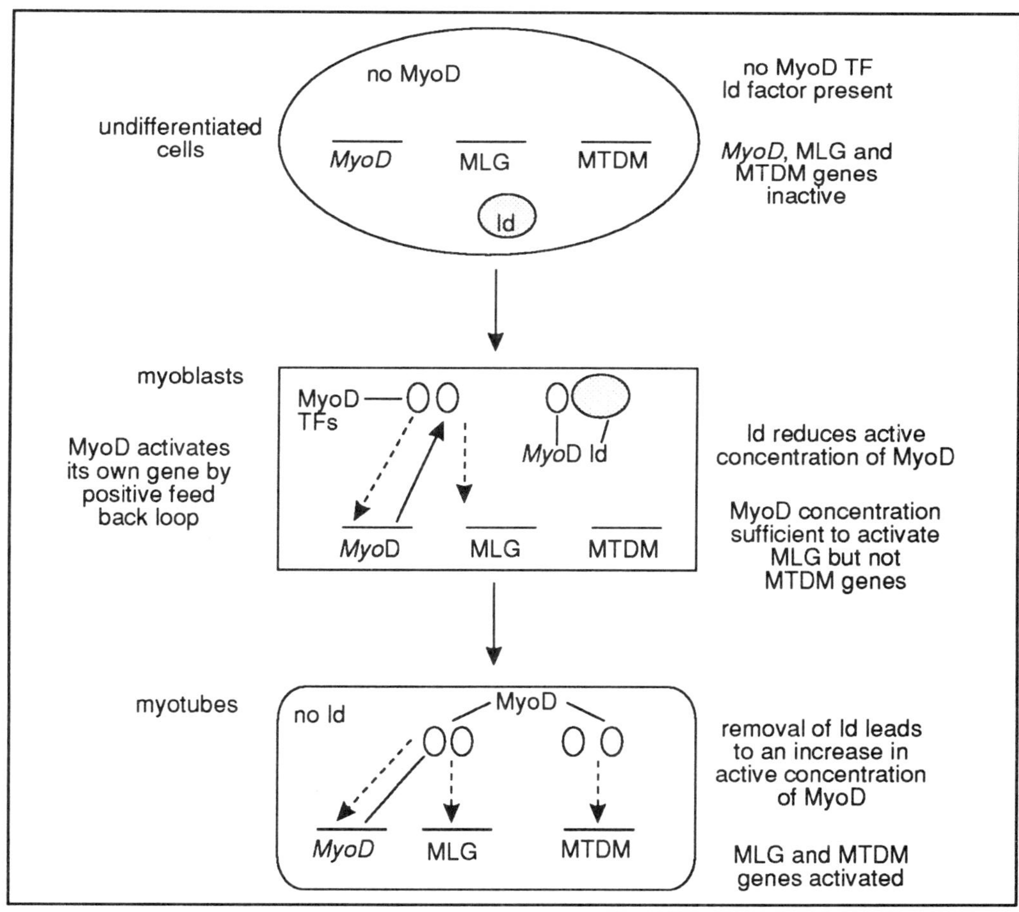

Figure 7.10 The involvement of MyoD in myoblast and myotube development. Only MLG is expressed in myoblasts, MLG amd MTDM are expressed in myotubes.

7.4.3 Transcription factors and complex systems

We should now consider how transcription factors can interact in complex co-ordinated systems. In this section, we will look at some of the mechanisms involved in transmitting signals in multicellular organisms, how these signals affect the activities of transcription factors and the consequences of this on gene expression. We will consider in detail two transcription factor systems, AP1 and the cAMP system and show how they can interact with a third system, the glucocorticoid receptor (GR).

The transcription factor AP1 is actually a composite of two proteins, which form either hetero- or homodimers. These are encoded by two cellular proto-oncogenes, *c-fos* and *c-jun*. (Actually both genes are members of gene families. However, as *c-fos* and *c-jun* are the best studied we will not consider any other examples here).

Fos and Jun are examples of BZIP proteins.

Π What does this mean?

It means they are proteins containing a basic DNA binding domain in association with a leucine zipper. It is this latter structure that is involved in dimer formation.

Fos/Fos dimers are not stable and cannot bind DNA. However both Jun/Jun and Jun/Fos dimers can be formed. While the former can bind AP1 or TRE consensus sequence elements weakly (see Table 7.2), the latter heterodimer forms a very strong complex which normally activates transcription.

(Note that TRE = TPA (12-O-tetradecanoyl phorbol-13-acetate) Response Element).

AP1 activity is induced by a variety of polypeptide hormones, growth factors, cytokines and neurotransmitters.

Π At what levels do you think AP1 activity is controlled?

AP1 activity regulated by transcription rates and post-translation modification of the component proteins

Both Fos and Jun levels are regulated at the level of transcription and also post translationally. The stimulation signals are received at the cell surface from where the information is transmitted to the interior of the cell by a signal transduction pathway. This either stimulates transcription of the TF genes or phosphorylation of the TF proteins. Transcription of *c-fos* normally occurs at a low level but after stimulation there is a transient rise in the expression of the gene. The signal transduction pathway that brings this about is unknown (although it is likely to be related to that which brings about phosphorylation) but the transcription factors and the binding site involved have been identified. The *c-fos* promoter contains a serum response element (SRE) which is recognised by the serum response factor (SRF). Binding of p67SRF (a serum response factor) to the SRE brings about the transient expression of the gene.

Π What simple mechanisms might be involved in controlling the period of expression of *c-fos*?

There are at least two. SRE activity is stimulated by cAMP (more of this later) and SRE activity is down regulated by the Fos protein. So an increase in the level of Fos leads to a reduction in the level of *c-fos* gene expression.

The increased amount of Fos protein brings about a stimulation in AP1 activity as it binds to pre-existing Jun proteins forming more stable heterodimers.

The Jun protein gene is inducible by the same signals that switch on *c-fos*. Induction is for a longer period because the Jun protein autoregulates *c-jun*. The promoter of *c-jun* contains a TRE to which Jun can bind, activating its own transcription. The expression of *c-jun* is inhibited by cAMP.

SAQ 7.5

1) Complete the labelling (a-c) of the regulation of transcription of *c-fos* and *c-jun* genes.

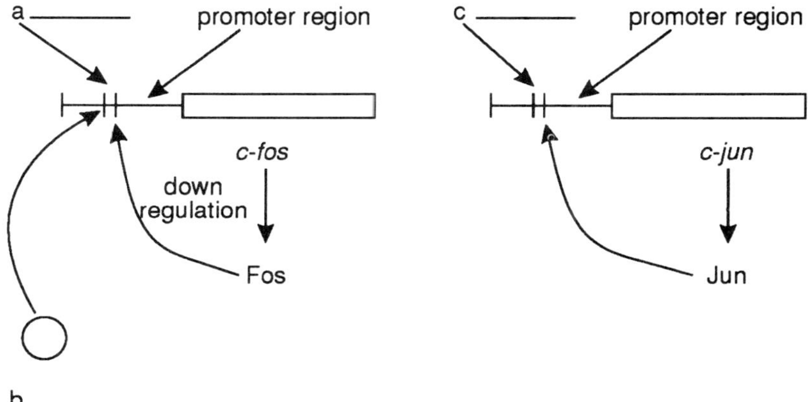

2) The expression of which gene (*c-fos* or *c-jun*) would you anticipate might be influenced by addition of 12-O-tetradecanoyl phorbol-13-acetate (TPA) to a cell?

Π What is the structure of the activation region in *c-jun*?

We mentioned this earlier, it contains a proline rich region.

Fos and Jun proteins are activated by phosphorylation. An outline of the phosphorylation of the protein coded by *c-jun* as shown in Figure 7.11

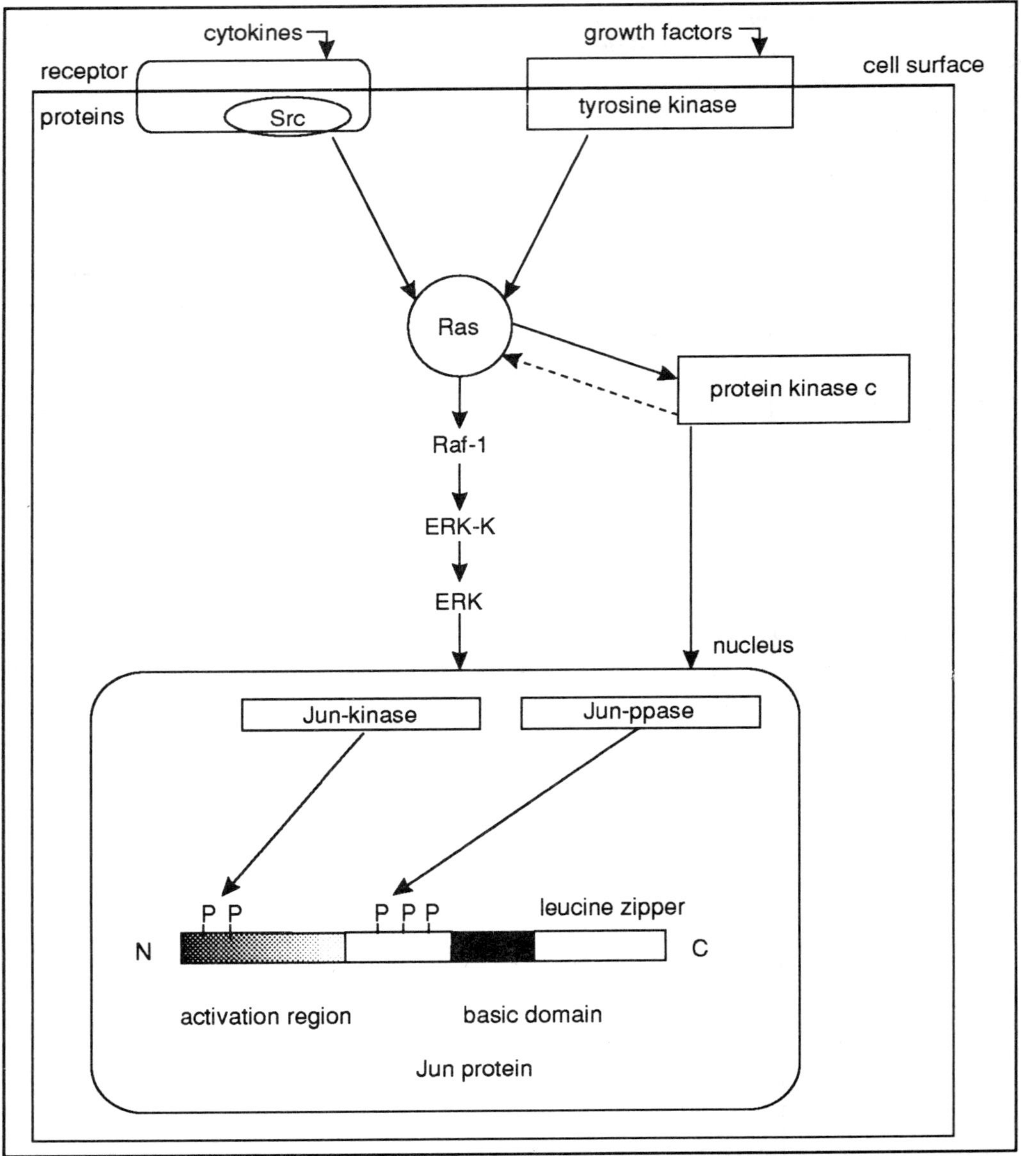

Figure 7.11 Post-translational activation of Jun protein. Extracellular signals stimulate signal transduction pathways that lead to the phosphorylation of the activation region and dephosphorylation near the DNA-binding domain.

Two regions of phosphorylation are indicated. Two phosphates are located in the amino-terminal activation domain and three clustered near to the DNA binding domain. The normal non-active condition of the protein is for the second region to be phosphorylated and the activation region to be under-phosphorylated. The presence of the phosphates near to the DNA binding region inhibits DNA binding.

Π What is likely to be the structure of the active protein?

influence of phosphorylation on TF activity

On activation phosphates are added to the activation region and removed from the DNA binding region. These actions are carried out by a Jun specific kinase and a Jun specific phosphatase respectively (Figure 7.11). Modification occurs in response to signals at the cell surface. Growth factors and other signals stimulate membrane associated tyrosine kinases (eg Src) or protein kinase C and these activate signal transduction pathways (Figure 7.11). In the case of the tyrosine kinases this involves a group of oncogene products such Ras, Raf and ERK. The Ras protein is a GTPase, Raf and ERK are protein kinases. The phosphorylation event increases activity 5 to 10 fold. Removal of the three phosphates from near the DNA binding region increases DNA binding.

CREB and CREM

The cAMP system is the second example we will look at in this section. It is a TF system that is influenced by a wide range of signals. At least two families of TFs are involved in controlling this system; they are CREB proteins (cAMP-responsive element binding protein) and CREM proteins (cAMP-responsive element modulator). Both of these proteins bind to CRE (cAMP response element) a sequence found in the promoters of many of genes. Both proteins are BZIP TFs and various forms of both proteins can act as either gene activators or repressors.

Π What is the structure of the CREB activation region?

Again we mentioned this earlier. CREB contains a glutamine rich activation region.

alternative splicing leads to two forms of CREB

The CREB protein seems to be constitutive and the level of protein is controlled at the level of post-transcriptional RNA processing and post-translational processing. Alternative splicing (see Chapter 5) has been observed for the mammalian CREB factor. Two forms of CREB have been detected, one that is complete and functional and a second that does not contain part of the activation region. As this protein can bind to CRE but does not activate RNA polymerase, it can act as a repressor.

Post-translational control by protein phosphorylation relies on a shorter pathway linking surface binding of the stimulus and the CREB protein. cAMP synthesis is stimulated at the cell surface and this directly stimulates protein kinase A (PKA) activity. This enzyme phosphorylates CREB at serine residue 133. This residue lies outside the activation region, but nevertheless, transcription activity is stimulated 10 to 20 times. Two other sites might be phosphorylated in this region but they have not been linked to transcriptional activity. CREB protein dephosphorylation is also believed to regulate activity.

CREM proteins have been identified as antagonists of CREB. They bind to the CRE, inhibiting the binding of CREB. Since they do not contain activation domains they block transcription. It is becoming clear that even this is not the whole story.

alternative forms of CREM produced during spermatogenesis

One recent example that has come to light is that during spermatogenesis there is a switch in CREM production that is related to RNA splicing. In pre-meiotic germ cells, CREM is expressed at low amounts in an antagonist form. From the pachytene spermatocyte stage a splicing event creates a new form of CREM. This form of CREM is an activator of transcription from the CRE. This protein contains two glutamine rich regions, analogous to activation domains in CREB, and Sp1. The precise role that the switch plays in the development of sperm in testes is yet to be elucidated but it is clearly an interesting example of the regulation of gene expression and a switch in regulation of transcription factors.

Some of the examples we have so far investigated may give the impression that TFs act independently. This is not the case and we have indicated the complex nature of gene regulation in Figure 7.1 and in the previous example.

Π How could several different regulatory systems interact to co-ordinate the expression of a gene?

Two simple mechanisms will be discussed here, one of which you have already met. Examine Figure 7.12, which shows part of the promoter region of a glycoprotein hormone α-subunit gene. cAMP stimulates the expression of this gene in placental cells by activating CREB factor binding to the two CRE, stimulating transcription. This gene is repressed by glucocorticoid hormones in placental cells. This phenomenon is regulated by GR, a TF that binds the hormones. A simple mechanism to explain this can be worked out from Figure 7.12.

Π What is it?

It has been suggested that GR binding sites overlap with CRE and that GR binding competes with the positive activator CREB for binding to the promoter. If GR factors binds to the DNA then the gene is repressed.

Π What other mechanisms could there be?

A second mechanism that could explain the phenomenon is that the two receptor proteins physically interact. Such is the case between AP1 (Fos/Jun) and GR. These two TFs can bind with one another, in effect reducing the active concentrations of the TFs. This could inhibit the expression of genes that are activated by either system, or it could derepress genes that are inhibited by the TFs. Such crosstalk, as it is known, can play a central role in co-ordinate regulation.

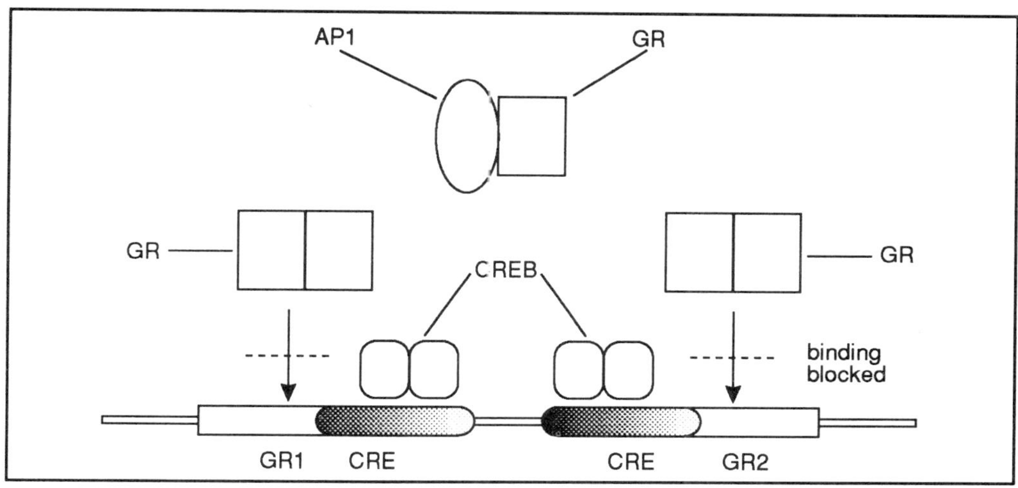

Figure 7.12 Part of the promoter of the glycoprotein hormone α-subunit gene. Two CRE sites overlap with two glucocorticoid receptor binding sites.

Oncogenes

Oncogenes are genes which are responsible for cellular transformation leading to cell proliferation and tumour formation. Some of the first oncogenes to be discovered were identified because they are carried by oncogenic viruses.

Viral oncogenes have subsequently been shown to have cellular counterparts encoded by the nuclear genome. Cell transformation may result either from infection by an oncogenic virus or by mutation or altered expression of a cellular oncogene.

Oncogenes identified in viruses are designated *v-onc* and the cellular equivalents are designated *c-onc*. The genes encoding the Fos and Jun TFs were originally identified as oncogenes and only later was their function as the AP1 transcription factor characterised, hence their designation as *c-fos* and *c-jun*.

changes in Fos and Jun production can lead to transformation

As we have already seen, Fos/Jun is stimulated by serum growth factor. This is a transient induction which brings about transient growth of the cell. If the *c-fos* or *c-jun* genes are mutated in such a way as to permanently switch them on, or so they no longer respond to negative control, then the AP1 transcription factor will remain in the cell at an elevated level. This in turn will maintain the activity of growth factor responsive genes leading to abnormal growth of the cell and potential transformation. Virus borne forms of the Jun protein exist that can also transform cells (Figure 7.13).

The Fos/Jun system demonstrates a mechanism whereby the continuous expression of certain genes brings about transformation. Another example we shall view here works in the opposite way in that genes which are 'switched off' can bring about cellular transformation.

Erb A transcription factors and erythroblast differentiation

The c-Erb A transcription factor is the cellular receptor for thyroid hormone. Following hormone binding the transcription factor: hormone complex binds to recognition sites in the DNA of thyroid responsive genes and activates their transcription. For example c-Erb A in the presence of thyroid hormone switches on differential gene expression which causes erythroblast cells to differentiate into non-dividing erythrocytes. A form of the gene v-*Erb* A is carried by avian erythroblastosis virus (AEV). It codes for an interesting protein because, like c-Erb A, it can specifically bind DNA, like c-Erb A, but it is mutated in the hormone binding region and as such is not converted to a form that activates transcription after hormone binding. This has certain consequences for the cell, as the virus protein acts as a dominant repressor of thyroid hormone responsive genes.

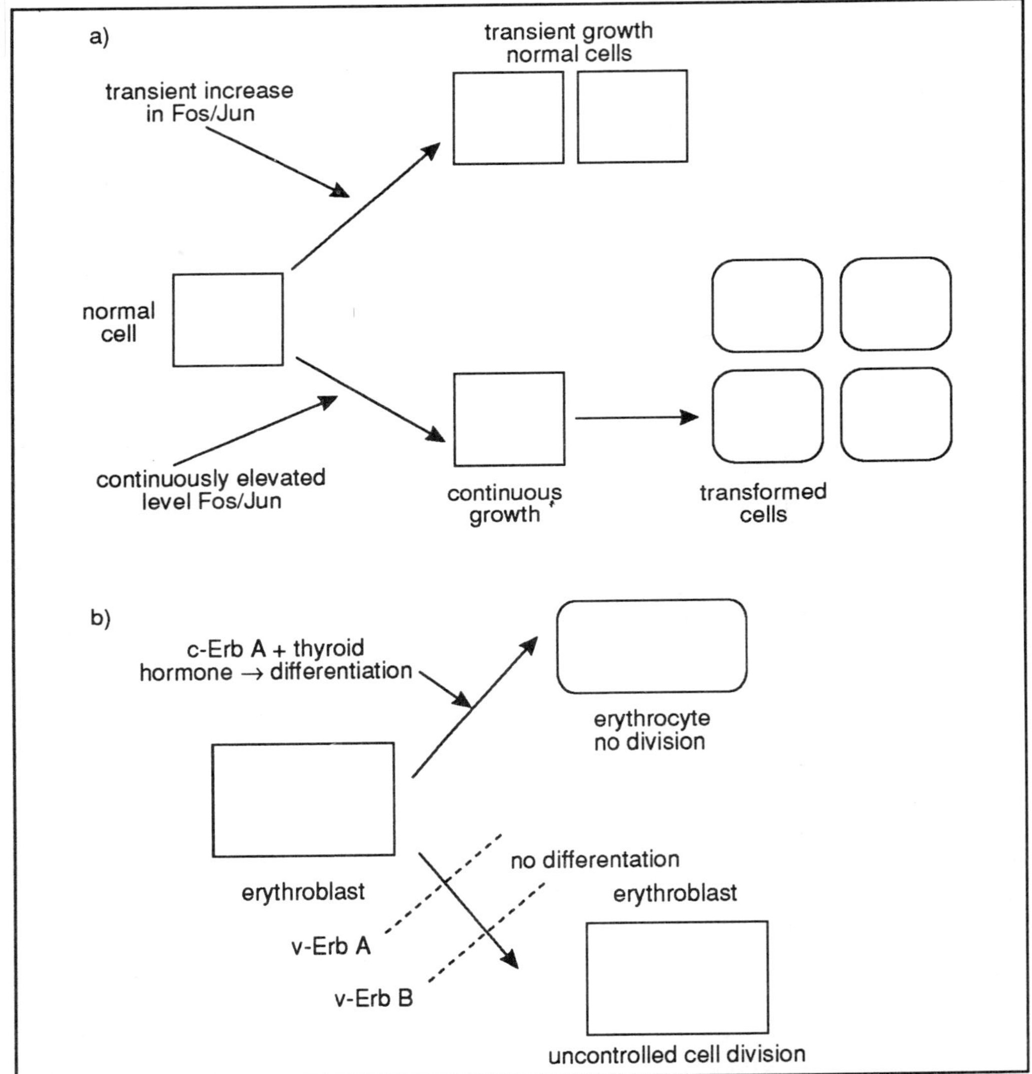

Figure 7.13 Two mechanisms by which cells can become oncogenic. In a), if the production of Fos/Jun is permanently switched on, cells continue to grow and are transformed. In b), c-Erb A activity is blocked by the presence of a non functional virus encoded protein (v-Erb A). No differentiation occurs and the cells proliferate as erythroblasts (see text for further details).

So as shown in Figure 7.13 the binding of hormone to the c-Erb A protein in erythroblast cells specific hormone stimulated gene expression which allows differentiation to occur, the cells becoming erythrocytes which no longer divide.

introduction of v-*Erb* B can lead to transformation

On infection by the virus v-*Erb* A is expressed. The protein inhibits erythrocyte gene expression and the cells proliferate indefinitely as erythroblasts. A second virus encoded *onc* gene, v-*Erb* B, once expressed makes the cell independent of external growth factors because it encodes a truncated form of the epidermal growth factor receptor and as such the cell becomes a tumour cell type. Many other examples of similar oncogenes have been discovered.

SAQ 7.6

A chimaeric gene has been produced in which the heat shock consensus has been inserted into the promoter region of the c-Erb A transcription factor gene. This constructed gene has been inserted into an erythroblast cell line where it has been shown to have replaced the normal c-Erb A gene. What might be the outcome if such cells were incubated in the presence of thyroid hormone?

7.4.4 Regulation of whole organ development

Up to now we have been considering how TFs can influence the growth of individual cells or tissues. Recently TFs have been linked with regulating differentiation on a greater scale. The development of whole organs and segments of bodies may be regulated by single transcription factors.

homeotic genes

The homeotic mutations of *Drosophila* have shown that single mutations can bring about radical re-organisation of the fly's body. For example, instead of an eye, a wing may be protruding from the head, or a leg may be substituted for an antenna. These effects have been linked to a small number of genes, the so called homeotic genes. Perhaps a 100 or so such genes exist in *Drosophila* and they may be regarded as master regulatory genes. Master regulatory genes are considered to operate by activating and/or repressing the transcription of a group of genes (as shown in Figure 7.7). If the products of some of these genes in turn regulate the activity of other sub-groups of genes, a cascade phenomenon will result, in which the activity of a large number of genes ultimately depends upon a single master regulatory TF. A mutation in the master regulatory gene will therefore have profound consequences, as demonstrated by the homeotic mutants. Proteins coded by such genes seem to all contain a motif we have mentioned earlier, the homeobox motif, and act as TFs.

bithorax and antennapedia complexes

The two best studied sets of genes are the Bithorax complex and Antennapedia complex. These are small groups of genes on *Drosphila* chromosome 3. The Bithorax complex is involved in regulating abdominal and thoracic segments while Antennapedia genes regulate thoracic and head segments. Each complex contains only 3 to 5 genes which appear to regulate their own expression in a complex fashion. These TF's can repress or activate other genes.

The mechanisms of gene regulation involved here are quite complex. The order of the genes in each cluster can influence expression which may relate to whether the chromatin in a particular region is open for transcription. RNA processing also plays a major role in regulating these genes. Perhaps the most exciting feature of these genes is that similar homeobox containing proteins have now been identified in mammalian systems. In mice, so called *Hox* genes have been found which are involved in segment formation in the developing central nervous system of embryos. These proteins are similar in many ways to homeobox proteins and even their genetic organisation seems to have been conserved.

Summary and objectives

In this chapter we have introduced you to the exciting area of study, eukaryotic gene transcription factors. This is an area of very active research and much progress has been made in the past few years but there is still much we do not know. We began the chapter by examining the general features of transcription factors. We learnt that they must consist of at least two parts, one which recognises particular nucleotide sequences and one which activates other (ultimately RNA polymerase) proteins. We also explained that the structures of the TFs which bind to DNA fall within a limited number of types such as homeoboxes, zinc fingers and basic domains. We also explained that TFs may bind to promoter and enhancer sequences and that multiple TFs may bind to these regions.

We also described the roles of transcription factors in the response to external stimuli (for example heat shock) and as master regulator genes. We described how the regulatory effects mediated by TFs may be controlled through the regulation of TF gene expression, by changes in RNA splicing and through post-translational modification of TFs. The role of TF genes as oncogenes was also described. In the final part of the chapter, we briefly described the role of TFs in the regulation of whole organ development.

Now that you have completed this chapter you should be able to:

- list the general features of transcription factors in terms of DNA binding motifs and activation domains;
- explain how transcription factors bind to particular (consensus) sequences in promoter and enhancer regions;
- draw models representing the structures of transcription factors;
- predict the outcome of introducing particular consensus sequences into promoters;
- explain how transcription factor genes may act as master regulatory genes in cellular and organ differentiation;
- describe, using suitable examples, how the activities of transcription factors may be influenced by changes in TF gene transcription, the processing of TF gene transcripts and by post-translational modifications.

Translation and post-translational modifications in eukaryotes

Translation and post-translational modifications in eukaryotes

8.1 Introduction

transcription, ribosomes, mRNA, tRNA, amino acids, ATP, GTP, proteins

Translation is the biosynthesis of polypeptides using the information stored in the base sequence of mRNA. The mRNA is complementary to a sequence of bases in DNA and its production (transcription) is described in Chapter 5. Translation is carried out on specialised structures called ribosomes. It requires mRNA, tRNA molecules, amino acids, ATP, GTP and a number of proteins. The reactions involved in translation in prokaryotes have been described in the BIOTOL text 'Genome Management in Prokaryotes' and will not be greatly elaborated upon here. It is assumed that the reader has an overview of translation in prokaryotes. These reactions are essentially the same in prokaryotes and eukaryotes although there are a number of differences. This chapter will briefly describe features of translation in eukaryotes and the major differences in between eukaryotes and prokaryote will be explained. Table 8.1 summarises the major differences between eukaryotic and prokaryotic protein synthesis. Read this table translation carefully and then attempt the following SAQ.

translation

	Eukaryotes	Prokaryotes
Synthetic apparatus		
	80S ribosome	70S ribosome, with 50% protein content of 80S ribosome
	mRNA codes for single gene product	mRNA codes for several gene products
	mRNA is capped and acquires a poly A tail	mRNA not capped or polyadenylated
Initiation		
	10 protein factors involved	3 protein factors involved
	Uses cap-binding proteins	Shine-Dalgarno sequence
Termination		
	Single protein factor involved	3 protein factors involved
	Single termination factor	Three termination factors

Table 8.1 Comparison of translation in prokaryotes and eukaryotes.

SAQ 8.1

Use Table 8.1 to fill in the missing spaces using the words listed below.

Polypeptides synthesis requires the participation of [ ], a source of metabolic [ ], mRNA and a number of [ ]. The [ ] ribosome has a [ ] of 80S and contains about twice as much protein as the smaller [ ] ribosome.

The genetic information is carried from the DNA to the ribosome by [ ]. In prokaryotes, the mRNA codes for several [ ], while [ ] is a copy of a [ ] gene. Initiation is more complex in eukaryotes than prokaryotes and requires the participation of a [ ]. In prokaryotes, a [ ] sequence is needed. Many proteins are required for initiation, at least [ ] in eukaryotes and three in prokaryotes. However, [ ] is an apparently simpler process in eukaryotes, requiring only a single protein factor compared to the [ ] needed by prokaryotic cells.

Choose from the following: termination, proteins, Shine-Dalgarno, polypeptides, sedimentation coefficient, three, ribosomes, mRNA, single, 7-methylguanine cap, 10, eukaryotic, energy, 60S prokaryotic, eukaryotic mRNA.

8.2 The role of mRNA

polycistronic

monocistronic

Messenger RNA carries the genetic information encoded in DNA to the cytoplasm. In prokaryotes mRNA is polycistronic, ie it is a copy of more than one gene and the processes of transcription and translation are coupled. As soon as a part of the mRNA molecule is synthesised translation begins. However, in eukaryotes mRNA is monocistronic, is synthesised in the nucleus, and must pass through the nuclear membrane to reach the ribosome. At the 5′ end of the mRNA a 7-methylguanine cap is added which appears to prevent degradation of the molecule by nucleases and phosphatases during its transport.

Π The structure and synthesis of the cap was covered in Chapter 5. See if you can remember the basic details of cap structure before continuing.

7-methylguanine cap

leader sequence

The 7-methylguanine cap is formed by the reaction of the terminal residue of the mRNA with GTP to form an unusual 5′-5′ triphosphate link (Figure 8.1). The guanine is methylated at N-7. Methylations may also occur at 2′ positions on the first two ribose units. Eukaryotic mRNA molecules also acquire a polyadenylate tail of upto 200 adenosine residues which is added after transcription (see also Chapter 5). The function of the poly A tail is not certain since if it is removed translation is not impeded. The 5′ end of the molecule also contains a variable number of bases termed the leader sequence which is not translated but is involved in initiation of protein synthesis.

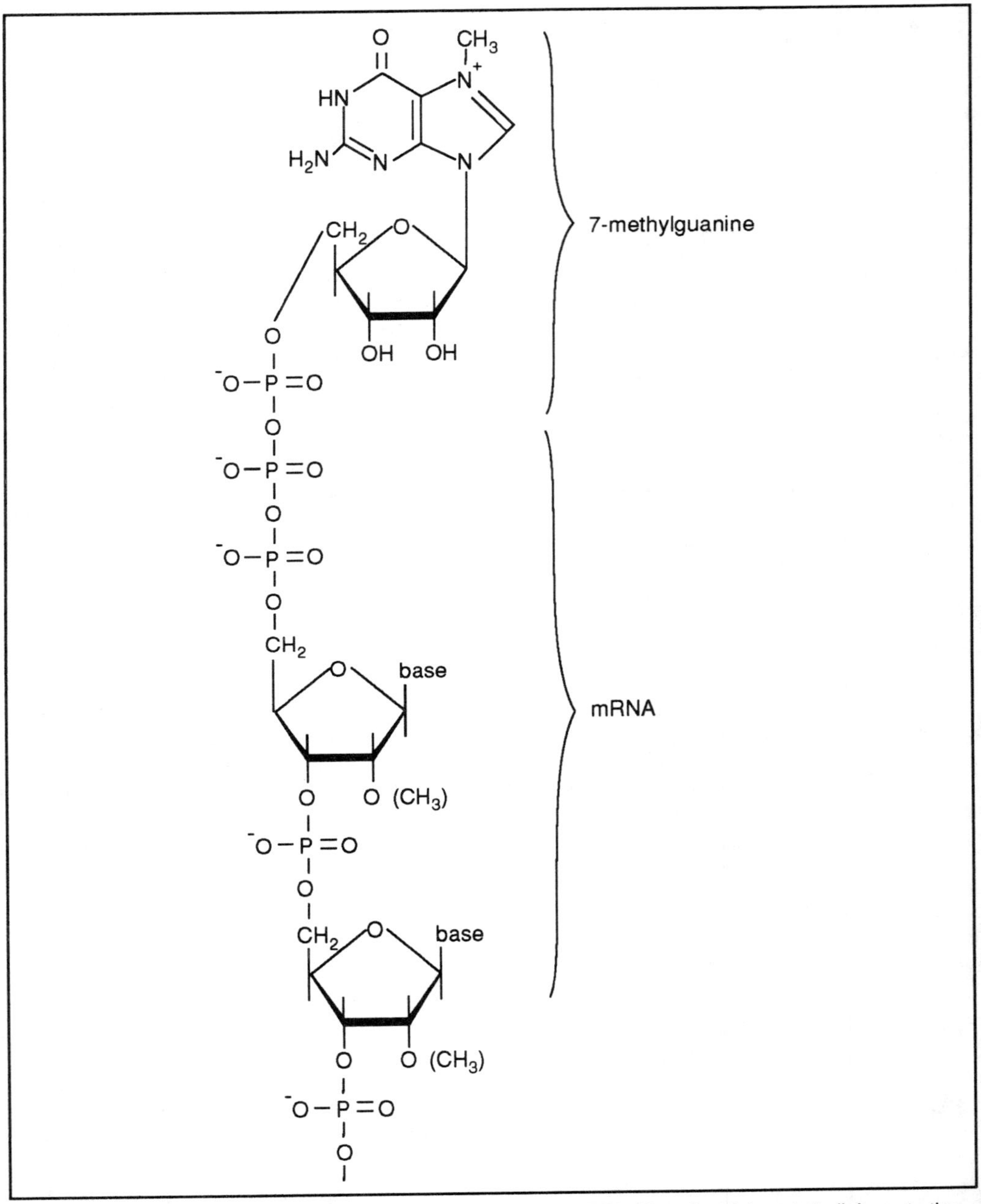

Figure 8.1 Structure of the 7-methylguanine cap of eukaryotic mRNA. Note the triphosphate linkage to the 5′ of the ribose.

8.3 Ribosomes

Ribosomes are ribonucleoprotein particles that contain the enzymes that synthesise polypeptides. Eukaryotic ribosomes have a sedimentation coefficient of 80S. They are composed of two subunits of sedimentation coefficient 60S and 40S respectively. They are, therefore, larger and composed of more proteins (ribosomal or r-proteins) and rRNA molecules than the prokaryotic type. Table 8.2 summarises the types of proteins and rRNAs found in eukaryotic and prokaryotic ribosomes.

	Prokaryotic	Eukaryotic
Whole ribosome	70S (mol wt 3 x 10^6)	80S (mol wt 4.5 x 10^6)
Large subunit		
	50S (mol wt 2 x 10^6)	60S (mol wt 3 x 10^6)
Composition	23S RNA (mol wt 1.2 x 10^6)	28S RNA (mol wt 1.9 x 10^6)
		5.8S RNA (mol wt 0.3 x 10^6)
	5S RNA (mol wt 0.25 x 10^6)	5S RNA (mol wt 0.25 x 10^6)
	34 r-proteins	50 r-proteins
Small subunit		
	30S (mol wt 0.9 x 10^6)	40S (mol wt 1.5 x 10^6)
Composition	16S RNA (mol wt 0.6 x 10^6)	18S RNA (mol wt 0.72 x 10^6)
	21 r-proteins	30 r-proteins

Table 8.2 Differences between eukaryotic and prokaryotic ribosomes. Note the molecular weight (masses) of RNA are reported in Daltons.

Π Calculate the approximate number of bases in each rRNA in prokaryotic and eukaryotic ribosomes. (Take the average molecular weight of each nucleotide moiety as 300 Daltons).

The molecular weight of 23S rRNA in prokaryotes is 1.2 x 10^6 Daltons. This is equivalent to 1.2 x $10^6 \div 300 = 4000$ bases. The other lengths can be calculated in exactly the same way.

polyribosome

During protein synthesis, a number of ribosomes bind to a single mRNA molecules so that large numbers of protein molecules may be produced simultaneously. This arrangement is called a polyribosome.

SAQ 8.2

^{14}C-labelled amino acids were added to a cell lysate under conditions that allow polypeptide synthesis. Free ribosomes were subsequently separated from polyribosomes by centrifugation. The amount of protein (absorbance at 280 nm) and the amount of radioactivity in each fraction were determined. The results are shown graphically:

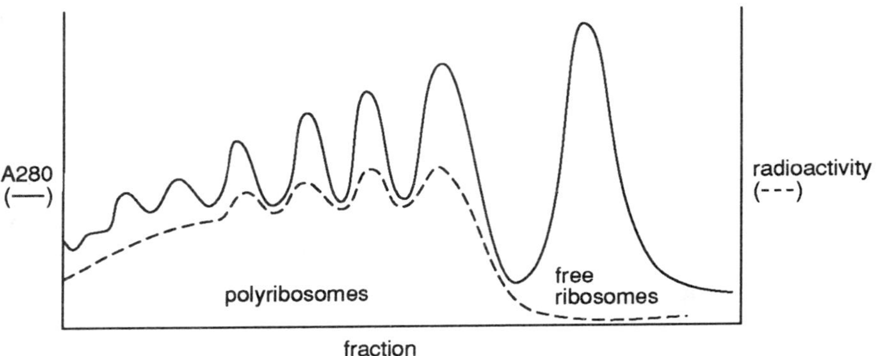

a) What do the peaks in the protein profile correspond to?

b) Account for the distribution of radioactivity.

8.4 Transfer RNA

Transfer RNA molecules carry amino acids to the ribosome for incorporation into the polypeptide. The tRNA molecules from eukaryotic and prokaryotic organisms have a similar structure. This is typically represented as a 'clover leaf' structure. In a stylised form it is often shown as:

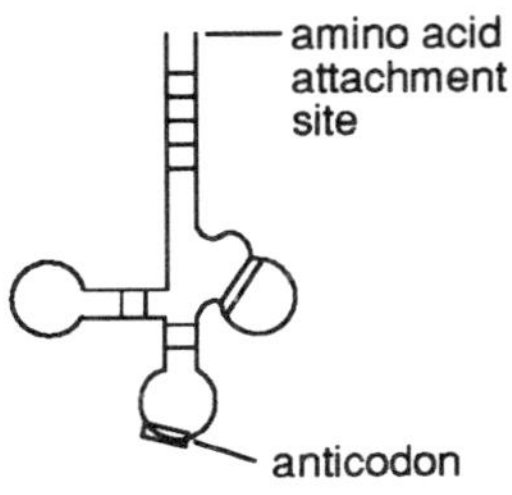

genetic code is degenerate

We will not elaborate on its structure here, except to remind you that each tRNA has a site which recognises an appropriate codon in the mRNA and another site to which the amino acid attaches (details of these molecules are given in the BIOTOL text 'Genome Management in Prokaryotes'). The genetic code is degenerate, therefore there may be as many as six codons for a particular amino acid and there may be a different tRNA corresponding to each of these codons. There appears to be a greater number of types of tRNA in eukaryotic cells although the various tRNA molecules do not markedly differ.

aminoacyl-tRNA synthetase

cognate tRNAs

The enzyme which links the tRNA molecule to its corresponding amino acid is an aminoacyl-tRNA synthetase. A single synthetase links an amino acid to all tRNA molecules which accept that amino acid. The tRNAs recognised by a synthetase are called cognate tRNAs. Not all codons for an amino acid are used with the same frequency and therefore there will not be equal amounts of cognate tRNAs in the cell. The linking of the amino acid to the terminal adenosine residue of the tRNA requires ATP. The reaction sequence is shown in Figure 8.2.

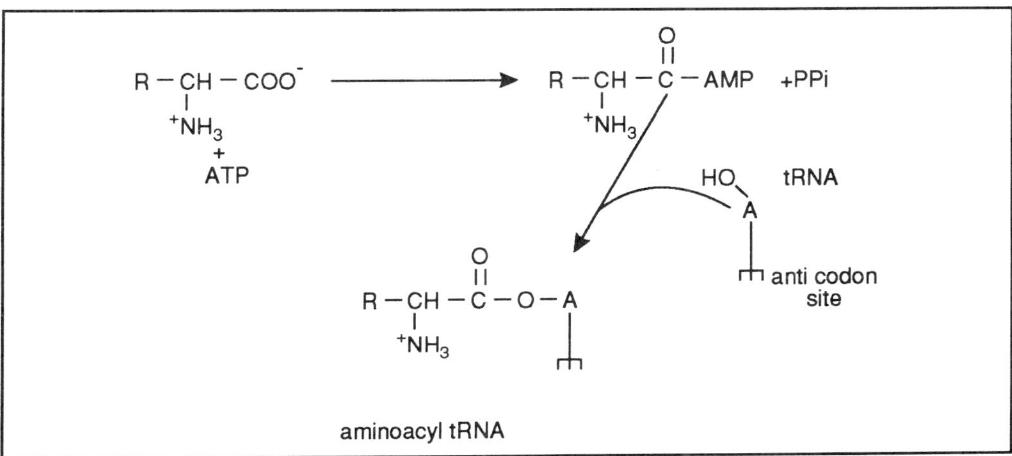

Figure 8.2 Addition of amino acid to its corresponding tRNA. Note that the structure of tRNA shown in this figure is highly stylised.

8.5 The mechanism of translation

The synthesis of a polypeptide in eukaryotic systems is essentially the same as the process found in prokaryotes. It consists of the three stages:

- initiation, (the start of translation);
- elongation, (the addition of specific amino acids residues to the growing polypeptide);
- termination, (the cessation of synthesis and release of the polypeptide from the ribosome).

Π Following its synthesis on the ribosome, most polypeptides appear to spontaneously fold into the native (active) conformation. Suggest why this situation has evolved, rather than an enzyme-catalysed folding of the polypeptide chain.

If folding were enzyme-catalysed, a separate enzyme would be required to 'fix' the position of each amino acid residue as it was added to each growing polypeptide. Clearly, an astronomically large number of enzymes would be required since prokaryote and eukaryote cells contain on average 20 000 and 30-50 000 proteins respectively! Also, since mistakes may occur in any biochemical reaction, the large number of possible reactions would make such a system error-prone. The folding of most proteins occurs spontaneously and the configuration which is taken up is the thermodynamically favoured form. This is dictated by the amino acid sequence of the protein and the environment in which the protein is synthesised.

folding of protein is usually spontaneous

It may be noted that some large proteins do require the 'assistance' of helper proteins to fold appropriately. These helper proteins are called chaperonins.

chaperonins

8.5.1 Initiation

Initiation is the binding of the mRNA to the ribosome and the beginning of polypeptide synthesis. An initiation complex is required to start translation and consists of the 40S and 60S ribosomal subunits, a molecule of mRNA, an initiator amino acyl-tRNA and at least ten protein initiation factors, considerably more than the three of prokaryotes. The prefix eIF is used to designate eukaryotic initiation factors and Table 8.3 lists their functions.

initiation factors

Eukaryotic	Prokaryotic	Functions
*NCF	IF1	Involved in forming initiation complex
eIF2	IF2	Involved in forming initiation complex
eIF3, eIF4C	IF3	Involved in forming initiation complex
CAPBPI	NCF	Involved in cap-binding in eukaryotes
eIF4A, B, F	NCF	Search for first AUG codon in eukaryotes
eIF5	NCF	Dissociates eIF2, eIF3 and eIF4C
eIF6	NCF	Dissociates 60S subunit from intact ribosome

Table 8.3 A comparison of the eukaryotic and prokaryotic initiation factors. *NCF = no corresponding factor.

Eukaryotic initiation uses met-tRNA but it is not formylated *in vivo* as in prokaryotes, although it may be *in vitro*. Only a single initiator sequence, AUG, is recognised. The Met-tRNA involved in initiation is designated t–RNA_f^{Met} or t–RNA_i^{Met} (f indicates that the Met can be formylated *in vitro* and i stands for initiation). The tRNA which decodes internal AUG codons is called t–RNA_M^{Met}.

Eukaryotes do not have a Shine-Dalgarno sequence to correctly align the ribosome. They establish the reading frame by using the 7-methylguanine cap and cap binding proteins (CBP) for this purpose. The first AUG on the mRNA serves as the initiator and this is usually the one nearest to the 5′ end of the mRNA. The 40S ribosomes attaches to the 5′ cap and moves along the mRNA hydrolysing ATP in the process. When AUG is encountered, the Met–$tRNA_f^{Met}$, pre-bound to the 40S subunit, attaches to the mRNA. Occasionally the 40S ribosome bypasses the first AUG. In this case the context of the AUG, ie the bases either side of it, appear to be important. The sequence CCA/GNNAUGG, where N can be any base, is the optimal initiation context sequence. If a pyrimidine appears at a position -3 or +1 to the A or the G respectively then the site

cap binding proteins

initiator

will be bypassed and the next appropriate AUG will be the initiation codon. Figure 8.3 summarises the process of initiation.

Π Examine Figure 8.3 carefully. It may be worthwhile to redraw this figure for yourself and write on it the information given in Table 8.3. Where is GTP involved in the initiation process?

You should have spotted that it involved in the binding of Met-tRNA to the initiation factor eIF2. It is subsequently hydrolysed as the initation complex is formed with the mRNA and the 43S ribosomal subunit.

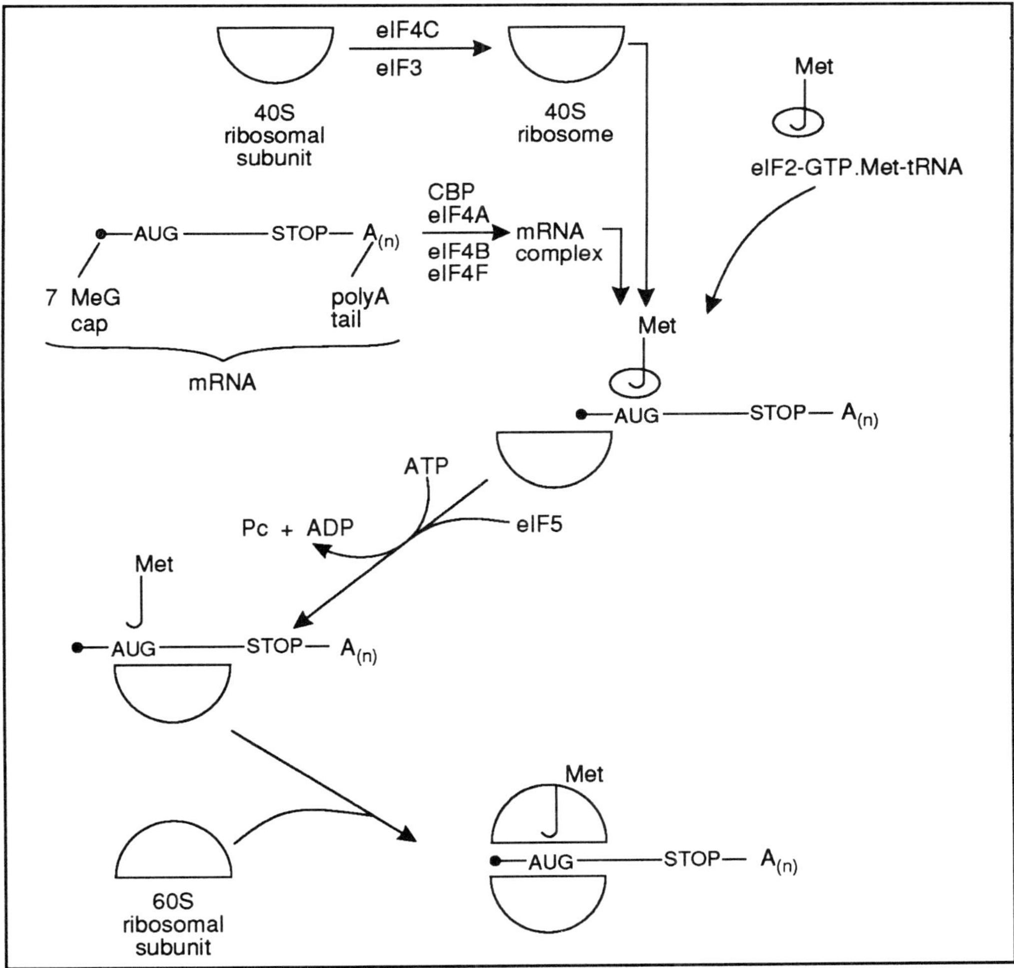

Figure 8.3 Overview of the initiation of translation in eukaryotes. This scheme is a simplification since some of the initiation factors involved in the process are omitted for clarity.

8.5.2 Elongation

elongation factors

Each round of the elongation cycle results in the addition of an amino acid residue to the growing polypeptide. The formation of a peptide bond, catalysed by peptidyl transferase, has been described in this series for prokaryotes and the mechanism is basically the same in eukaryotes. The protein elongation factors (eEF) differ between prokaryotes and eukaryotes (Table 8.4).

Eukaryotic	Prokaryotic	Functions
eEFIα	EF-Tu	Delivers aminoacyl-tRNA to small ribosomal subunit
eEFIβγ	EF-Ts	Recycles eEF1α and EF-Tu
eEF2	EF-G	Involved in translocation

Table 8.4 Comparison of eukaryotic and prokaryotic elongation factors.

GTP is hydrolysed

eEF1$_{\alpha}$ corresponds to EF-Tu in prokaryotes. It forms a ternary complex with GTP and an aminoacyl-tRNA. GTP is hydrolysed to GDP in the process. eEF1$_{\beta\gamma}$ assists in the recycling of eEF1$_{\alpha}$. It catalyses the release of GDP and binding of GTP (GDP-GTP exchange) and allows eEF1$_{\alpha}$ to bind another aminoacyl-tRNA. eEF2 corresponds to prokaryotic EF-G, and promotes translocation of the aminoacyl-tRNA from the A (amino acid site) to the P (polypeptide) site of the ribosome. Translocation requires the hydrolysis of GTP. EF-G and eEF2 are examples of GTP-binding proteins.

SAQ 8.3

In a study of the biosynthesis of haemoglobin, bone marrow cells were incubated with ^{14}C-labelled amino acids at a reduced temperature to slow the rate of synthesis and make it more amenable to study. The radioactive amino acids were added to the medium for short periods of time (from four to twenty minutes) and the globin chains synthesised isolated. The data is presented diagrammatically below.

After 4 minutes exposure to radioactive amino acids, the radioactivity in completed globin chains could be traced to amino acid g. After 7 minutes exposure the radioactivity in completed globin chains was traced to amino acids f and g. After 10 minutes exposure, the radioactivity in completed globin chains was traced to amino acids e, f and g. These, and the data for longer exposure to radioactive amino acids, are represented in the figure below. Given that the amino acids a-g are in the sequence in globin: amino terminus a- b- c- d- e- f- g- carboxy terminus, deduce whether protein synthesis begins at the amino or the carboxy terminus of the globin molecule.

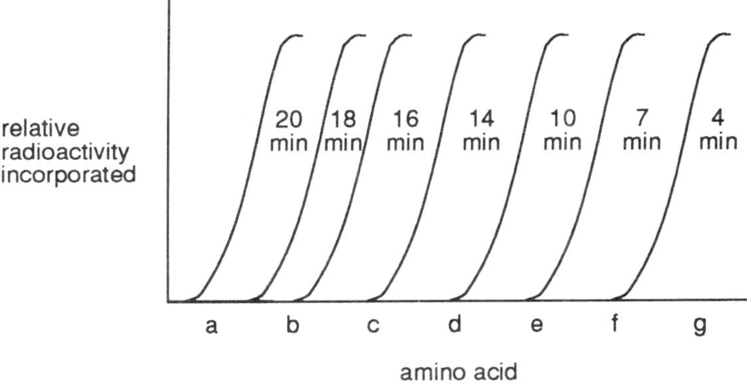

8.5.3 Termination

Elongation occurs in a continuous cycle until the complete polypeptide is synthesised. Termination is the hydrolysis of the completed polypeptide from the tRNA on the P site and occurs when the ribosome reaches one of three termination (nonsense or STOP) codons UAA, UAG or UGA which do not have a corresponding tRNA.

termination factor

Termination in eukaryotes require a single protein termination factor compared to the three used by prokaryotes (Table 8.5). A consequence of termination is that the ribosome dissociates into its subunits. Peptidyl transferase is also involved in termination since it must allow water to substitute for the α-amino group of the amino acid in a nucleophilic attack on the peptidyl-tRNA.

Eukaryote	Prokaryote	Functions
eRF	RF1, RF2 and RF3	Promote release of completed polypeptide

Table 8.5 Comparison between eukaryotic and prokaryotic termination factors.

SAQ 8.4

Identify which of the following are true and which are false.

For protein synthesis to occur in eukaryotes:

1) only ribosomes, tRNA molecules, mRNA, GTP, and amino acids are required.

2) UTP and ATP are required as free energy sources.

3) the absence of one of the twenty amino acids will stop the process.

4) the polypeptide produced depends only on the type of mRNA used.

5) a sequence of bases on the rRNA complementary to part of the mRNA is required to align the ribosome correctly.

8.6 Fidelity of translation

Approximately one amino acid residue in 10^4 is inappropriately incorporated into a polypeptide during its biosynthesis. Two main proofreading processes ensure that this low level of errors is maintained during translation:

- discrimination by synthetase enzymes;
- discrimination at the ribosomal level.

We will examine these in turn.

8.6.1 Discrimination by aminoacyl-tRNA synthetases

The first level of proofreading ensures that an appropriate amino acid is attached to its corresponding tRNA. Some groups of amino acids have similar structures.

Π For example, how does isoleucine differ from valine?

It differs by only a single methylene group. Other large hydrophobic amino acids such as leucine and phenylalanine also resemble valine. However, synthetases are amongst the most specific of enzymes and are able to make such fine distinctions.

double sieve mechanism

The ability of the valyl-tRNA synthetase to distinguish between these different amino acids arises from the enzyme having two adjacent active sites. One site catalyses the addition of valine to $tRNA^{Val}$, the other hydrolyses the amino acyl-tRNA to the free amino acid and tRNA if an amino acid, other than valine, is attached to the tRNA. Amino acids larger than valine cannot enter the site which catalyses addition to the tRNA and so are discriminated against. Amino acids smaller than valine can enter the synthetic active site and be linked to $tRNA^{Val}$. However, they are small enough to enter the second, hydrolytic site and be cleaved from the tRNA. The process of having two recognition steps is sometimes referred to as a double sieve mechanism. Having two steps leads to a considerable lowering in the rate of incorporation of 'incorrect' amino acids.

Π If each discrete step resulted in the formation of one inappropriate aminoacyl-tRNA in 10^2, what would the error in the overall process be?

It would be one in 10^4 that is ($10^2 \times 10^2$).

8.6.2 Discrimination at the ribosomal level

Perhaps you might find it surprising that the accuracy of codon-anticodon interactions is less than the discrimination achieved during translation. This enhancement is obtained by having accuracy (proofreading) built into the system at several steps. Each step has only a modest level of fidelity, but taken all together, they amplify the accuracy in selecting the required aminoacyl-tRNA for the codon in question. However, such a scheme can only work if, after the first discriminatory step, subsequent stages are driven by an input of free energy. The stages in ribosomal proofreading appear to occur at the initial binding of the aminoacyl-tRNA and then at a so-called 'kinetic' proofreading stage.

ribosomal discrimination

The specificity of codon-anticodon interactions depends, in part, upon the structure of the ribosome and on structural features of the tRNA molecule, additional to its anticodon, which influence the mutual binding of the two. This ribosomal discrimination helps ensure that when an appropriate aminoacyl-tRNA binds then a peptide bond is formed. However, when an 'incorrect' aminoacyl-tRNA complexes, it is discriminated against and leaves the binding site. This selection depends upon two competing tendencies: the strength of binding of the tRNA to the ribosome at sites other than through the anticodon; and the relative stability of the correct codon-anticodon versus an incorrect codon-anticodon interaction. The stability of tRNA-ribosome binding, other than codon-anticodon interactions, influences specificity. Thus, for example, if the general ribosomal affinity is too large it will stabilise tRNA binding and mask unfavourable codon-anticodon interactions. If, however, this general affinity is too weak, then even appropriate aminoacyl-tRNA molecules will be retained with only a low efficiency.

A decoding region of the ribosome can discriminate between the different aminoacyl-tRNAs. Three proteins in the small subunit (mol wt. of 12 000, 5000 and 4000 Daltons) of prokaryotic ribosomes appear to be associated with a built in low level misreading since mutations in the genes coding for these proteins increases the errors in translation. It is uncertain if similar proteins serve the same function in eukaryotes but proteins of the same size are found in the 40S subunit. It also appears that rRNA is important in ribosomal-level discrimination.

kinetic proofreading

Kinetic proofreading constitutes the accuracy-amplifying step in ribosomal discrimination of aminoacyl-tRNAs. This step is little understood, but appears to be associated in prokaryotes with an EF-Tu-dependent flux of GTP over the ribosome. The free energy for the discrimination is thus supplied by the hydrolysis of the GTP and proofreading is driven by the production of GDP.

SAQ 8.5

mRNA is transcribed from a template DNA by RNA polymerase. This enzyme produces mRNA with an error of only one incorrect ribonucleotide in 10^5. In translation, one inappropriate amino acid residue is incorporated into a polypeptide per 10^4. If mRNA is produced at a rate of 50 nucleotides a second and polypeptides at 15 amino acid residues a second:

1) How long does it take before:

   a) a defective mRNA;

   b) a defective polypeptide, is produce?

2) What proportion of:

   a) mRNA;

   b) protein molecules;

will be defective assuming the average length of an mRNA molecule to be 1kb?

8.7 Antibiotics and protein synthesis

antibiotics are often selective

Many antibiotics inhibit protein synthesis in prokaryotes without affecting the equivalent eukaryotic processes ie they are selectively toxic (Table 8.6). This difference is the basis of their therapeutic use. The selectivity may arise through differences in the mechanism of protein synthesis or through differences in the uptake of the antibiotic by the cell. Antibiotics may be classified according to the stage of protein synthesis which is blocked. This can be illustrated by reference to a number of exemplary antibiotics.

Antibiotic	*Cell type affected	Site of action	Effect
Puromycin	E, P	Peptidyl transferase	Premature release of polypeptide
Tetracycline	P	Recognition	Inhibits binding at A site
Streptomycin	P	Recognition	Prevents movement of the initiation complex
Chloramphenicol	P	Peptidyl transferase	Blocks binding at P site
Erythromycin	P	Translocation	Inhibits translocation
Fusidic acid	E, P	Translocation	Prevents release of EF-Tu and GDP
Cycloheximide	E	Translocation	Inhibits peptidyl transferase

Table 8.6 Selected antibiotics and their sites of action. *E, eukaryotic; P, prokaryotic.

8.7.1 Antibiotics acting on recognition

Π Puromycin has a similar structure to the terminal aminoacyl-adenosine portion of tRNA (Figure 8.4). Which charged tRNA does puromycin most resemble?

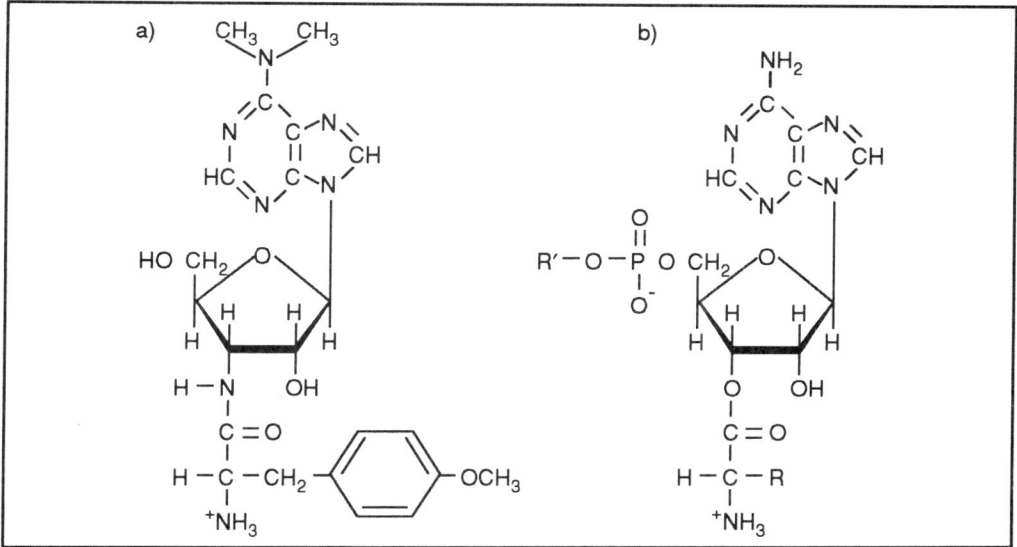

Figure 8.4 The structures a) puromycin and b) an amino acyl-tRNA showing the terminal amino acyl-adenosine residue. Note their similarity in structure. R is the sidechain of the amino acid residue carried by the tRNA. R′ is the remainder of the tRNA molecule.

The sidechain on puromycin is similar in structure to tyrosine. Puromycin binds to the A site of the ribosome and can accept the polypeptide from the P site. However, the binding is reversible and an incomplete polypeptide carrying a puromycin residue at its C terminal is released.

We can represent this process in the following way.

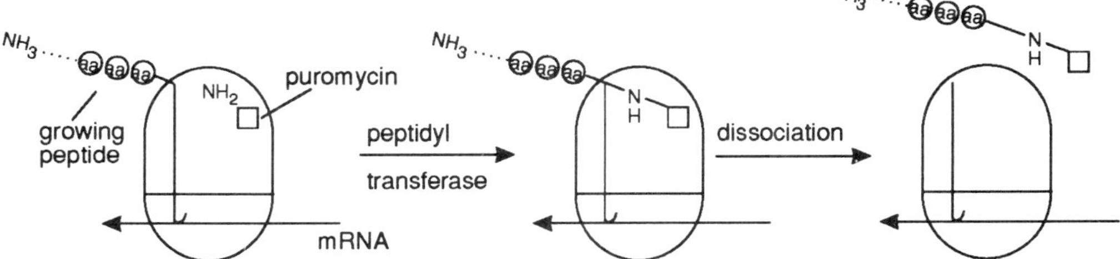

In this situation the ribosomes fail to recognise that the puromycin is not an aminoacyl-tRNA. The puromycin enters the ribosomal A site irrespective of the nature of the next anticodon on the mRNA.

Tetracycline (Figure 8.5) binds to the small subunit of the ribosome and blocks binding of the aminoacyl-tRNA to the A site. Tetracycline can block eukaryotic protein synthesis but only *in vitro* as it is not able to cross the eukaryotic cell membrane.

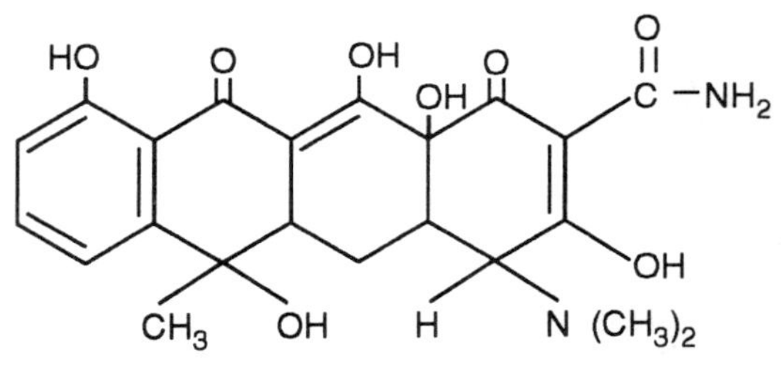

Figure 8.5 Tetracycline.

8.7.2 Antibiotics acting on peptidyl transferase

chloramphenicol

Chloramphenicol (Figure 8.6) binds to the large subunit of the prokaryote ribosome and prevents peptidyl transfer. The blockage may be due to distortion of the ribosome rather than action on the enzyme.

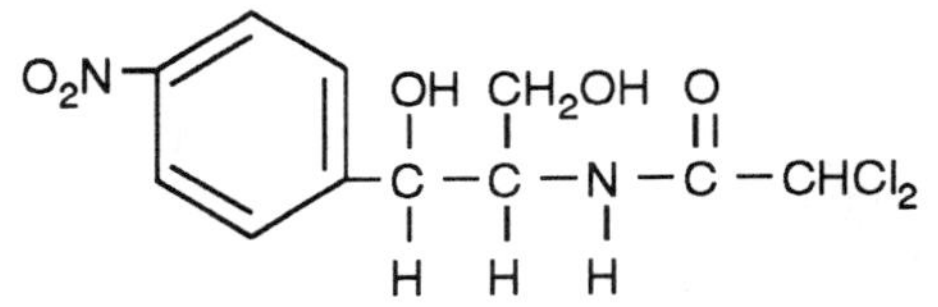

Figure 8.6 Chloramphenicol.

8.7.3 Action on translocation

erythromycin

Erythromycin (Figure 8.7) binds reversibly to free ribosomes but not to polyribosomes. Erythromycin allows the production of short polypeptides but prevents extended synthesis since the ribosomal attachment to mRNA is unstable in the presence of the drug.

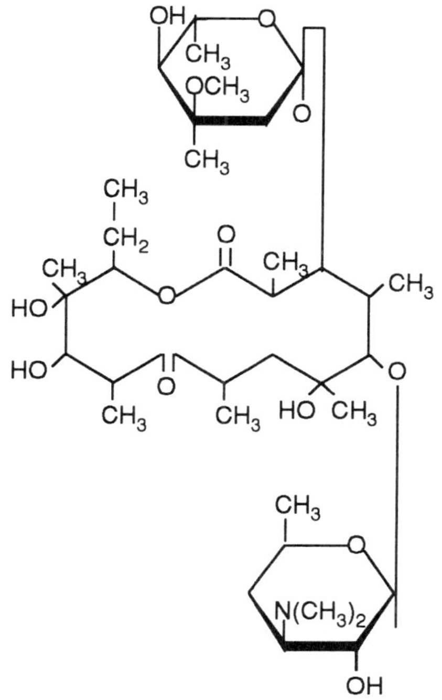

Figure 8.7 Erythromycin. Note that we have used a simplified representation of this molecule by omitting C-H bonds from the various rings.

8.7.4 Antibiotics and other compounds which block eukaryotic protein synthesis

cycloheximide

A small number of compounds block only eukaryotic protein synthesis. Cycloheximide (Figure 8.8) is an antibiotic which blocks translocation and is used experimentally to prevent protein synthesis. Diphtheria toxin is an enzyme which catalyses the addition of ADP-ribose from NAD^+ to a modified histidine residue of eIF2. The process by which this occurs is represented in Figure 8.9. This results in the irreversible inactivation of eIF2. Diphtheria toxin is one of the most deadly toxins known, such that one molecule may be sufficient to kill a cell.

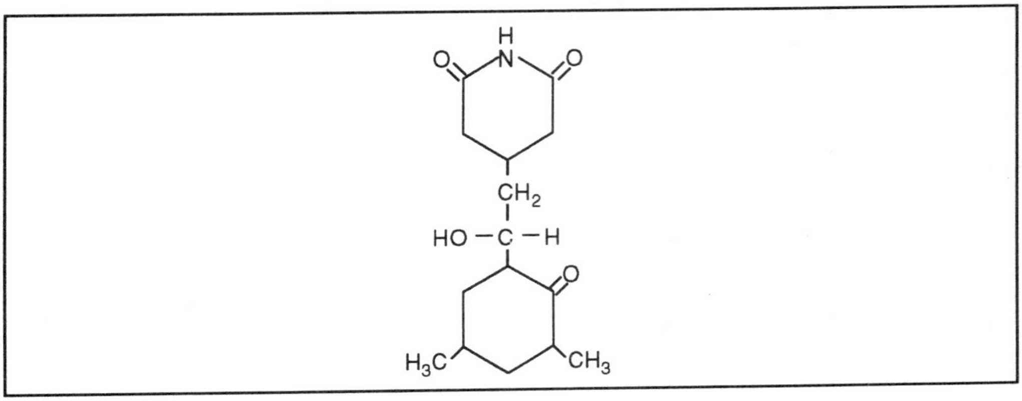

Figure 8.8 Cycloheximide.

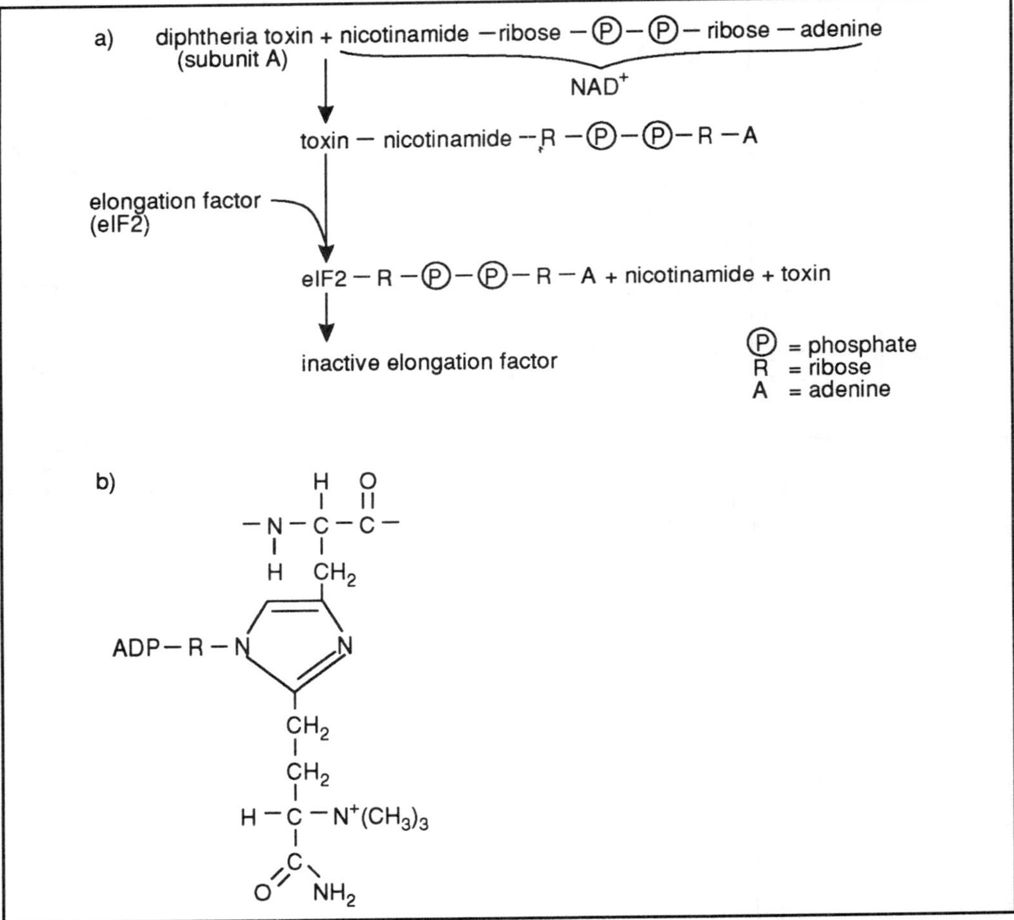

Figure 8.9 Diptheria toxin and the inactivation of eIF2. a) Overall reaction sequence. b) The structure of the ADP-ribosylated modified histidine residue of eIF2.

SAQ 8.6

State which of the following are true and which are false.

1) Some antibiotics are selective because they take advantage of the differences between eukaryotic and prokaryotic protein synthesis.

2) Puromycin has a similar structure to the 3′ end of an aminoacyl-tRNA.

3) Chloramphenicol blocks binding of an aminoacyl-tRNA to the A site of the ribosome.

4) Tetracycline binds to the large subunit of the ribosome and inhibits the binding of the aminoacyl-tRNA.

5) Cycloheximide prevents translation in both prokaryotes and eukaryotes.

8.8 Signal peptides and secretion

two populations of ribosomes in eukaryotes

Ribosomes in eukaryotic cells comprise two populations: those found free in the cytosol and those attached to the membrane of the rough endoplasmic reticulum (RER). The ribosomes of both populations are identical and in fact this segregation only occurs after translation of the mRNA begins. Free ribosomes translate mRNA which codes for proteins which will generally function in the cytosol. Ribosomes of the RER produce proteins which will become associated with the plasma membrane or be secreted from the cell.

signal peptide

The production of polypeptides by cytosolic ribosomes proceeds as described previously. However, translation of mRNA associated with secretory and membrane proteins produces an ER-directing signal peptide. The signal peptide consists of the first 25-30 amino terminal amino acid residues. Signal sequences differ between species and, indeed, between different proteins in the same species. Some examples are shown in Table 8.7.

Bovine proalbumin	Met-Lys$^+$-Trp-**Val**-Thr-**Phe-Ile**-Ser-**Leu-Leu-Leu-Phe**-Ser-Ser-**Ala**-Tyr-Ser-
Mouse antibody H chain	Met-Lys$^+$-**Leu**-Ser-**Leu-Leu**-Tyr-**Leu-Leu**-Thr-**Ala-Ile-Pro**-His-**Ile-Met**-Ser-
Human proinsulin	Met-**Ala-Leu**-Trp-**Met**-Arg$^+$-**Leu-Leu-Pro-Leu-Leu-Ala-Leu-Leu-Ala-Leu**-Trp-Gly-**Pro**-Asp-**Pro-Ala-Ala-Ala-**

Table 8.7 Examples of signal peptides. Basic residues are indicated by + and hydrophobic residues are shown in bold.

Π From the examples given in Table 8.7, derive a common structure for signal sequences.

The basic structure for signal sequence is shown in Figure 8.10. Their mechanism of action also appears to be the same irrespective of species; in fact prokaryotic signal peptides, which direct the secretion of proteins from bacterial cells, also promote the export of proteins from eukaryotic cells!

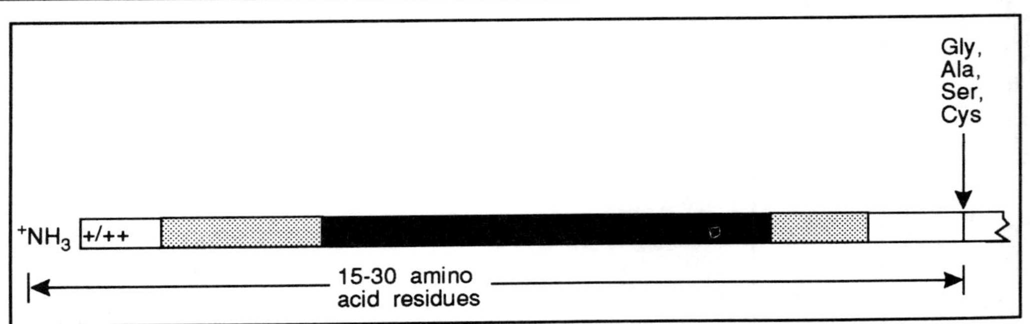

Figure 8.10 Overall structure of a signal peptide. Note the initial basic residues (+), uncharged regions (cross hatched), the continuous hydrophobic stretches (shaded) and the small terminal residue.

signal recognition particle (SRP)

SRP receptors

signal peptidase

The mechanism of action of signal-directed segregation is summarised in Figure 8.11. Translation proceeds normally on free ribosomes until about 70-80 amino acid residues have been polymerised. This means that the signal sequence protrudes from the ribosome. The signal peptide is recognised by a signal recognition particle (SRP) which binds to it. Binding of the SRP stops elongation. The complex then docks with the ER membrane by attaching to an SRP receptor protein which is an integral ER membrane protein. Once docked, the ribosome is then held by ribosome binding proteins and the SRP is released. Translation now resumes, however, the growing polypeptide is translocated through the ER membrane (Figure 8.11). A signal peptidase (SPase) within the ER lumen hydrolytically removes the signal peptide, seemingly before translation is complete. Secretory proteins pass completely through the ER membrane into the lumen, are transferred to the Golgi apparatus and eventually exported from the cell.

Membrane proteins are not completely translocated through the ER membrane but remain associated with the biomembranes during their transfer from the ER to the Golgi apparatus to the plasma membrane.

SAQ 8.7

Recombinant DNA techniques ('genetic engineering') allow chimeric proteins (that is novel proteins composed of portions of different proteins) to be produced. If chimeric mRNA consisting of the codons for a signal sequence was appropriately combined with those for a protein which is normally formed on free ribosomes and is functional in the cytosol, what would be the expected outcome of its translation?

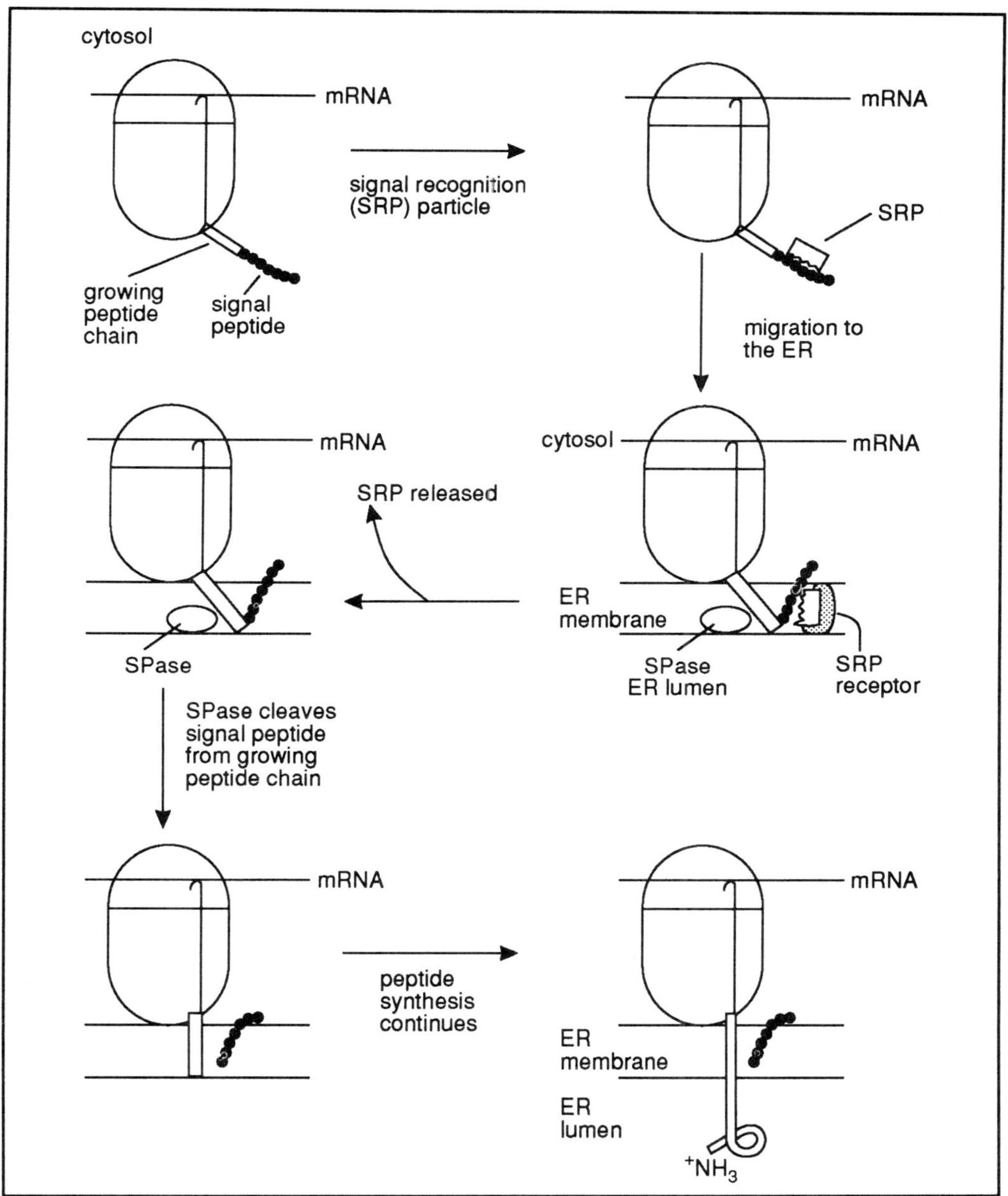

Figure 8.11 Outline of the mechanism of translocation of polypeptides across the ER membrane. SP = signal peptide, SRP = signal recognition particle and SPase, signal peptidase. Not all steps have been illustrated. See text for details.

8.9 Post-translational modifications

co- and post-translational modifications

Translation, in itself, rarely produces a fully functional protein. The polypeptide formed is usually subjected to a number of changes, both during and after its initial synthesis on the ribosome. These changes are called co- and post-translational modifications respectively. However, the distinction between the two types is often blurred and post-translational modification is often used as a general term. Although some such changes do occur in prokaryotes, they are much more extensive in eukaryotic systems.

SAQ 8.8

Is the removal of an amino terminal methionine residue during translation a co- or a post-translational event?

8.9.1 Types of post-translational modifications

Most post-translational modifications can be classified into one of three major types:

- modifications to the amino or carboxyl termini;
- modifications to specific amino acid residue sidechains;
- limited, specific proteolysis.

These divisions are rather arbitrary and, indeed, a given protein may be subjected to more than one type of modification.

The biological sites of post-translational modifications vary. For examples, limited proteolysis can occur intra- and/or extracellularly. Most intracellular modifications involve the cytosol, endoplasmic reticulum and the Golgi apparatus and its associated secretory vesicles.

Given the diverse nature of post-translational activities this chapter will be restricted to those which are the most representative and about which most is known! Table 8.8 summarises the major types of post-translational modifications.

Modification	Sites of change	Examples of proteins affected
Acetylation	Amino terminus	Many cell proteins
ADP-ribosylation	Carboxy terminus	Histone H1
Amidation	Carboxyl terminus	Many proteins
Phosphorylation	Ser, Thr and Tyr residues	Glycogen phosphorylase
Sulphation	Tyr residues	Number of secretory and membrane proteins
Oxidation	Cys residues	Extracellular proteins
Acylation	N-Gly, Cys and Ser residues	Membrane and cytosolic proteins
Glycosylation	Ser, Thr and Asn residues	Most secreted and membrane proteins
Limited proteolysis	Polypeptide chain	Many secreted enzymes and hormones

Table 8.8 Summary of major types of post-translational modifications.

8.9.2 Modifications of amino and carboxyl termini

The major modifications are:

$^+NH_3$-	formylation,
	acetylation;
$-COO^-$	ADP-ribosylation,
	formation of a substituted amide,
	formation of a terminal amide.

Amino terminus modifications

N-acetylation

N-acetyltransferase

Modifications to the amino terminus, such as the formylation of the amino terminal Met residue is typical in prokaryotic system. However, N-acetylation is the most common amino terminal modification of eukaryotic proteins. Over 80% of cytosolic proteins in some cell types show this addition. The acetyl group is added from acetyl CoA by N-acetyltransferase (AcT):

$$^+NH_3\text{-R} + \text{Acetyl CoA} \xrightarrow{(AcT)} CH_3CO\text{-NH-R} + \text{CoASH} + H^+$$

The role of acetylation is unknown. Amino termini may also be modified by the addition of long chain fatty acids (see acylation below).

Carboxyl terminal modifications

ADP-ribosylation

amidation

ADP-ribosylation of the carboxy terminus occurs at the terminal lysine residue of the protein histone H1, while a substituted amide is formed by the addition of a tyrosine residue to tubulin following its release from the ribosome. Carboxyl amidation is, however, much commoner and amidated peptides are found in many types of organisms suggesting it arose early in evolution. About 50% of the bioactive peptides in the nervous and endocrine systems are amidated, as are a number of neurotransmitters and growth factors. In general, this amidation is essential for biological activity.

Amidation occurs by modification of a glycine residue at the C-terminus of the peptide. Ascorbate (vitamin C), and Cu^{2+} are necessary co-factors:

$$\underset{\text{peptidyl-glycine}}{\text{R-Gly}} + O_2 + \text{ascorbate} \xrightarrow{Cu^{2+}} \text{R-}NH_2 + \text{glyoxylate} + \text{dehydroascorbate} + H_2O$$

Recent studies suggest that amidation is a two step process, with an hydroxylated intermediate being formed. The second step converts the hydroxylated intermediate to the peptide amide, and releases glyoxylate. Although the second step can occur spontaneously, *in vivo* it appears to be enzyme-catalysed:

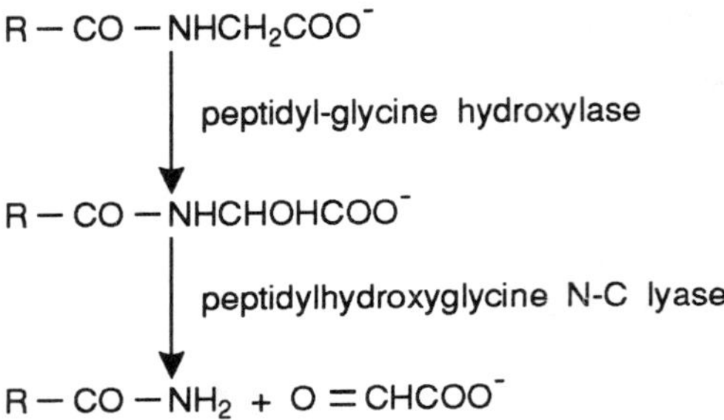

8.9.3 Modifications to specific amino acid sidechains

Changes to specific amino acid sidechains are the most diverse of all post-translational modifications. Examples of secondarily modified amino acids are given in Table 8.9. These changes may be comparatively simple, such as the additions of phosphate to serine, threonine or tyrosine residues (phosphorylation), sulphate to tyrosine residues (sulphation), hydroxyl groups to lysine or proline residues (hydroxylation), palmitate to cysteine residues (acylation) or the oxidation of pairs of cysteine residues (disulphide bond formation).

N-Acetylaspartate	π-Methylhistidine
O^{β}-Mannosylation	O-Adenosyltyrosine
N-Acetylalanine	γ-Carboxyglutamate
β-Hydroxyphenylalanine	S-Galactosylcysteine
ε-N-Methyllysine	Glycinamide
4-Hydroxyproline	Hydroxylysine
O^{β}-Phosphoserine	O^{4}-Phosphotyrosine
3-Chlorotyrosine	O^{4}-Sulphotyrosine

Table 8.9 Examples of amino acids formed by post-translational modifications.

prolyl hydroxylase

Collagen is a widely distributed connective tissue protein, which consists largely of the amino acid residues proline, glycine and lysine. Some of the proline residues are post-translationally hydroxylated to 4-hydroxyproline by prolyl hydroxylase, an enzyme with a ferrous ion at its active site. Hydroxylation is dependent upon the presence of O_2, α-oxoglutarate and ascorbate (vitamin C), (a reducing agent necessary to maintain the iron in a ferrous state).

We can represent the reaction in the following way:

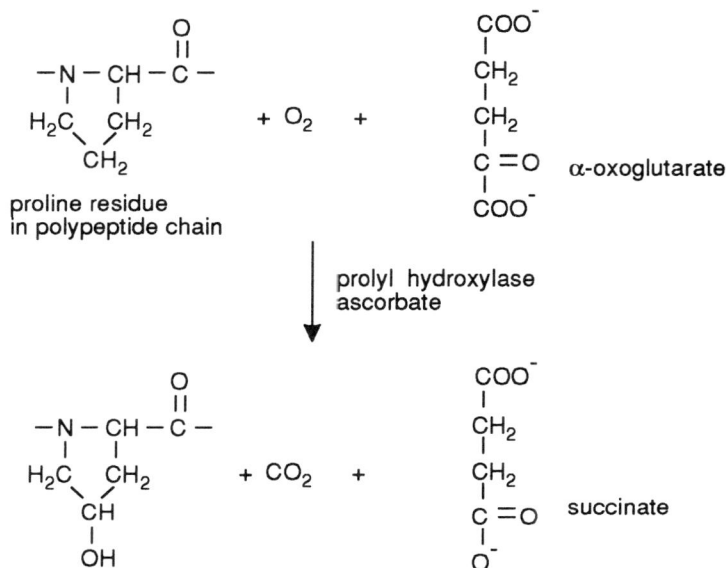

$$\underset{\text{proline residue in polypeptide chain}}{-N-CH(-CH_2-CH_2-CH_2-)-C(=O)-} + O_2 + \underset{\alpha\text{-oxoglutarate}}{{}^{-}OOC-CH_2-CH_2-C(=O)-COO^{-}}$$

$$\downarrow \text{prolyl hydroxylase, ascorbate}$$

$$-N-CH(-CH_2-CH(OH)-CH_2-)-C(=O)- + CO_2 + \underset{\text{succinate}}{{}^{-}OOC-CH_2-CH_2-C(=O)-O^{-}}$$

lysyl hydroxylase

A small number of proline residues are also hydroxylated at C-3 but by a different enzyme. A comparatively small proportion of the lysine residues are also hydroxylated, to hydroxylysine, in a reaction catalysed by lysyl hydroxylase. This reaction also requires the presence of O_2, α-oxoglutarate and ascorbate and the overall reaction can be written as:

$$^{+}NH_3-CH_2\,CH_2\,CH_2\,CH_2-\overset{|}{\underset{H}{C}}(NH-)-\overset{O}{\overset{\|}{C}}- \longrightarrow {}^{+}NH_3-CH_2-\overset{OH}{\overset{|}{CH}}-CH_2-CH_2-\overset{|}{\underset{H}{C}}(NH-)-\overset{O}{\overset{\|}{C}}-$$

Native collagen is a triple stranded helical molecule, the integrity of which is markedly stabilised by hydroxylation of proline residues. Collagen fibres in connective tissue are formed by linking the triple helical molecules together by covalent cross links. These covalent bonds are formed by joining two hydroxylysine and one lysine of different helices together.

Π Use the information provided above to predict the major clinical symptoms one would expect to be associated with a deficiency of ascorbate.

scurvy

Prolyl and lysyl hydroxylases will have reduced activity, therefore the collagen produced will be incompletely hydroxylated. The lack of hydroxyproline means defective triple helix formation and symptoms typical of scurvy will result eg skin lesions, blood vessel fragility, poor wound healing and abnormal growth.

The lack of hydroxylysine means triple helices are defectively cross-linked giving rise to the symptoms of, for example, type IV Ehlers-Danlos syndrome; musculoskeletal abnormalities, poor wound healing and increased extensibility of skin and joints.

Other modification to amino acid residues may be more extensive. For example the formation of large oligosaccharides at specific serine, threonine or asparagine residues. Thus although only 20 amino acids are specified by the genetic code the hydrolysis of all the proteins of an organism results in the release of upto 200 different types of amino acids.

SAQ 8.9

1) Use Tables 8.9 and your knowledge of the genetic code to determine which of the following amino acid residues are produced as a result of post-translational modifications:

   a) 3- Chlorotyrosine

   b) S-Galactosylcysteine

   c) Glycinamide

   d) Glutamine

   e) N-Acetylalanine.

2) State the 'parent' amino acid residue for each of the post-translationally modified types given in 1).

3) Prothrombin has 10 γ-carboxyglutamate residues (see figure below) which are necessary for Ca^{2+}-binding. The prothrombin-Ca^{2+} complex is essential for appropriate blood clotting. The γ-carboxyglutamate residues are formed by post-translational modification of glutamate residues during the intracellular biosynthesis of prothrombin. Vitamin K is essential for this processing:

$$\begin{matrix} | \\ O=C \\ | \\ HC-CH_2-CH_2-COO^- \\ | \\ N-H \\ | \end{matrix} + CO_2 \xrightarrow[K]{\text{vitamin}} \begin{matrix} | \\ O=C \\ | \\ HC-CH_2-CH \begin{matrix} \diagup COO^- \\ \diagdown COO^- \end{matrix} \\ | \\ N-H \\ | \end{matrix}$$

Suggest the most likely outcome of a dietary deficiency of vitamin K.

Phosphorylation

glycogen phosphorylase

The enzymes which control the breakdown and formation of glycogen are probably the best known examples of proteins which are subject to phosphorylation (and dephosphorylation). Glycogen phosphorylase catalyses the breakdown of glycogen by promoting its phosphorylation with the release of glucose 1-phosphate:

$$\underset{(\text{glucose})_n}{\textbf{glycogen}} + \textbf{P}_i \xrightarrow{\text{phosphorylase}} \underset{(\text{glucose})_{n-1}}{\textbf{glycogen}} + \textbf{glucose 1-phosphate}$$

Skeletal muscle phosphorylase occurs in one of two interconvertible forms: phosphorylase a (which is active) and phosphorylase b (which is generally inactive).

The enzyme is dimeric, each subunit having a molecular mass of 97 000 Daltons. Phosphorylase b is converted to the a form by being phosphorylated itself at residue Ser-14 in each subunit:

phosphorylase kinase

○–○ + 2 ATP ⟶ P–□–□–P + 2 ADP

(phosphorylase b) (phosphorylase a)

phosphatase

The enzyme can be converted back to the b form by the action of a phosphatase which catalyses the dephosphorylation of the serine residues.

This is a somewhat simplified explanation because the activity of both forms of phosphorylase are affected by the relative concentrations of AMP, ATP, glucose 6-phosphate and glucose. The control of this process of phosphorylation and dephosphorylation of this enzyme is best considered in the context of metabolism. If you wish to follow up this aspect, we recommend the BIOTOL text 'Biosynthesis and the Integration of Metabolism'.

Sulphation

tyrosines may be sulphated

Sulphation occurs with a number of proteins and neuropeptides. Most sulphated proteins are secretory or membrane proteins, all of which are produced on ribosomes of the rough endoplasmic reticulum (Section 8.8). Tyrosine residues appear to be the main sites of attachment, and upto 1% of tyrosine residues in the total protein may be sulphated.

Sulphation occurs in the Golgi apparatus. Phosphoadenosine phosphosulphate, (PAPS) (Figure 8.12) is the sulphate donor:

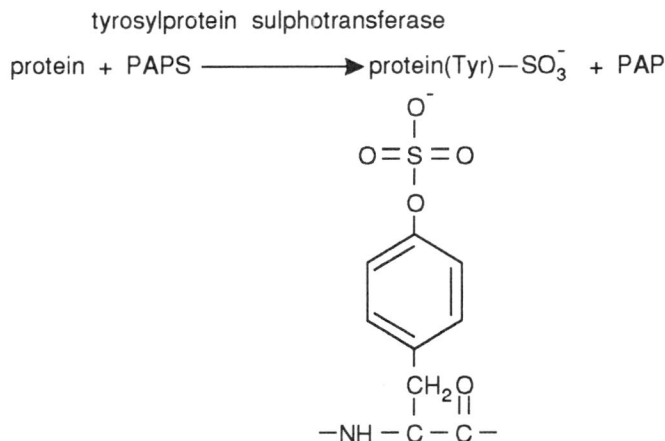

tyrosylprotein sulphotransferase

$$\text{protein} + \text{PAPS} \longrightarrow \text{protein(Tyr)}-SO_3^- + \text{PAP}$$

O^-
|
O=S=O
|
O
|
(benzene ring)
|
CH_2 O
| ||
–NH – C – C –

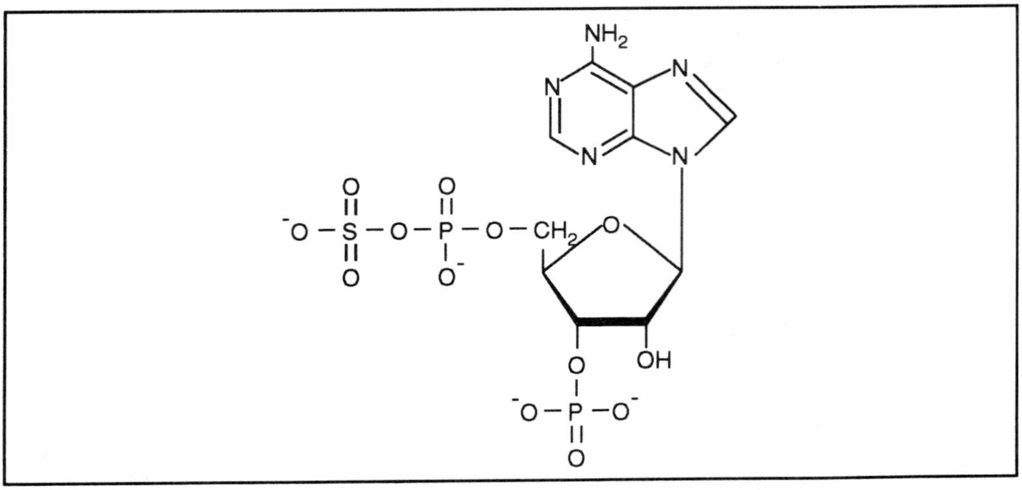

Figure 8.12 Phosphoadenosine phosphosulphate (PAPS).

A biological role for sulphation has been established in only few cases. For example, hormonal activity of sulphated cholecystokinin is 260 times more active than the nonsulphated form.

Oxidation

The oxidation of adjacent cysteine residues to form disulphide bonds occurs in relatively few proteins, and these are largely restricted to secreted or cell-surface proteins. Good examples of proteins containing disulphide bonds are antibodies.

Π Explain why you would not expect to find disulphide bonds in proteins of the cytosol of a cell. Hint: think about the reducing/oxidising potential in the cytosol.

The cytosol of the cell is highly reducing, therefore disulphide bonds would be reduced to two cysteine residues.

formation of disulphide bonds

Disulphide bonds are the only type of 'cross-link' commonly found in globular proteins. For a disulphide bond to form, the folding of the protein must already have occurred to bring the two cysteine residues to a distance of 0.4-0.9 nm apart and in an appropriate orientation. Little is known of the nature of the oxidising agent (hydrogen acceptor). *In vitro*, O_2 is a common agent. Recent evidence indicates that a pool of oxidised glutathione (Figure 8.13) present in the lumen of the ER (in other words, separated from the cytosol) is the most likely hydrogen acceptor for *in vivo* disulphide bond formation.

$$2\ R{-}SH + \text{oxidised glutathione} \longrightarrow R{-}S{-}S{-}R + 2\ \text{reduced glutathione}$$

a)

$^{+}NH_3-CH(COO^-)-CH_2-CH_2-C(=O)-NH-CH(CH_2-S-S-CH_2)-C(=O)-NH-CH_2-COO^-$

$^{+}NH_3-CH(COO^-)-CH_2-CH_2-C(=O)-NH-CH(CH_2-)-C(=O)-NH-CH_2-COO^-$

γ glutamate cysteine glycine

b)

$^{+}NH_3-CH(COO^-)-CH_2-CH_2-C(=O)-NH-CH(CH_2SH)-C(=O)-NH-CH_2-COO^-$

Figure 8.13 a) Oxidised and b) reduced forms of glutathione.

disulphide bond and stabilisation of structure

Disulphide bonds form in a protein only after it has folded into its active conformation. Thus they do not contribute to the folding to form the final stable state. However, disulphide bonds may stabilise a folded conformation that is no longer stable in their absence, because of other subsequent covalent modifications such as proteolysis (Section 8.9.4). This is the case with, for example, insulin which is synthesised as a comparatively large, inactive precursor polypeptide. This folds, and disulphide bonds form. The hormone is then activated by specific proteolytic processing and is stabilised by the disulphide bridges under most conditions. We can represent this in the following way:

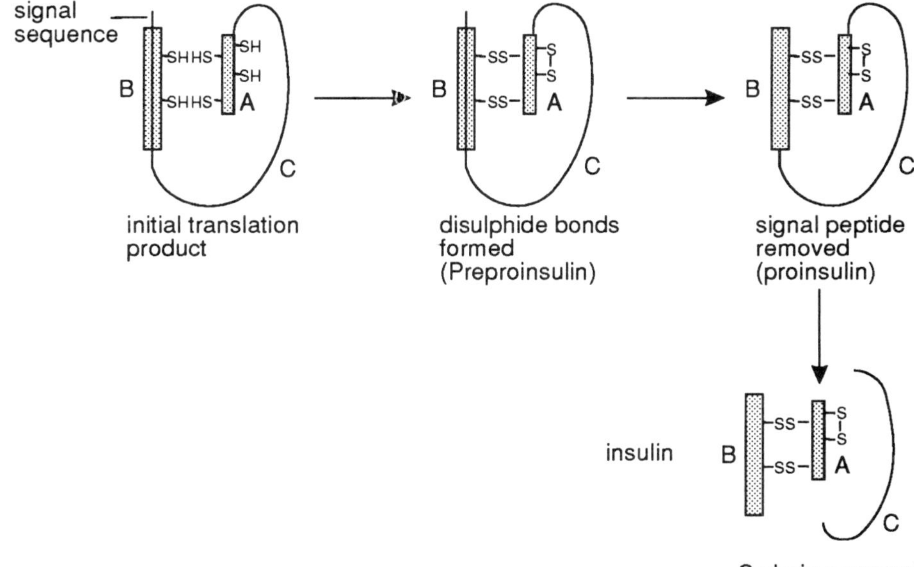

If, however, disulphide exchange is permitted, rearrangement of the disulphide bonds occurs, the hormone denatures and biological activity is lost. In this, and similar instances, disulphide bonds are probably stabilising an inherently unstable conformation.

Acylation

myristate

palmitate

stearate

Acylation is the covalent attachment of long chain fatty acids to proteins. It is apparently ubiquitous in eukaryotes of all types. The commonest fatty acyl groups added are myristate ($CH_3(CH_2)_{12}COO^-$) and palmitate ($CH_3(CH_2)_{14}COO^-$), although stearate ($CH_3(CH_2)_{16}COO^-$) may also be added. In all cases the acyl group is added from the corresponding coenzyme A derivative (Figure 8.14).

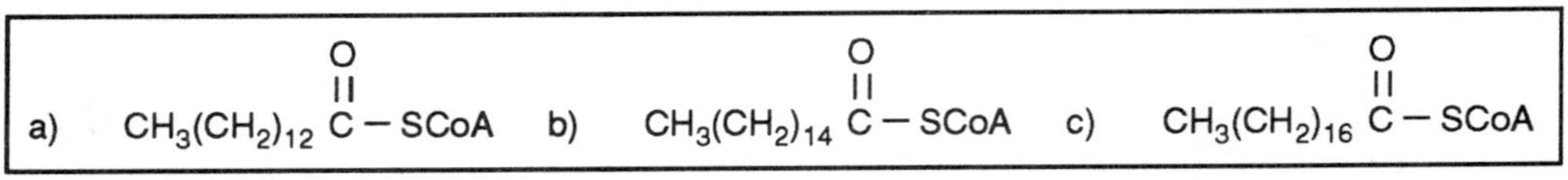

a) $CH_3(CH_2)_{12}\overset{O}{\overset{\|}{C}}-SCoA$ b) $CH_3(CH_2)_{14}\overset{O}{\overset{\|}{C}}-SCoA$ c) $CH_3(CH_2)_{16}\overset{O}{\overset{\|}{C}}-SCoA$

Figure 8.14 Coenzyme A derivatives of a) myristate, b) palmitate and c) stearate.

myristate is an amino terminal modification

Myristate is commonly added to an amino terminal glycine residue before translation is completed (Figure 8.15a). It is, therefore, also an amino terminal modification (Section 8.9.2). The link is a biologically stable amide bond. In contrast, palmitate is normally added to internal cysteine residues (Figure 8.15b) when the polypeptide is complete. Serine residues may also by acylated. The thioester (or ester, if serine is involved) link formed is biologically labile and the palmitate may have a shorter half life than the protein carrying it.

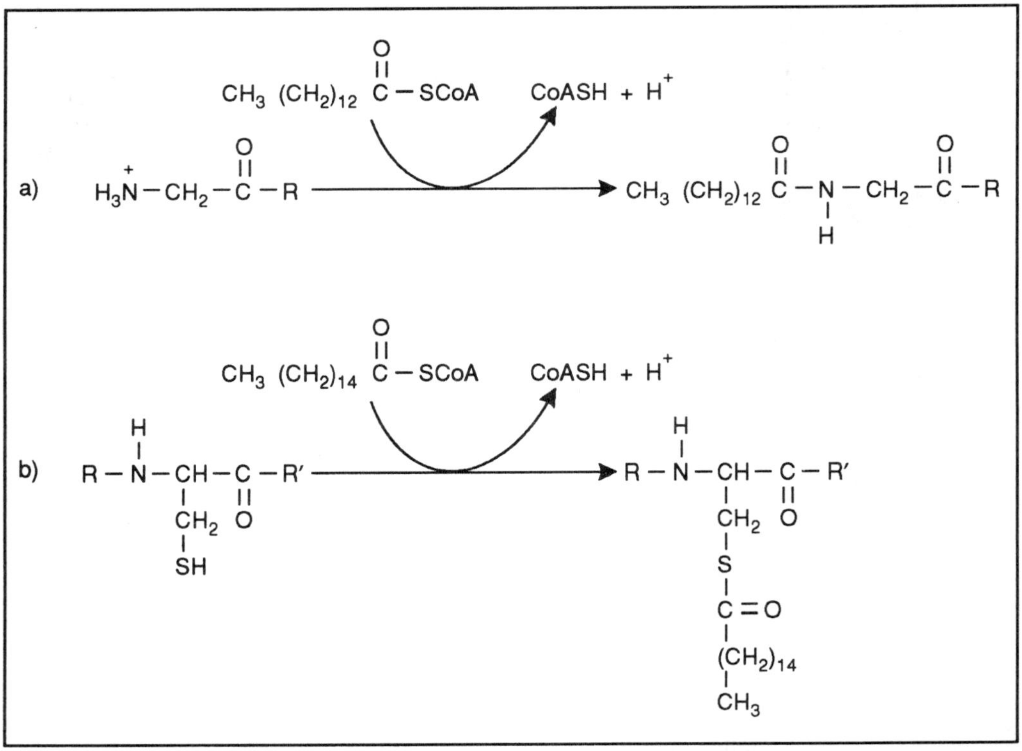

a) $\overset{+}{H_3N}-CH_2-\overset{O}{\overset{\|}{C}}-R \xrightarrow[\text{CoASH} + H^+]{CH_3(CH_2)_{12}\overset{O}{\overset{\|}{C}}-SCoA} CH_3(CH_2)_{12}\overset{O}{\overset{\|}{C}}-\underset{H}{\underset{|}{N}}-CH_2-\overset{O}{\overset{\|}{C}}-R$

b) $R-\overset{H}{\overset{|}{N}}-\underset{CH_2-SH}{\underset{|}{CH}}-\overset{O}{\overset{\|}{C}}-R' \xrightarrow[\text{CoASH} + H^+]{CH_3(CH_2)_{14}\overset{O}{\overset{\|}{C}}-SCoA} R-\overset{H}{\overset{|}{N}}-\underset{CH_2-S-\overset{O}{\overset{\|}{C}}-(CH_2)_{14}-CH_3}{\underset{|}{CH}}-\overset{O}{\overset{\|}{C}}-R'$

Figure 8.15 a) N-acylation of glycine residue with myrisrate. b) Addition of palmitate to a cysteine residue.

The biological roles of acylation are uncertain. Palmitate is found associated with membrane bound proteins and helps anchor them in the bilayer by forming stable hydrophobic interactions (Figure 8.16). However, myristoylation is found in both membrane and cytosolic proteins.

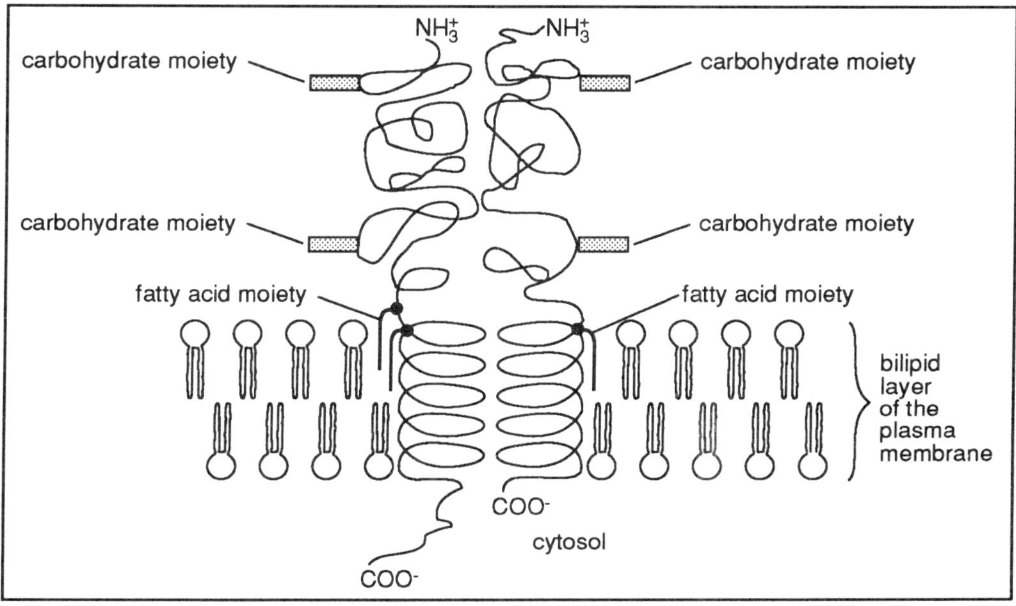

Figure 8.16 Schematic showing insertion of a protein residue into a lipid bilayer. The fatty acyl chains assisting in this are shown in bold. Note the carbohydrate which has also been added to the protein post-translationally (see below). Redrawn from Schlesinger, M.J. (1981) Ann. Rev. Biochem., 50, 193-206.

SAQ 8.10

Complete the following table to contrast myristoylation and palmitoylation.

Feature	Myristate	Palmitate
Chemical link		
Amino acid residue involved		
Half life		
Post- or co-translational		
Biological site of protein		

Glycosylation

glycoproteins

Glycosylation is the enzyme-catalysed addition of sugar residues to proteins forming glycoproteins. Glycoproteins are common, occurring on cell surfaces, in secretions, as part of the extracellular matrix and in viral envelopes. Their biological roles are diverse (Table 8.10).

Glycoprotein	Source	Carbohydrate content (%)*	Class/Function
Ribonuclease B	Bovine	10.7	Enzyme
Glucose oxidase	*Aspergillus niger*	16.2	Enzyme
Thyroglobulin	Human	10.6	Hormone
Chorionic gonadotrophin	Human	33.9	Hormone
IgA	Horse	2.4	Protection
α-Acid glycoprotein	Bovine	36.5-42.7	Serum protein, released following stress
Prothrombin	Human	8.9	Clotting
Fibrinogen	Human	9.5	Clotting
Transferrin	Human	11.3	Serum transport of iron
Interferon-γ	Human T-lymphocytes	15-32	Antiviral activity, inhibitor of cell division
Interleukin-2	Human T-lymphocytes	21-31.8	Promotes proliferation of lymphocytes
Tumour necrosis factor B	Human T-lymphocytes	7.5-26	Cytotoxic effects with some tumours
Anion exchanger	Human erythrocyte	7.9	Transmembrane transport
Lipopolysaccharide receptor	Human erythrocyte	37.7	Membrane receptor
Glycophorin	Human erythrocyte	59.0	?
Collagen	Bovine achilles tendon	1.5	Locomotion
Proteoglycan	Rat cartilage	80-90	Articulation, resistance to compression
Mucins	Human cervical fluid	89.9	Protection, lubrication
Mucins	Loach skin	41.6	Protection
Glycopeptides from extensin	Plant cell walls	70.5	Structural
Lectin A	*Lotus tetragonolobulus*	9.4	Protection
Agglutinin	*Ricinus communis*	10.6	Protection

Table 8.10 Examples of glycoproteins. Reproduced with permission from Cole, C.R. and Smith, C.A. (1989) Glycoprotein biochemistry (structure and function) - a vehicle for teaching many aspects of biochemistry and molecular biology. Biochemical Education, 17, 179-189 (* % by weight).

SAQ 8.11

Use Table 8.10 to determine an approximate range of the carbohydrate content in globular glycoproteins.

role of carbohydrates in glycoproteins

The biological roles of the carbohydrate of glycoproteins is not well understood. It appears to be necessary for appropriate folding of some proteins into their native conformation, to stabilise some glycoproteins against thermal denaturation and proteolysis, to increase general solubility and be necessary for the insertion of some glycoproteins into membranes and for others to be secreted. It has been suggested that in some cases it acts as a 'molecular address' allowing some glycoproteins to be targeted to appropriate regions of the cell. The extensively glycosylated glycoproteins of the extracellular matrix are able to bind large quantities of water due to their high carbohydrate content. It is likely that the carbohydrate of glycoproteins therefore has different roles in different glycoproteins.

glycans

The carbohydrate portions of glycoproteins are often called glycans, and generally occur as oligosaccharides. The oligosaccharides usually contain mannose (Man), galactose (Gal), glucosamine (GlcNAc) but less often glucose (Glc), galactosamine (GalNAc), L-fucose (L-Fuc) and N-acetylneuraminate (NeuNAc), a sialic acid represented generally as Sia. The glycans are generally branched and can contain upto 15-20 sugar residues. Each glycan is attached to the glycoprotein to a specific amino acid residue by a single sugar residue. The two commonest links are O-glycosidic bonds involving serine or threonine residues and N-acetylgalactosamine, and N-glycosidic bonds between the amide nitrogen of asparagine and N-acetylglucosamine (Figure 8.17).

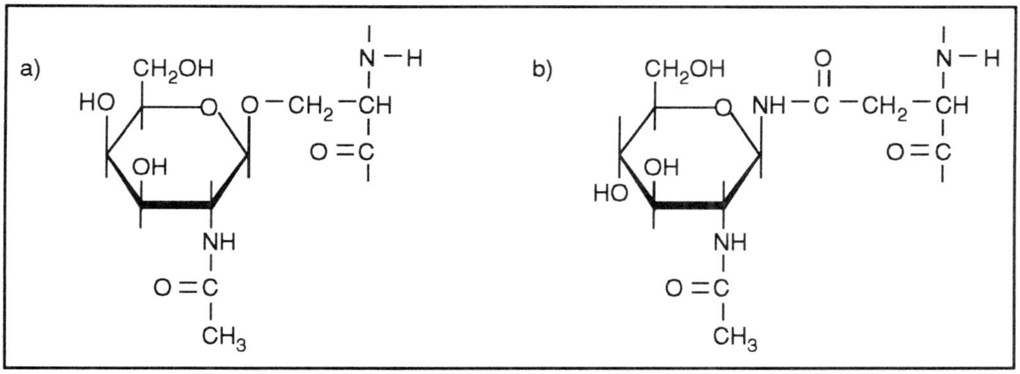

Figure 8.17 Structures of O- and N-linked glycosidic bonds. a) GalNAc linked to a serine residue and b) GlcNAc to an asparagine residue.

core structure

O-linked glycans generally have simpler, if more diverse, structures than N-linked types (Figure 8.18a). N-Linked glycans can be subdivided into three basic types: high-mannose; hybrid and complex (Figure 8.18b). All three types have a common core structure:

```
Man
   \
    Man—GlcNAc—GlcNAc—Asn
   /
Man
```

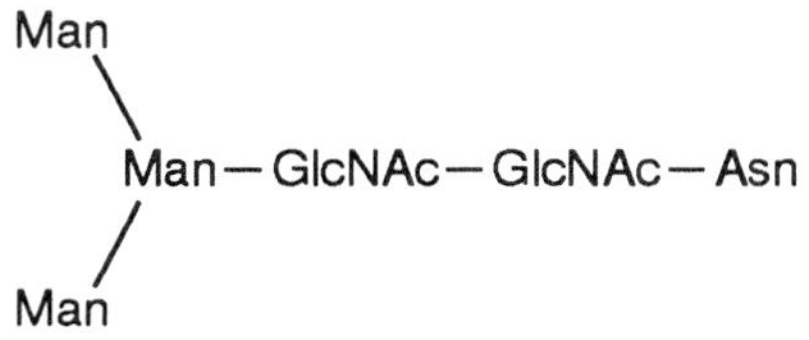

High-mannose types contain only mannose and the core GlcNAc residues (Figure 8.18b i). Hybrid types have a high-mannose extension attached to one of the two terminal mannoses of the core, the other has substituents of the complex types (Figure 8.18b ii). Complex type N-linked glycans contain a variable number of branches, each branch consisting of at least a GlcNAC residue and often being extended by the addition of Gal, NeuNAc or Fuc residues (Figure 8.18b iii).

Π In order to gain a greater understanding of these various structures, use Figure 8.18 to draw out the structures of these oligosaccharides. We have drawn the compound represented by Figure 8.18b i) for you as an example.

```
                              Man 1 —α— 2 Man 1
                                             |α
                                             6
                                   Man 1 —α— 3 Man 1
                                                  |α
                                                  6
Man 1 —α— 2 Man 1 —α— 2 Man 1 —α— 3 Man 1 —β— 4 GlcNAc 1 —β— 4 GlcNAc 1 —β— N
```

You might even try to draw them out structurally!

```
a)  i)    Galβ1-3GalNAcβ1-O

    ii)   GlcNAcβ1-6
                     GalNAcβ1-O
             Galβ1-3

    iii)  NeuNAα1-3Galβ1-4GlcNAcβ1-6
                                     GalNAcβ1-O
                 NeuNAcα1-3Galβ1-3

b)  i)    Manα1-2Manα1-6
                        Manα1-6
                 Manα1-3
                                 Manβ1-4GlcNAcβ1-4GlcNAcβ1-N
          Manα1-2Manα1-2Manα1-3

    ii)   Manα1-6
                 Manα1-6
          Manα1-3
                          Manβ1-4GlcNAcβ1-4GlcNAcβ1-N
        GlcNAcβ1-2Manα1-3

    iii)  Siaα2-3Galβ1-4GlcNAcβ1-2Manα1-3
                                           Manβ1-4GlcNAcβ1-4GlcNAcβ1-N
          Siaα2-3Galβ1-4GlcNAcβ1-2Manα1-6
```

Figure 8.18 Examples of a) O-linked glycans and b) i) high mannose, ii) hybrid and iii) complex N-linked glycans of glycoproteins.

Biosynthesis of glycans

glycosyl-transferases, glycosidases

The biosynthesis of glycans involves two general types of enzymes, glycosyltransferases, which add sugar residues and glycosidases which are necessary to remove them during the production of some types of glycans.

Glycosyltransferases (Gly-Ts) catalyse the transfer of a sugar residue from a donor to an acceptor substrate:

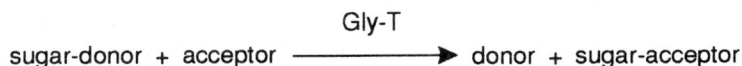

$$\text{sugar-donor} + \text{acceptor} \xrightarrow{\text{Gly-T}} \text{donor} + \text{sugar-acceptor}$$

The sugar donor is usually a nucleoside diphosphate sugar (Figure 8.19). UDP is the donor for Glc, Gal, GlcNAc and GalNAc, whereas GDP donates Man and Fuc and CMP donates NeuNAc. Lipids (dolicholphosphates) are the direct donors in the transfer of some glycosyl residues during the synthesis of N-linked glycans.

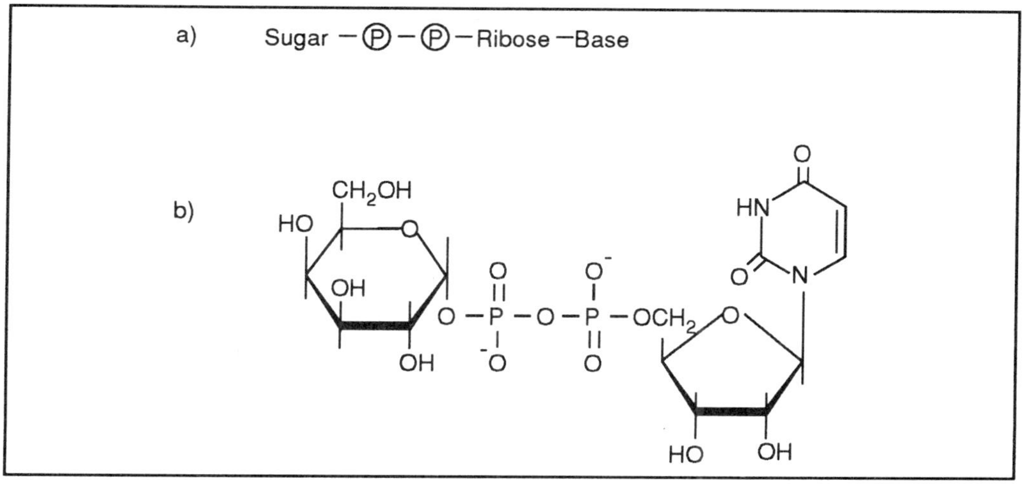

a) Sugar —Ⓟ—Ⓟ—Ribose—Base

b)

Figure 8.19 a) Generalised structure of a nucleoside diphosphate sugar (NDP-sugar). b) The specific NDP-sugar, uridine diphosphate galactose (UDP-Gal).

Π If glucosyltransferase catalyses the reaction:

UDP—Glc + acceptor ⟶ UDP + Glc—acceptor

Name the glycosyltransferases which catalyse the following reactions:

UDP—Gal + acceptor ⟶ UDP + Gal—acceptor

UDP—GlcNAc + acceptor ⟶ UDP + GlcNAc—acceptor

GDP—Man + acceptor ⟶ GDP + Man—acceptor

GDP—L—Fuc + acceptor ⟶ GDP + L—Fuc—acceptor

CMP—NeuNAc + acceptor ⟶ CMP + NeuNAc—acceptor

You should have given these enzymes the following names.

Galactosyltransferase; N-acetylglucosyltransferase; mannosyltransferase; fucosyltransferase and N-acetylneuraminyltransferase (a sialyltransferase).

Glycosyltransferases are remarkably specific for the sugar being transferred, the base component of the nucleotide, the acceptor substrate and the glycosidic bond formed. A separate Gly-T is required to form each of the glycosidic bonds found in a glycoprotein.

SAQ 8.12 Determine the number of Gly-Ts required to form each of the glycans shown in Figure 8.18a i)-iii) and in Figure 8.18b i-iii).

The biosynthesis of O-linked glycans proceeds by the stepwise addition of sugar residues to a protein core. Once the first sugar has been added to the polypeptide, the assembly of the glycan is determined by the substrate specificities of the rest of the

Gly-Ts and glycosylation will continue until a sialic acid residue is added. Sialic acid residues are signals which dictate the termination of glycosylation. O-Glycosylation can be explained by referring to the formation of mucins by submaxillary glands. A number of mucins are known, but synthesis starts with the addition of GalNAc to a serine (or threonine) residue using the galactosyltransferase UDP-GalNAc:polypeptide transferase.

$$\text{UDP}-\text{GalNAc} + \text{Ser/Thr(-protein)} \longrightarrow \text{GalNAc}-\text{Ser/Thr(-protein)} + \text{UDP}$$

Once synthesis is initiated in this fashion, Gal, Fuc, further GalNAc or NeuNAc may be added depending upon the enzyme complement of the species.

Π Given that ovine (sheep) submaxillary glands are rich in sialyltransferase activity, and that porcine (pig) glands are relatively deficient in sialyl- but contain appreciable galactosyl-, fucosyl- and N-acetylgalactosyltransferase activities, determine the most probable structures of the glycans in the mucins produced by these two species. (Try to draw out a reaction scheme before reading on).

A figure of the accompanying type would be suitable.

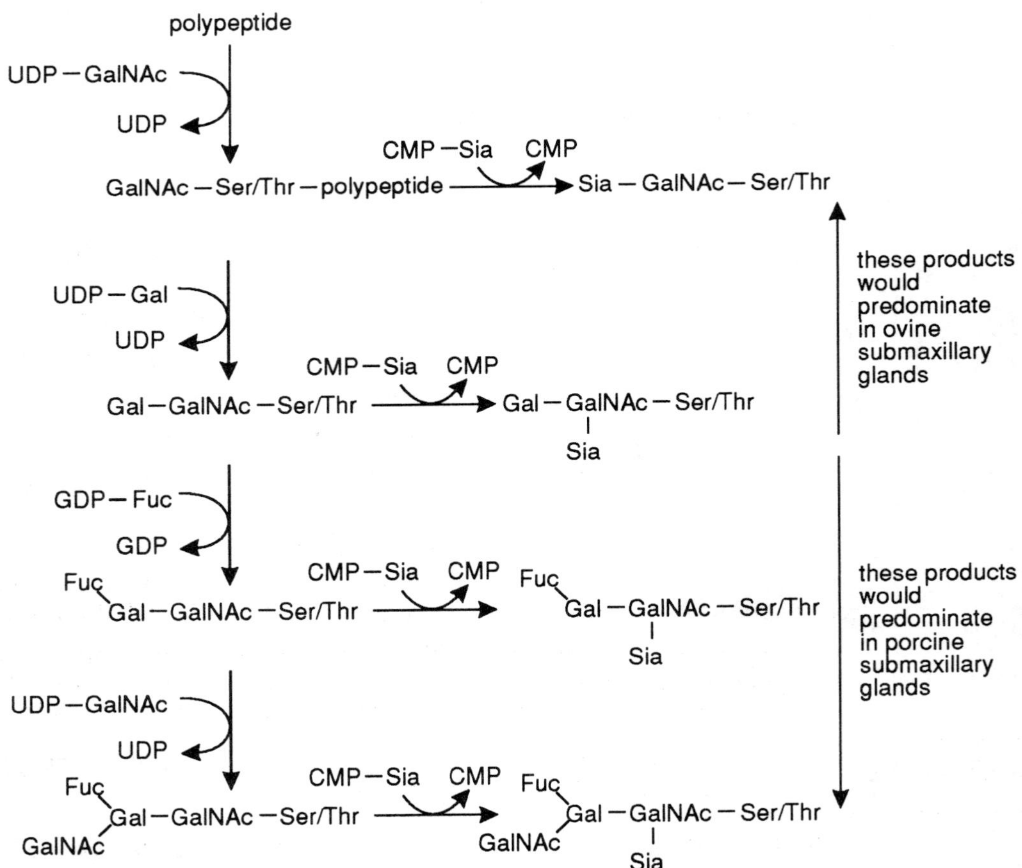

Do not worry too much if your diagram is not like this, it is a difficult question!

The point we are trying to establish is that in ovine submaxillary glands with high sialyltransferase activity, glycan synthesis is soon terminated. In porcine submaxillary glands, with low sialyltransferase activity, glycan synthesis can be quite extensive. The exact order in which the carbohydrate moieties are added depends upon the substrate specificities of the transferases present.

The production of O-linked glycans appears to occur exclusively in the Golgi apparatus which contains the necessary enzymes.

N-linked glycans made via common intermediate

The biosynthesis of N-linked glycans is more complex than that of O-linked types. Several subcellular sites are involved and a common, lipid-linked intermediate glycan is first produced which is then added to specific Asn residues of the polypeptide. The initial oligosaccharide is then processed: a procedure involving its shortening ('trimming') by specific glycosidases and elongation by glycosyltransferases in a variety of ways to give high mannose, hybrid or complex types.

The structure of the common intermediate is shown in Figure 8.20.

```
                              ╱Man—Man—Man—Glc—Glc—Glc
— GlcNac — GlcNAc — Man
                              ╲            ╱Man—Man
                               Man
                                           ╲Man—Man
```

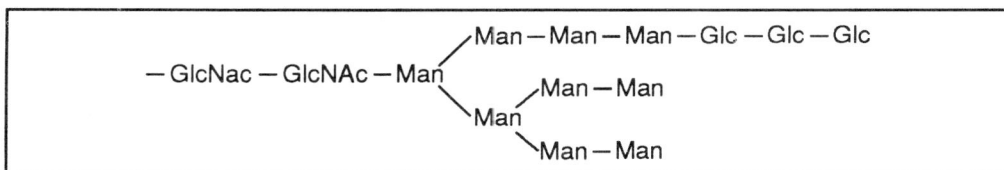

Figure 8.20 Structure of the common intermediate formed during the biosynthesis of N-linked glycans.

dolichol pyrophosphate

This intermediate is assembled by the stepwise addition of sugar residues to a dolichol pyrophosphate carrier (Figure 8.21). Note the hydrophobic nature of this molecule. It is made up of isoprenoid units.

$$CH_3-\underset{CH_3}{C}=CH-CH_2\left[CH_2-\underset{CH_3}{C}=CH\ CH_2\right]_{15\text{-}18}CH_2-\overset{CH_3}{CH}-CH_2-CH_2-O-\underset{O^-}{\overset{OH}{\mathbf{P}}}-\mathbf{O}-\underset{OH}{\overset{O^-}{\mathbf{P}}}=\mathbf{O}$$

Figure 8.21 Dolichol pyrophosphate. The pyrophosphate group is shown in bold.

oligosaccharyl-transferase

Nucleotide sugars are the direct donors of the first seven sugar residues (two GlcNAc and five Man), the remaining seven residues (four Man and three Glc) are added via dolichol phosphate intermediates (Figure 8.22). Begin reading this figure at the top (Dol-P) and, using the description given below, trace the arrows on the figure to follow what happens during the production of N-linked glycans. Synthesis starts with the addition of GlcNAc residues to a dolichol phosphate on the luminal surface of the endoplasmic reticulum. The structure is then translocated across the ER membrane to the cytosol side and five Man residues added. A second translocation returns the lipid-glycan intermediate to the lumen where further Man and Glc residues are added to give the completed $Glc_3Man_9GlcNac_2$-PP_i-dolichol. The glycan portion is then transferred by an oligosaccharyltransferase to specific Asn residues of polypeptides entering the endoplasmic reticulum. Appropriate Asn residues are identified by the enzyme because they constitute part of the sequence Asn-X-Ser or Thr, where X can be any amino acid although proline is uncommon.

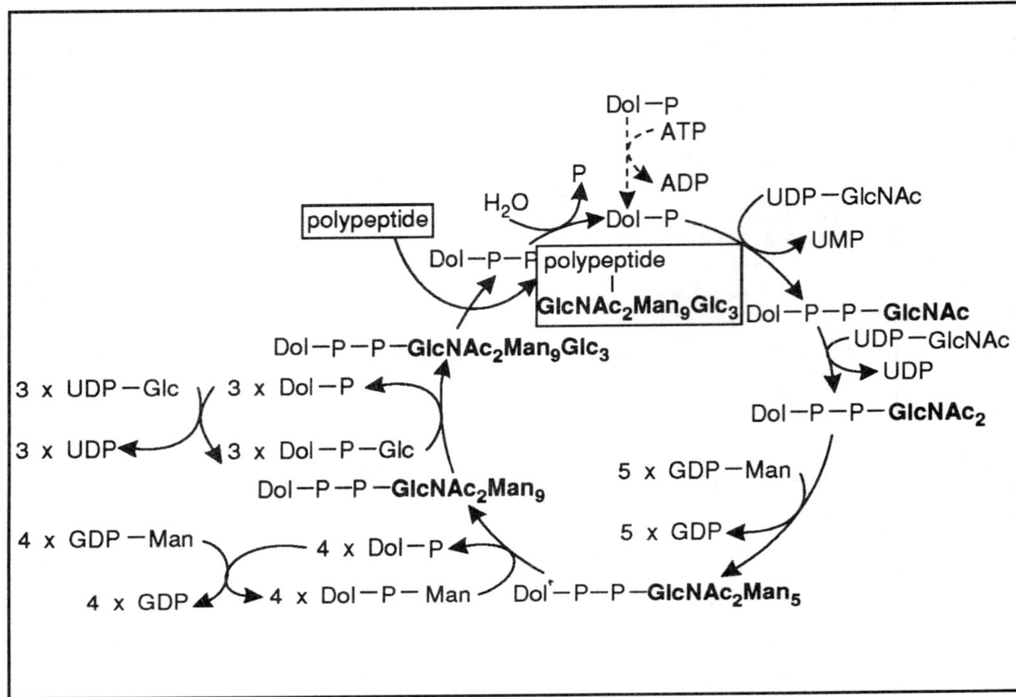

Figure 8.22 Biosynthesis of the common intermediate of N-linked glycans. See text for details.

The common intermediate is then trimmed by specific glycosidases in the ER and Golgi apparatus and elongated by Gly-Ts in the Golgi apparatus to give a variety of different types of glycans. An example of a sequence of trimming and elaboration of glycans is given in Figure 8.23.

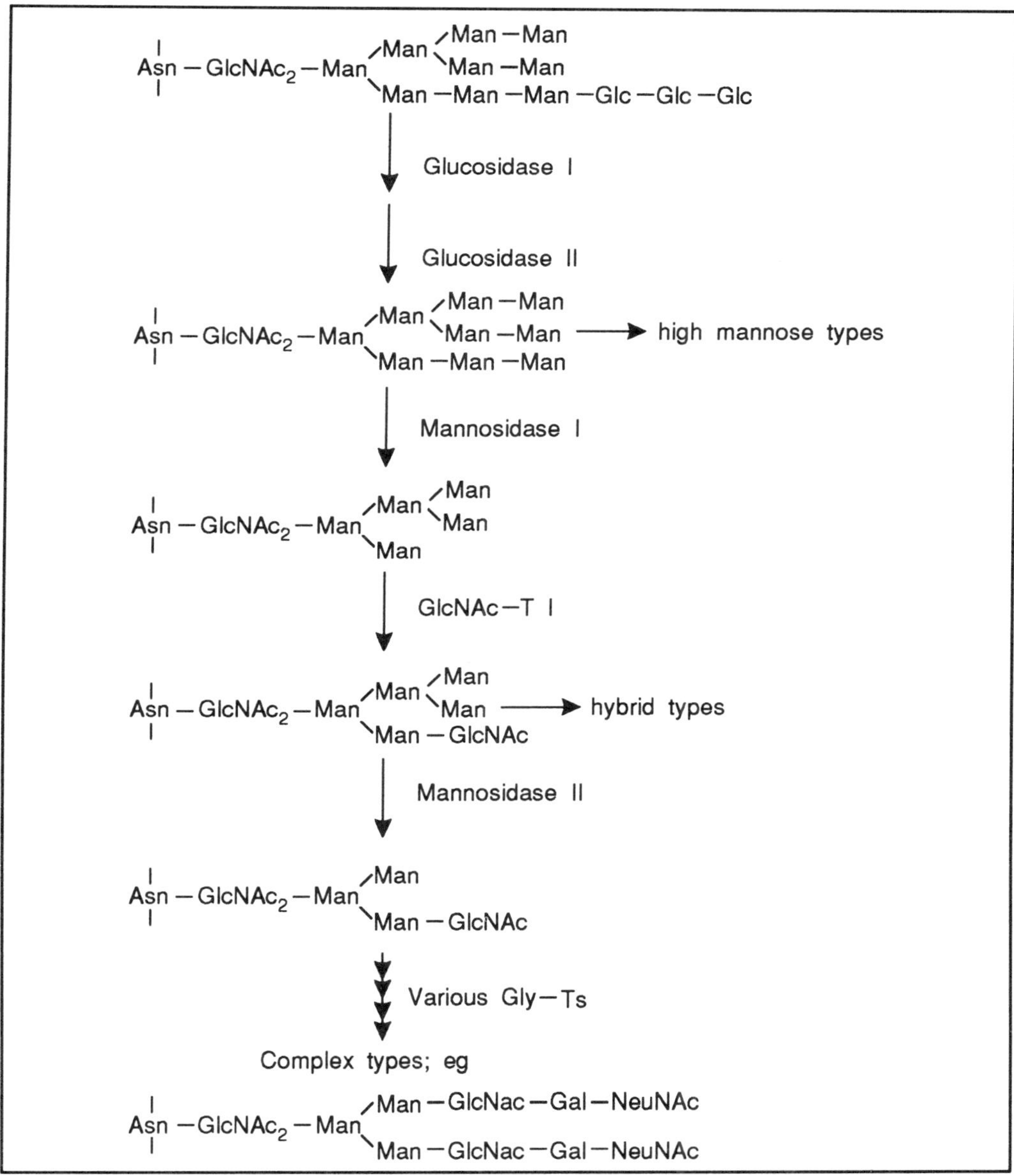

Figure 8.23 Simplified representation of the processing pathway of N-linked glycans. Redrawn from Elbein, A.D. (1991) The role of N-linked oligosaccharides in glycoprotein function. Trends in biotechnology, 9, 346-352.

Π Knowledge of glycosylation has been advanced by the use of several compounds which inhibit specific enzymes involved in the biosynthesis of glycoproteins. 1) Tunicamycin inhibits the enzyme UDP-GlcNAc:dolichol GlcNAc-1-phosphate transferase, the first enzyme involved in adding a sugar residue to form the common intermediate of N-glycosylation synthesis. What will be the most likely effects of this antibiotic on the biosynthesis of glycoproteins? 2) 1-Deoxymannojirimycin has the structure:

1-Deoxymannojirimycin

It is an inhibitor of one of the glycosidases of the Golgi apparatus involved in processing the common intermediate in N-glycosylation. Suggest a) the type of glycosidase involved and b) the general effect on the glycans produced following its uptake by the cell. (Look back at Figure 8.23 for clues).

Your answers should be:

1) Glycoproteins deficient or lacking in N-linked glycans will be produced.

2a) A mannosidase (specifically an α1-2 mannosidase) will be inhibited. Note the resemblance of 1-deoxymannojirimycin to mannose:

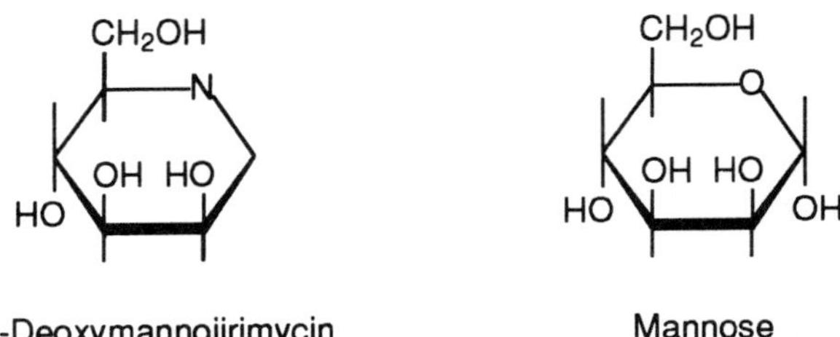

1-Deoxymannojirimycin Mannose

2b) Since 'trimming' is affected the cell will produce glycoproteins with altered N-linked glycans.

8.9.4 Limited proteolysis

Specific, limited proteolysis occurs in the formation and activation of a number of proteins and peptides. These hydrolyses may involve the removal of discrete portions of a polypeptide to form a single active protein or may produce a number of peptides, each distinctively active, from a single large polypeptide.

The removal of the initial amino terminal methionine residue and the signal sequences of secretory and membrane proteins described earlier (Section 8.8) are examples of such processing. Digestive enzymes (zymogens) eg pepsinogen and trypsinogen, and blood clotting and complement proteins of the immune system are well known examples of proteins which are activated by specific hydrolytic events.

sequence of proteolysis of preproinsulin

Specific proteolysis can be conveniently divided into intracellular and extracellular processing. A number of proteins eg zymogens are subject to both, others eg insulin and other hormones only the former. For example, insulin is produced on ribosomes of the RER in islet of Langerhans cells in the pancreas. Look back to the section entitled 'Oxidation' in Section 8.9.3 for drawings of the overall post-translational processing of insulin. The initial amino terminal methionine of preproinsulin is removed, then the signal peptide is cleaved to give proinsulin. Removal of the C-peptide region of the molecule then generates active insulin. The C-peptide is removed in the secretory vesicles in several stages. A protease with trypsin-like specificity hydrolyses the molecule after two pairs of basic residues (amino acids numbers 31,32; 59,60) to leave separate A and B chains joined by disulphide bonds. A carboxypeptidase B-like activity then removes the terminal basic, 31 and 32 residues to give the final active form of insulin. It might be helpful for you to draw out your own scheme showing these modifications.

SAQ 8.13

Summarise all the proteolytic steps associated with the maturation of insulin. State whether each step is a co- or post-translational event

proopio-melanocortin can give rise to ACTH or α-MSH

Proteolytic processing can be versatile. Thus the precursor, proopiomelanocortin is produced in the pituitary gland but differentially processed in the anterior and intermediate lobes to give, for example ACTH in the former and α-MSH in the latter (Figure 8.24). We would not expect you to remember all of the details of Figure 8.24. However, you should remember that by the differential processing of translation products may produce molecules with quite different properties. Another way of thinking about this is that the activity of a gene products is not merely a reflection of the primary sequence of amino acids coded for by the gene, but reflects the way the primary translation product is modified. Such modification is, of course, dependent upon the environment (cell type) in which the nucleotide sequence of mRNA is translated.

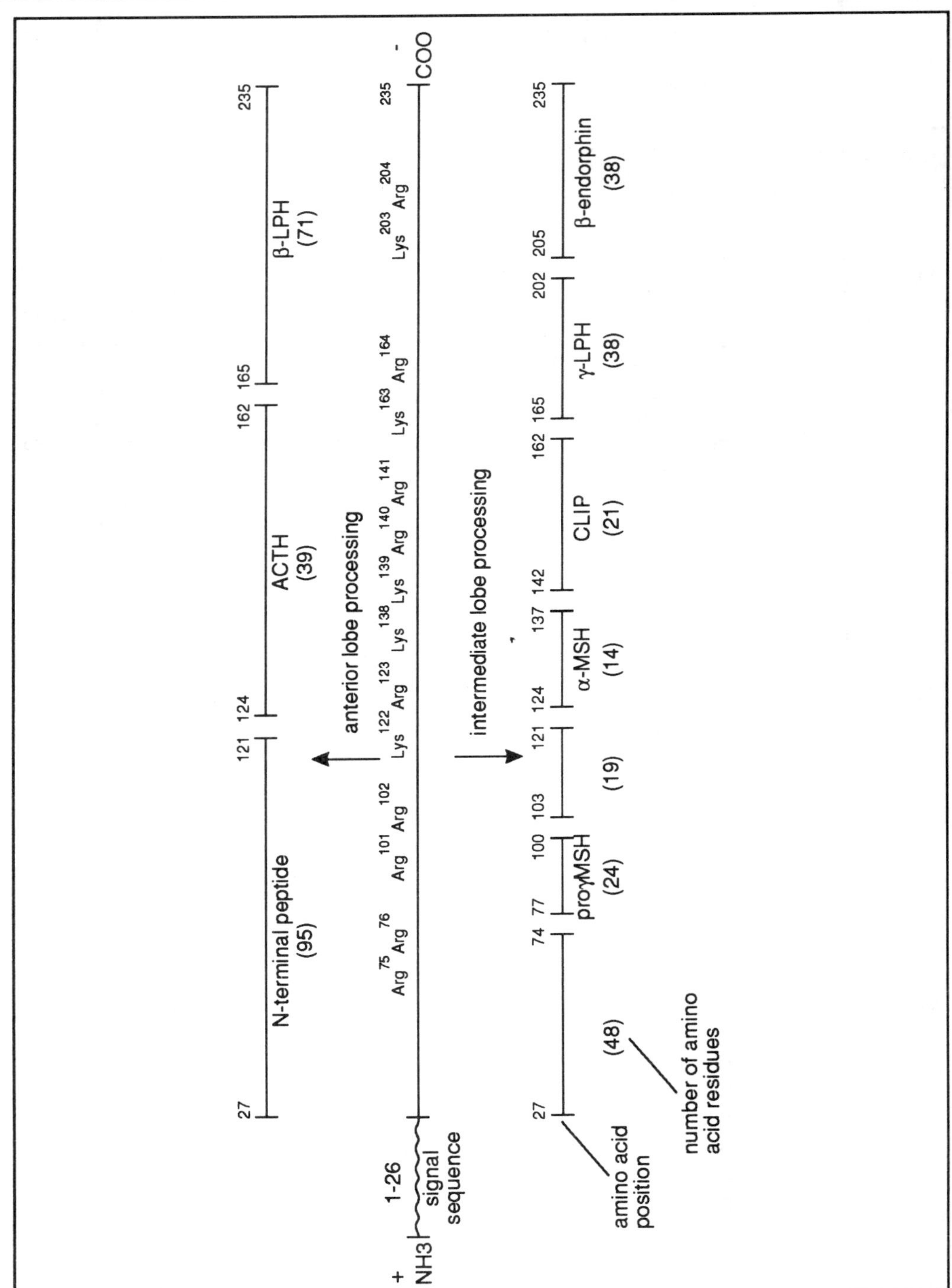

Figure 8.25 Differential proteolytic processing of proopiomelanocortin in the pituitary gland. Redrawn from Turner, A.J. (1986) Processing of neuropeptides. Essays in Biochemistry, 22, 69-119. We have not included the definitions of the abbreviations of the products derived from anterior and posterior lobe processing as they are not essential to the discussion here. You might, however, recognise some of these from your knowledge of physiology.

Summary and objectives

Translation in eukaryotes proceeds by steps essentially similar to those in prokaryotes. There are, however, a number of differences. Eukaryotic ribosomes are larger than those in prokaryotes; their mRNA is monocistronic; a larger number of protein factors are involved in initiation, and the factors in elongation and termination differ between the two types of cells. These differences form the basis of the selective action of a number of antibiotics.

Eukaryotic cells have two populations of ribosomes: free and membrane-bound. Free ribosomes produce polypeptides for internal use by the cell. Ribosomes bound to the RER synthesise polypeptides for secretion or for integration into the plasma membrane. This segregation is made possible by the production of a signal peptide and its recognition by a signal recognition particle.

Many polypeptides require co- and post-translational modifications to convert them to biologically active proteins or peptides. There are many types of such modifications although the most numerous occur in eukaryotic systems. Post-translational processing includes the chemical modification of amino and carboxyl termini, addition/cleavage reactions at specific residues and limited, specific proteolysis of the polypeptide backbone. A given protein may be subjected to several different types of modifications.

Now you have completed this chapter you should be able to:

- describe the basic mechanisms of translation in eukaryotes;
- identify the mechanisms of translation which ensure high fidelity of gene expression;
- distinguish, in broad terms only, between translation in eukaryotes and in prokaryotes;
- outline how differences in translation in eukaryotes and in prokaryotes form the basis of action of some antibiotics;
- distinguish between protein synthesis in the production of cystolic and secretory/integral membrane proteins;
- recognise the role of post-translational modifications in producing biologically active proteins;
- list and classify the major types of post-translational modifications;
- describe addition/cleavage reactions, limited (specific) proteolysis, glycosylation and acylation as examples of post-translational modifications using specific examples where appropriate.

The mitochondrial and chloroplast genome

The mitochondrial and chloroplast genome

9.1 Introduction

Mitochondria are typically rod-shaped organelles about 2µm in length and 0.5mm in diameter, although there is a large variation in their size and shape. For example some plant mitochondria are cup-shaped. Mitochondria can be isolated from cells by a process of cellular fractionation and preparative centrifugation. Examination by electron microscopy shows that mitochondria have an outer membrane and a highly folded inner membrane which encloses the mitochondrial matrix. Electron transport and oxidative phosphorylation occur on the inner membrane while the reactions of the TCA cycle occur in the matrix. The matrix also contains the mitochondrial genome.

mitochondria genome

photosynthesis

Chloroplasts are the site of photosynthesis. They are oval structures and are typically about 10 µm by 5 µm. The chloroplast has an outer and inner membrane. The inner membrane surrounds a stroma which contains a system of membranes called the thylakoids, soluble enzymes and the chloroplast genome. Both mitochondria and chloroplasts contain circular double stranded DNA molecules. The DNA is transcribed, translated and replicated independently of nuclear DNA. In protists, the mitochondrial DNA may be linear. In all cases the organelle genome codes for only a small number of proteins, but does so for all of the tRNA and rRNA molecules required for protein synthesis within the organelle. The majority of the organelle proteins are coded by nuclear DNA, and are synthesised in the cytosol and imported into the organelle.

thylakoids

endosymbiotic origin of mitochondria and chloroplasts

The probable endosymbiotic origins of these two organelles is now well established (we will discuss this at the end of this chapter) and the genetic information carried in these organelles are probably relics of the genomes of the endosymbionts. Nevertheless the persistence through evolution of genes within these organelles indicates that they perform vital functions within the cell and genetic analysis indicates that they may exert considerable influence on the phenotypic properties of organisms. It is therefore fitting that we examine the features and functions of these genes.

We will begin by examining the structure of mitochondrial DNA in terms of its size and nucleotide composition. Then we will examine the genes that are carried by the human (mammalian) mitochondrial genome and the functions of their products. We will then turn our attention to plant mitochondrial DNA and to the mitochondrial DNA of trypanosomes. We will also examine some of the features of translation within these organelles. Subsequently, we will turn our attention to chloroplast genomes, examining both their organisation and expression.

In the final parts of the chapter, we will examine the replication of organelle genomes and discuss the orgins of these organelles.

9.2 The structure of mitochondrial DNA

The smallest mitochondrial DNA molecules are found in vertebrates and invertebrates and are a circular duplex 5-6 μm in circumference containing 14-40 kbp. Figure 9.1 shows an electron micrograph of mitochondrial DNA and Table 9.1 summarises the properties of these molecules.

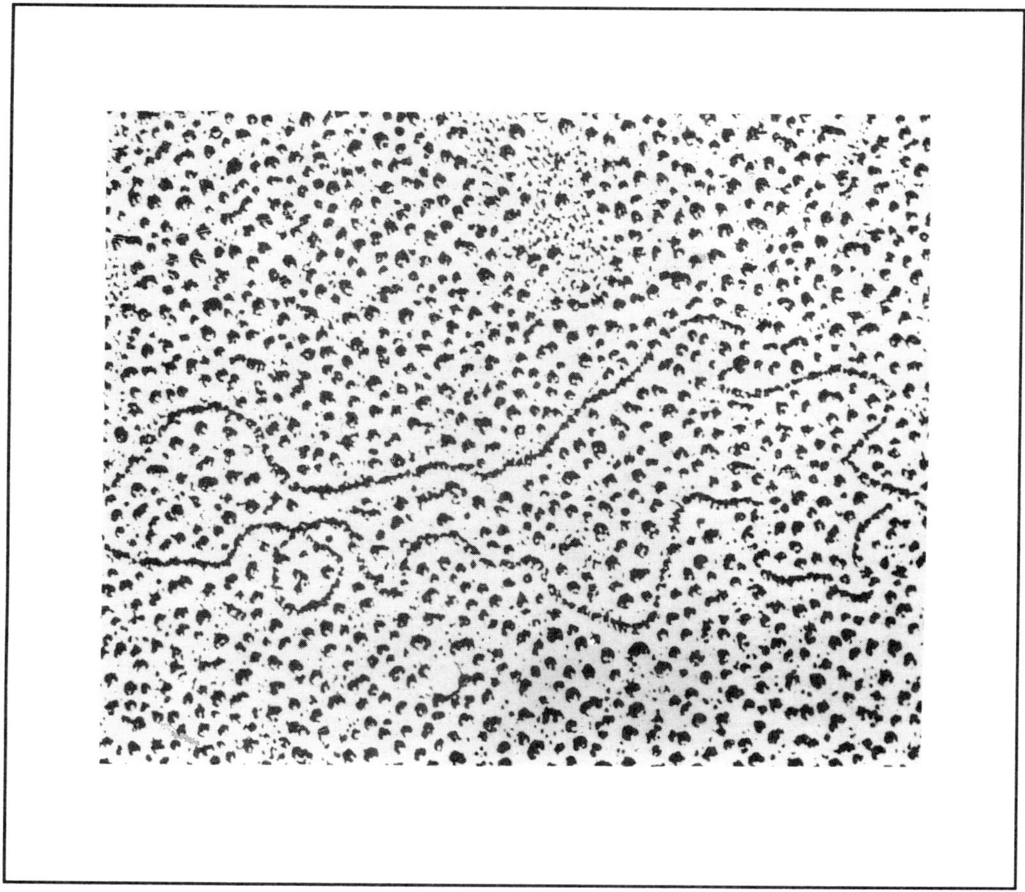

Figure 9.1 An electron micrograph (magnification x 10 000) of mitochondrial DNA from a tumour cell line. Courtesy of Dr P. Kumar, Department of Biological Sciences, Manchester Metropolitan University.

SAQ 9.1

Assume that the average molecular weight of a basepair in DNA is 642 Da and has a length of 0.34 nm. Use this information to complete Table 9.1.

Organism	length (μm)	mol wt (x 10^6 Da)	kbp
Protozoa			
Tetrahymena*	15	35	
Paramecium*	14	30-35	
Algae			
Chlamydomonas	4-5	9.8	
Fungi			
Saccharomyces	25	50	
Neurospora	19-26	40	
Higher plants			
Pisum (Pea)	30		360
Phaseolus (Bean)	20		
Zea (Maize)			570
Platyhelminthes			
Hymenolepis	4.8		
Nematoda			
Ascaris	4.8		14.2
Annelida			
Urechis	5.9		
Arthropoda			
Drosophilia	6.2	12.4	
Eschinodermata			
Echinoidea	4.6-4.9		
Chordata			
Fish	5.4		
Amphibia	4.9-5.8		
Birds	5.1-5.4		
Mammals	4.7-5.6		
Human	4.7	9.8	

Table 9.1 The properties of mitochondrial DNA. * indicates linear DNA. Adapted from Mitochondria, Chloroplasts and Bacterial Membranes, 1981, J.N. Prebble, p128, Longman, England.

Plant and protist mitochondrial DNA is larger than that of animal mitochondria with a circumference of 15-30 μm. Plant mitochondrial DNA is 200-600 kbp and is approximately the same size as the *Escherichia coli* genome.

heavy and light strands

The two strands of the mitochondrial DNA molecule are often different in density. This reflects the different base composition of each strand, particularly the different proportions of thymine in the two strands. The density difference can be marked, ranging from 5 mg cm^{-3} in the sea urchin to 44mg cm^{-3} in the chick. The two strands are called heavy (H) and light (L). Table 9.2 shows the base composition of a single strand of DNA from Metazoan species.

Species	kbp	Nucleotide composition (%)			
		A	G	C	T
Ascaris	14.28	22.2	7.6	20.4	49.8
Drosophilia	16.01	39.4	12.2	9.3	39.1
Strongylocentrotus	15.65	28.7	22.7	19.4	30.2
Paracentrotus	15.7	30.8	22.5	17.2	29.5
Xenopus	17.55	33.0	23.5	13.5	30.0
Mus	16.295	34.5	24.4	12.4	28.7
Rattus	16.29	34.1	26.2	12.5	27.2
Bos	16.34	33.4	25.9	13.4	27.3
Homo sapiens	16.57	31.0	31.2	13.1	24.7

Table 9.2 The base composition of single strands mitochondrial DNA from Metazoans (also see SAQ 9.2).

SAQ 9.2

Calculate the base composition for the strand complementary to those shown in Table 9.2.

absence of introns

The human mitochondrial genome has a compact organisation composed of 16 569 basepairs with no introns (intervening sequences).

Figure 9.2 shows a map of the human mitochondrial genome. The genome contains genes coding for 13 polypeptides, 22 tRNA and two rRNA molecules.

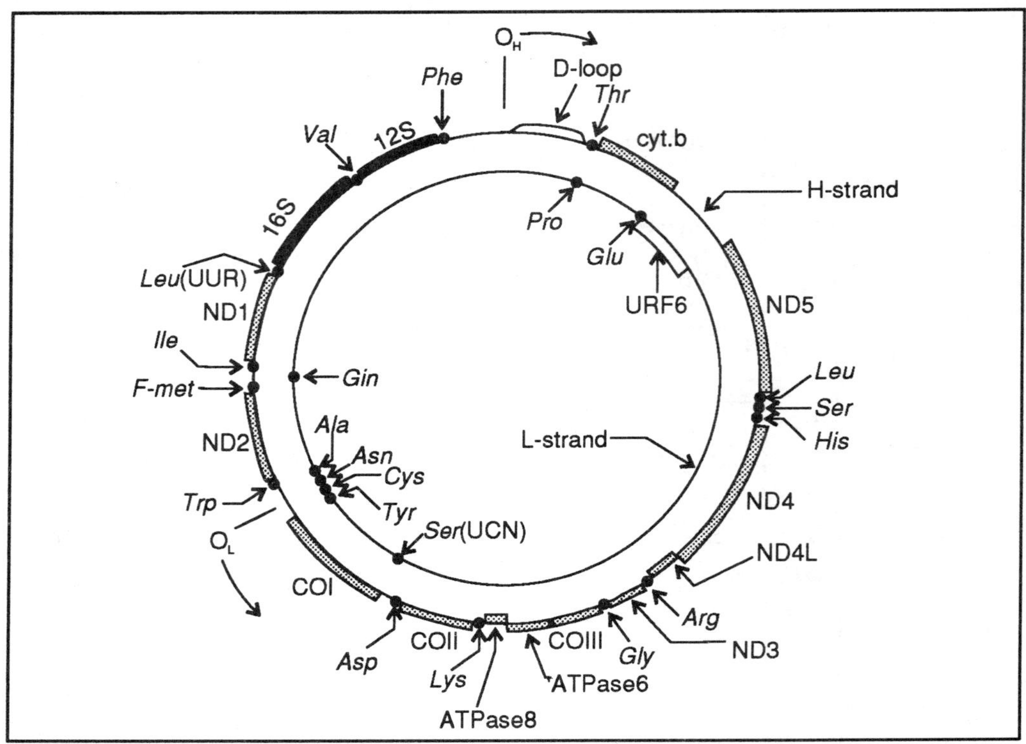

Figure 9.2 A genetic map of the mammalian mitochondrial genome. The outer circle is the H-strand, the inner the L-strand. The abbreviations refer to genes listed in Table 9.3. Spaces on one strand lie opposite genes on the complementary strand. tRNAs are designated by the amino acids for which they code. The codons recognised by tRNAs are shown for some molecules, (R = purine and N = any base, URF6 = unidentified reading frame). Redrawn from Watson *et al* Molecular Biology of the Gene Vol. 1 (1987), p 454, Benjamin Cummings, California (see text for further details).

Π Our representation of human mitochondrial DNA is quite complex. The polypeptide genes are represented by stippled chords, the two ribosomal genes by filled cords and the tRNAs by filled circles. It would help you unravel this complex genome by drawing it out for yourself. First draw a circle. Then put on the origin of replication (O_H) and then draw on the ribosomal genes. Then you can add the polypeptide genes and finally the tRNA genes. What is immediately apparent about the tRNA genes?

You probably noticed that they are scattered throughout the genome and are present on both DNA strands. The code abbreviations for the polypeptide genes are given in Table 9.3.

	Human	Yeast	Plants
Ribosomal RNAs			
large subunit	16S	21S	26S
small subunit	12S	15S	18S
5S	-	-	+
Transfer RNAs	22	25	a
Protein components of the respiratory chain			
NADH dehydrogenase			
ND1	+	-	+
ND2	+	-	+
ND3	+	-	+
ND4	+	-	
ND4L	+	-	
ND5	+	-	+
ND6	+	-	
Ubiquinol-cytochcrome c oxidoreductase	+	+	+
Cytochrome c oxidase			
COI	+	+	+
COII	+	+	+
COIII	+	+	+
ATP synthetase			
ATP6	+	+	+
ATP8	+	+	-
ATP9	-	+	+
ATPA	-	-	+

Table 9.3 Molecules coded by mitochondrial DNA. a - The number of plant tRNA genes is not known. + indicates the gene has been identified in the organism and - that it has not. Adapted from 'Plant physiology, biochemistry and molecular biology', 1990, p148. Edited by D. Dennis and D. Turpin, Longman, UK.

Π What cellular activities are coded for by these genes?

You should have concluded that they are concerned with making the components of the respiratory electron transport chain and with oxidative phosphorylation.

ribonucleotides

An unusual feature of the mitochondrial genome is the presence of ribonucleotides interspersed between the deoxyribonucleotides. In the human mitochondrial genome there are 18 of these and they may represent initiation points for replication of DNA.

Π Ribonucleotides may be detected in DNA because they make the DNA susceptible to degradation by alkali. Explain why ribonucleotides and not deoxyribonucleotides are hydrolysed by alkali.

Ribonucleotides contain the sugar ribose which has a hydroxyl group on C-2. This is susceptible to attack by free OH^- groups. Deoxyribonucleotides which lack this hydroxyl group are resistant to degradation by alkali.

9.2.1 Protein coding genes

long open reading frames

The human mitochondrial genome contains 13 deoxyribonucleotide sequences called long open reading frames which code for mitochondrial proteins. Three of these sequences code for subunits of the cytochrome oxidase complex (COI, COII and COIII; see Table 9.3), one for cytochrome b, two for subunits of the ATP synthetase (also called ATPase) and seven for subunits of NAD^+ dehydrogenase. In some cases, the mRNA copy of these sequences may contain the information for two polypeptides. In vertebrates, the information to synthesise 13 proteins is encoded on 11 mRNA molecules. The ND4 and ND4L sequences and the ATP6 and ATP8 sequences are overlapping and carried on single mRNA molecules.

There is a reading frame designated URF6 (for unidentified open reading frame), whose protein product has yet to be identified, but which represents a gene coding for a mitochondrial component.

Genes can be transcribed from either the H or the L-strand and the genes coding for proteins are separated from each other by genes for tRNA molecules. It is thought that the mRNA transcript of the L-strand is less stable than that of the H-strand.

9.2.2 tRNA genes

limited number of tRNAs made

Genes coding for 22 tRNA molecules are found in mitochondrial DNA of mammals compared with the minimum of 32 needed to translate the nuclear genetic code. Yeast mitochondrial DNA codes for 25 tRNA molecules. The situation in plants is not known.

There are many structural differences between tRNAs from the mitochondria and those coded by nuclear DNA but there are also similarities. The base preceding the anticodon is U (T in cytosolic tRNA), a pyrimidine is always in the first position of the anticodon loop and a purine is the first base after the anticodon.

An unusual feature of gene organisation in the mitochondrion is the presence of tRNA genes at the beginning and end of non-tRNA genes.

9.2.3 rRNA molecules

Ribosomal rRNA molecules are components of the mitochondrial ribosomes and are associated with the large and small subunit. Two genes coding for 16S and 12S rRNA molecules are found on the H strand. The human 12S rRNA has some resemblance (similar nucleotide sequence) to the 16S rRNA of *E. coli.*

9.3 Variation in genome organisation

The structure of the mitochondrial genome is similar in relatively closely related species for example human, mouse and frog, but there are large variations between vertebrate and invertebrates and plant mitochondria. Human mitochondrial DNA has genes joined end-to-end with very few non-coding regions or introns. The rRNA genes are smaller than their bacterial counterpart.

introns present in yeast mitochondrial DNA

In contrast, in yeast, genes are separated by AT-rich intervening sequences which comprise about half of the genome. Introns interrupt the coding sequences and there is considerable variation in gene order.

9.3.1 Non-coding regions

All mitochondrial DNA molecules contain a variable region of up to 1000 bp which is not involved in coding for a product (Table 9.4).

Species	Length (bp)
Ascaris suum	899
Drosophilia yakuba	1077
Paracentrotus lividus	132
Xenopus laevia	2135
Mus musculus	879
Rattus norvegicus	898
Bos taurus	910
Homo sapiens	1122

Table 9.4 The size of the non-coding regions of mitochondrial DNA. Redrawn from Organelles in Eukaryotic Cells (1989), p136, edited by J. Tager *et al*, Plenum Press.

D-loop

This region contains regulators of replication and transcription. In metazoans the non-coding region is called the D-loop (D for displacement) and contains the origin for DNA replication. In vertebrates, the D-loop is between the genes coding for $tRNA^{Pro}$ and $tRNA^{Phe}$ and contains the replication origin of the H strand and a promoter for L strand RNA synthesis. Figure 9.3 illustrates the organisation of the D-loop in vertebrates. The D-loop has a high AT and low GC content in all species that have been studied. In echinoids, the non-coding region is small, 100-200 bp long and is located among the genes coding for tRNAs. It also contains the replication origin.

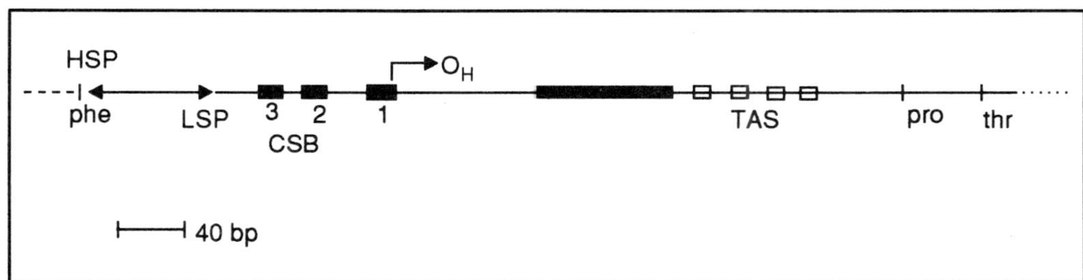

Figure 9.3 The structural organisation of the D-loop region in vertebrates between the genes for $tRNA^{Phe}$ and $tRNA^{Pro}$ molecules. The central part (dark line) is the most conserved region. HSP, heavy strand promoter; LSP, light strand promoter; CSB, conserved sequence blocks; TAS, termination associated sequences.

SAQ 9.3

Identify which of the following are true.

1) All of the open reading frames which code for mitochondrial proteins in mammals are found on the heavy strand of mitochondrial DNA.

2) The non-coding region of *Xenopus* mitochondrial DNA is almost twice as long as that found in human mitochondrial DNA.

3) The 13 peptides encoded by vertebrate mitochondrial DNA are produced by translation of 13 separate mRNAs.

4) The DNA segments of mammalian mitochondrial DNA coding for tRNA are all separated from each other by segments coding for proteins or rRNA.

5) The D-loop regions of different metazoans show many similarities.

9.4 Plant mitochondrial DNA

plant mitochondrial DNA larger than that from animals

There is relatively little information available on the structure of plant mitochondrial DNA because of the difficulty of isolating it in an undamaged form. It is, however, substantially larger than animal mitochondrial DNA (Table 9.1). If mitochondrial DNA is isolated from tissue cultured plant cells it is circular while that isolated from non-tissue cultured sources is linear. The difference may be due to the more rigorous isolation procedure required in the latter which probably shears the molecule. Another difference from animal molecules is that plant mitochondrial DNA is non-homogenous and several molecules of differing sizes may be isolated. The existence of these molecules has been explained by recombination between repeated sequences in the molecule as shown in Figure 9.4. It is possible that the larger molecules may not survive isolation and give rise to the linear molecules commonly isolated from plants.

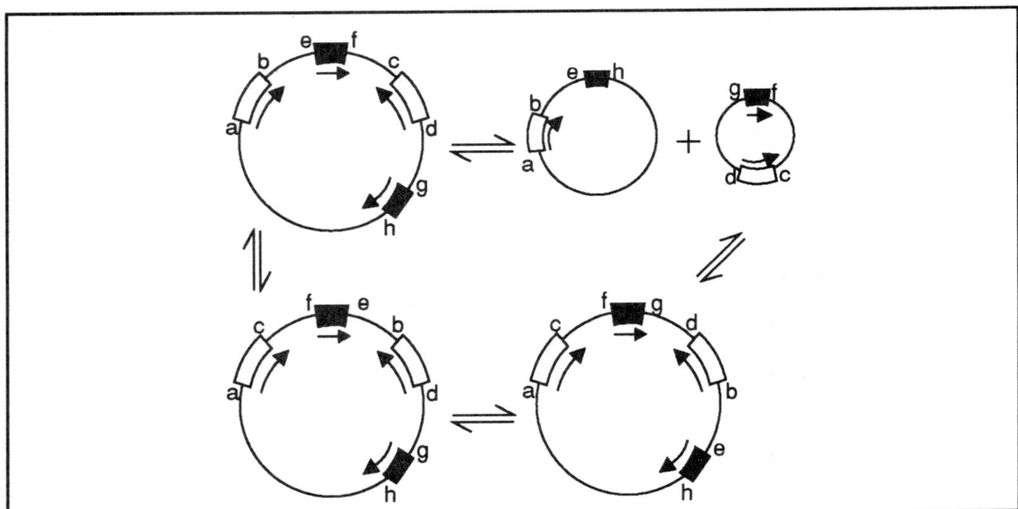

Figure 9.4 Circular DNA molecules of different sizes can be generated by recombination of repeated sequences in the initial parental molecule.

promiscuous DNA

Plant mitochondrial DNA can acquire DNA sequences from other DNA sources. These sequences are called promiscuous DNA. An example is a 12 kbp sequence found in both the mitochondrial and chloroplast genome of maize which has been shown to have originated from the chloroplast genome. The structure of the mitochondrial genome of maize is shown in Figure 9.5. The maize *COII* gene has a 794 bp intron. The equivalent gene in animals and yeast mitochondrial is not split by such an intron. Other plant mitochondrial genes do not appear to contain introns.

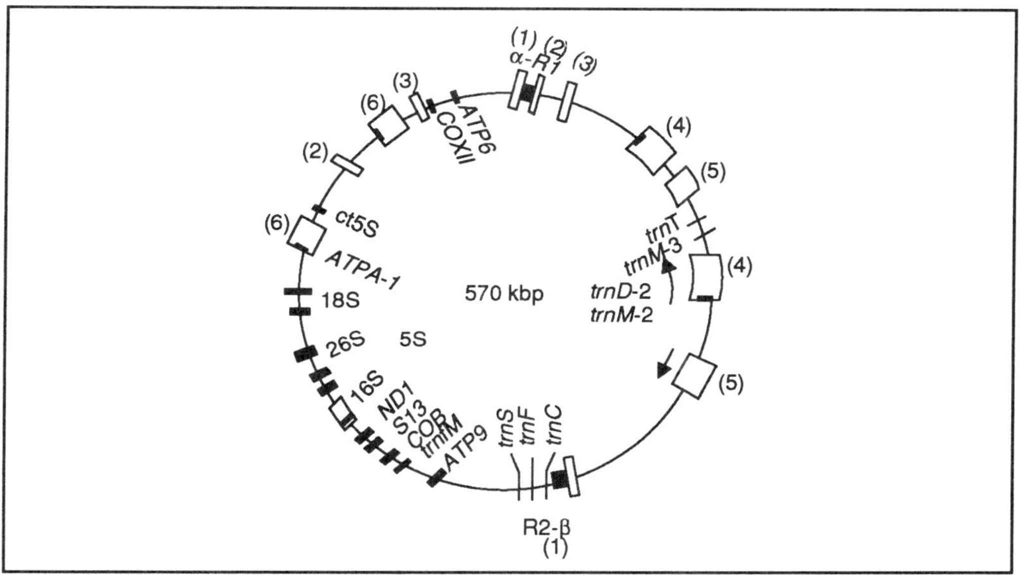

Figure 9.5 The structure of the 570 kbp mitochondrial genome of maize. The darker boxes represent sequences whose function has been identified. *trn* are tRNA genes for cysteine (C), aspartate (D), phenylalanine (F), initiator methionine (fM), elongator methionine (M), serine (S), and tyrosine (T). The numerical suffix indicates more than one member of the gene family. The abbreviation of protein and rRNA genes is as in Table 9.1. Redrawn from Plant Physiology, Biochemistry and Molecular Biology, edited by D.T. Dennis and D.H. Turpin, Longman.

9.5 Kinetoplast DNA

catenated

minicircles

maxicircles

Kinetoplast DNA is the mitochondrial DNA of trypanosomes. Trypanosomes are parasitic protozoa some of which cause important tropical diseases such as sleeping sickness. Each trypanosome contains a single mitochondrion whose DNA has a contour length of 0.45 μm (Table 9.10), but has an unusual structure consisting of thousands of interlocked circles. This type of DNA is called catenated or minicircle DNA (Figure 9.6). As well as highly catenated DNA, about 5% of the DNA is in the form of large circles of 20 to 38 kbp called maxicircles. The maxicircles are comparable to other metazoan mitochondrial DNA and have a similar gene content. The function of the minicircles is not known. The maxicircles of *Leishmania tarentalae* carries genes for the 9S and 12S rRNAs, cytochrome oxidase subunits I, II and III, ATP synthetase subunits 6 and 9 and cytochrome b. The DNA of kinetoplasts has a high AT content.

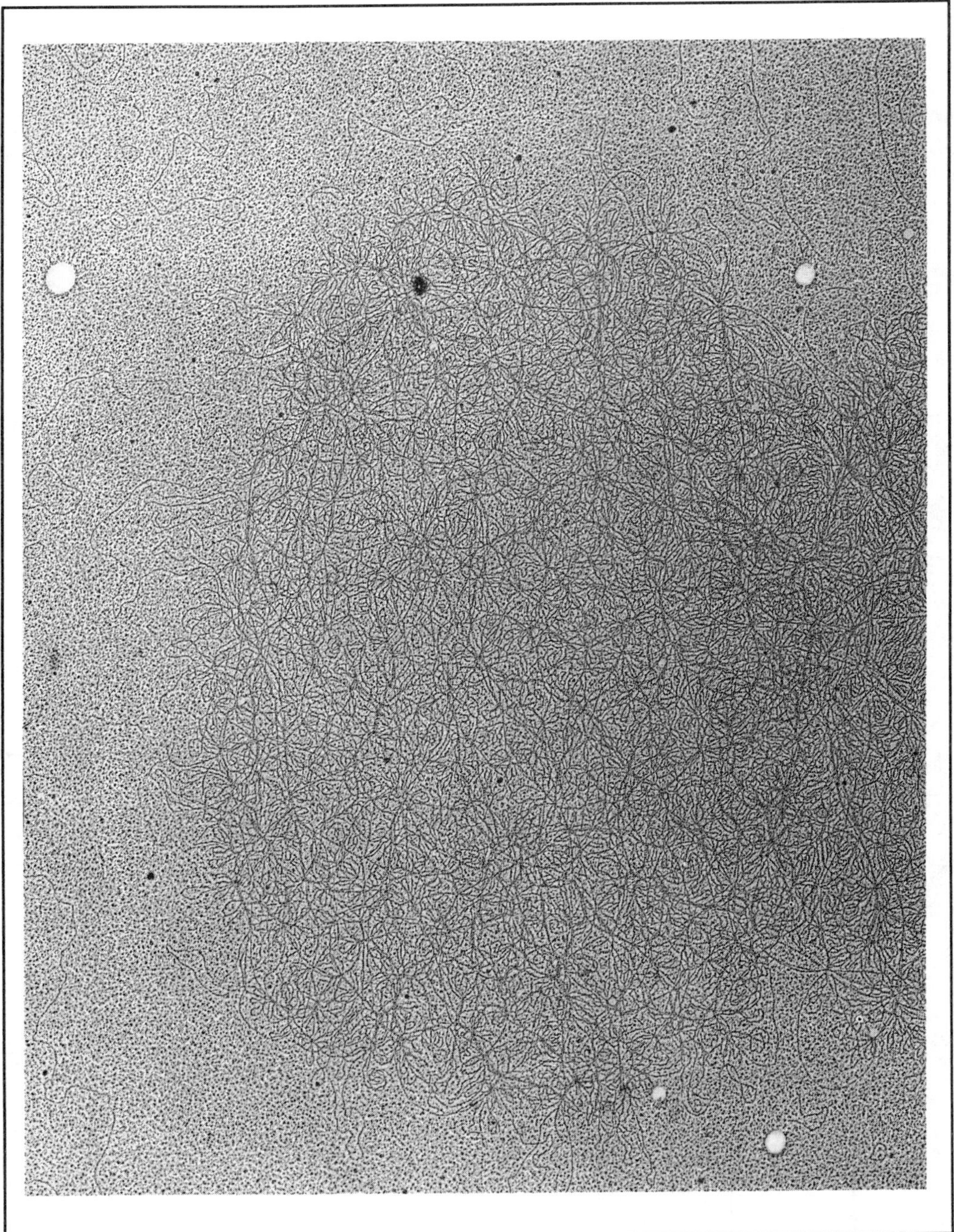

Figure 9.6 Electron micrograph of DNA released from a trypanosome. The DNA consists of thousands of interlocking circles called catenated or minicircular DNA.

SAQ 9.4

Complete the following by inserting the appropriate missing words or numbers using the list provided below.

The human mitochondrial genome contains [ ] bp and is a [ ] stranded, circular molecule. One strand (the [ ]) is heavier than the other (the [ ]) due to a difference in [ ] content. The DNA molecule contains 13 deoxyribonucleotide sequences called [ ] frames. These code for proteins such as [ ], [ ] and [ ]. The human mitochondrial genome also codes for two [ ]RNA molecules and 22 [ ]. Human mitochondrial DNA lacks [ ] but other species such as yeast have extensive [ ] rich sequences.

All mitochondrial DNA has [ ] regions which contain the D-loop, the point of [ ] of DNA synthesis.

Plant mitochondrial DNA is substantially [ ] than animal DNA and is heterogenous with a number of smaller [ ] forms. It may also contain [ ] sequences derived from other sources. [ ] DNA is derived from trypanosomes and consists of maxicircle and [ ] DNA. The latter is [ ]. The DNA of kinetoplasts has a high [ ] content.

Word and number list

larger, cytochrome b, initiation, introns, long open reading, L-strand, minicircle, tRNA, 22, catenated, non-coding, kinetoplast, 16 594, ATP synthetase, promiscuous, plant, AT (twice), maxicircle, NADH dehydrogenase, circular, double stranded, H-strand, rRNA, thymidine, RNAs, ribosomal, tRNAs.

9.6 Protein synthesis in mitochondria

involves both mitochondrial and cytosl components

Mitochondria contain all the components required to translate mitochondrial mRNA but co-operation with proteins produced in the cytosol is necessary to produce the protein synthesising apparatus. However, mitochondria can only synthesise a small fraction of their total proteins and the rest are imported from the cytosol. There are a number of differences between protein synthesis in the cytosol and that in the mitochondria. Mitochondrial RNA is resistant to RNase degradation, and protein synthesis is not inhibited by cycloheximide (which inhibits cytoplasmic synthesis) but it is inhibited by the antibiotics chloramphenicol, lincomycin and erythromycin (see Chapter 8).

9.6.1 Synthesis of mitochondrial mRNA

Mitochondrial mRNA has a different buoyant density and different base composition to cytosolic mRNA. There is no hybridisation between mitochondrial mRNA molecules and nuclear DNA indicating that mRNA is not imported into the mitochondria.

SAQ 9.5

Hybridisation experiments are performed by heating DNA molecules to temperatures above 80°C to encourage the double helix to separate into two single strands. If a molecule with a complementary base sequence is added and the solution cooled to 60°C then hybrids of single stranded DNA and the foreign molecule may form. Describe how this procedure may be used to determine whether mitochondrial mRNA is transcribed from mitochondrial or nuclear DNA?

mitochondrial transcriptase similar to the prokaryotic enzyme

Transcription in mitochondria and prokaryotes shows a number of similarities. The DNA-dependent RNA polymerase which transcribes the DNA in mitochondria is synthesised in the cytosol and coded for by nuclear DNA. The polymerase differs from other eukaryotic polymerases in that it is inhibited by rifampicin but not by α-amanitin. These properties are also shown by prokaryotic polymerases.

human mitochondrial mRNAs have tails

However, the production of mature mRNA molecules in human mitochondria differs markedly from prokaryotes, in that a single promoter on each strand of the mitochondrial genome is used to produce a full length primary transcript of the genome. This is subsequently cleaved to produce individual mRNAs, tRNAs and rRNAs. Some of these mature mRNAs remain polycistronic, ie they contain more than one translated coding sequence. Mitochondrial mRNA from humans, but not from yeast, contains a poly(A) tail of about 20 bases, which is shorter than the 50-200 base poly(A) tail of nuclear-derived mRNA. The coding regions of the DNA do not contain stop signals (UAA in the mRNA transcript) but one is generated after transcription by cleaving off the tRNA at the 3′ end of the transcript and adding the poly(A) tail. The process is shown in Figure 9.7.

Less is known about the transcription of plant mitochondrial DNA. *Brassica campestris* has a 214 kbp mitochondrial genome which codes for about 20 polypeptides. It is transcribed into 24 mRNAs of > 500 bases. Plant mitochondrial mRNAs contain non-coding regions at the 3′ and 5′ ends. These may be several hundred nucleotides in length and do not have a poly(A) tail but have the repeating sequence AAGTGAGG.

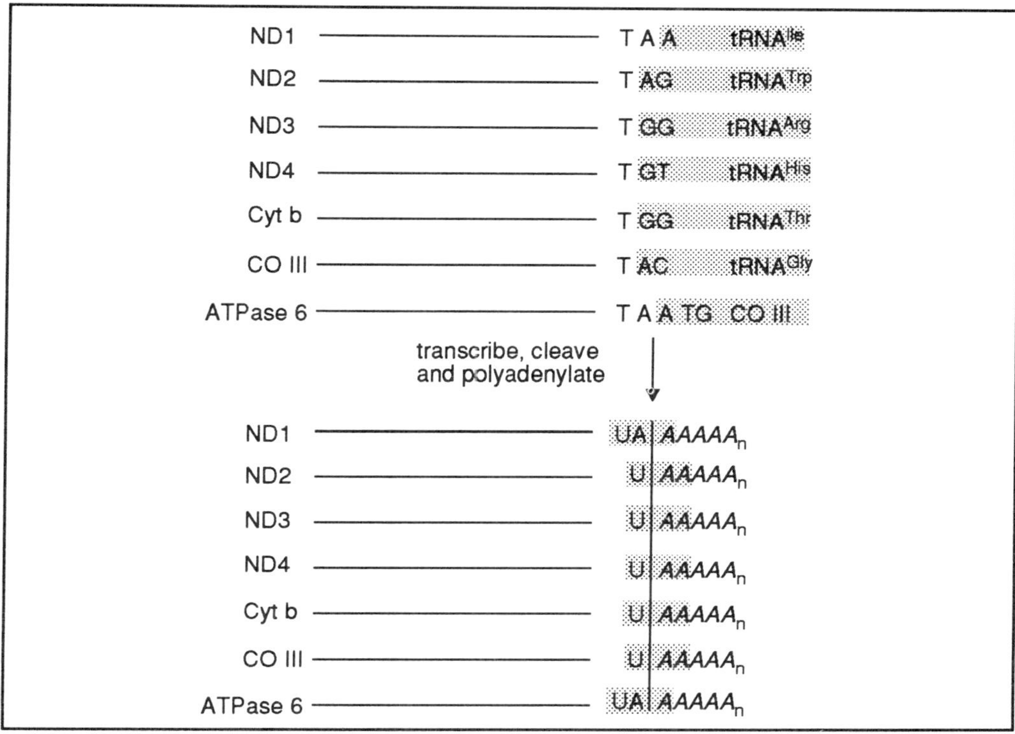

Figure 9.7 The production of stop signals to terminate transcription at the end of coding regions. The top diagram shows the protein-coding regions adjacent to tRNA genes. The boxes show the portion that is cleaved from the 3′ end of the transcripts made from their genes. The lower diagram shows the sequences that are generated after transcription, cleavage at the point marked by an arrow and addition of A residues. The stop signal UAA is produced. Redrawn from Watson *et al* Molecular Biology of the Gene, Vol. 1 (1987), p456, Benjamin Cummings, California.

9.6.2 Mitochondrial ribosomes

Ribosomes of two sizes are found in mitochondria (Table 9.5). Higher plant and animal mitochondria contain smaller ribosomes with a sedimentation coefficient of 55-60S, while the mitochondria of lower organisms contain 70-80S ribosomes. The ribosomes are usually attached to the inner mitochondrial membrane and are arranged in rows of up to a dozen. The ribosomes can be dissociated into two subunits. In mammals, these are of 39S (mol wt. 1.6 x 10^6 Daltons) and 28S (mol wt 0.86 x 10^6 Daltons) respectively. The larger subunit contains the 16S rRNA and the smaller 12S rRNA. About 40 proteins are associated with each subunit. The majority of these proteins are synthesised in the cytosol and imported into the mitochondria.

Organism	Ribosome	Subunits		rRNA	
Tetrahymena	80	55		21	14
Euglena	72	50	32	21	16
Yeast	72	50	38	23	16
Neurospora	73	51	37	23	16
Higher plants	77-78	60	44	26	18
Locust	60	40	25	16	12
Mammals	55	39	28	16	12

Table 9.5 The sedimentation coefficients in Svedberg (S) of mitochondrial ribosome, subunits and rRNA. Adapted from Mitochondria, Chloroplasts and Bacterial Membranes, 1981, J.N. Prebble, p129, Longman, England.

Π The antibiotics chloramphenicol, lincomycin and erythromycin (which all bind to the large subunit of ribosomes) inhibit protein synthesis in mitochondria, chloroplasts and prokaryotes but not in the cytosol or on the RER of eukaryotes. Explain this specificity.

The three antibiotics bind to 50S ribosomal subunits which are not found in the cytoplasmic ribosomes of eukaryotes. The comparable subunit in the cytoplasm of eukaryotes is 60S. Cytoplasmic eukaryotic protein synthesis can be inhibited by emetine or cycloheximide.

9.6.3 Soluble factors

aminoacyl tRNA synthestase

N-formyl-methionine

Aminoacyl tRNA synthetases (the enzymes which catalyse the addition of amino acids to their appropriate tRNAs) for at least 15 different amino acids have been identified in the mitochondria *Neurospora crassa*. These enzymes show a high degree of specificity for mitochondrial tRNA, showing little activity with cytosol tRNA molecules. As in prokaryotes, initiation of protein synthesis is by N-formylmethionine. The initiation, elongation and termination factors of mitochondria are interchangeable with those of bacteria. (These are described in the BIOTOL text 'Genome Management in Prokaryotes'.

Π Briefly, what are the functions of initiation, elongation and termination factors in protein synthesis?

initiation, elongation and termination factors

Initiation factors, of which there are three in prokaryotes, are required to stabilise the association of ribosomal subunits and mRNA in order that protein synthesis may be initiated. Elongation factors aid in the binding of charged tRNAs to the ribosome and in the translocation of the peptide chain. Termination factors terminate protein synthesis by dissociating the completed polypeptide from the ribosome.

9.6.4 The mitochondrial genetic code

The mitochondrial genetic code is shown in Table 9.6.

Π Compare this with the 'universal' code shown in Chapter 3. What are the major differences?

first position (5′ end)	second position: U	second position: C	second position: A	second position: G	third position (3′ end)
	U	C	A	G	
U	UUU] Phe (GAA) UUC] UUA] Leu (UAA) UUG]	UCU] UCC] Ser (UGA) UCA] UCG]	UAU] Tyr (GUA) UAC] UAA stop UAG stop	UGU] Cys (GCA) UGC] UGA] Trp (UCA) UGG]	U C A G
C	CUU] CUC] Leu (UAG) CUA] CUG]	CCU] CCC] Pro (UGG) CCA] CCG]	CAU] His (GUG) CAC] CAA] Gln (UUG) CAG]	CGU] CGC] Arg (UCG) CGA] CGG]	U C A G
A	AUU] Ile (GAU) AUC] AUA] Met (CAU) AUG]	ACU] ACC] Thr (UGU) ACA] ACG]	AAU] Asn (GUU) AAC] AAA] Lys (UUU) AAG]	AGU] Ser (GCU) AGC] AGA stop AGG stop	U C A G
C	GUU] GUC] Val (UAC) GUA] GUG]	GCU] GCC] Ala (UGC) GCA] GCG]	GAU] Asp (GUC) GAC] GAA] Glu (UUC) GAG]	GGU] GGC] Gly (UCC) GGA] GGG]	U C A G

Table 9.6 Genetic code of mammalian mitochondria. Note that the bracketed groups of codons are read by a single tRNA. The anticodon sequence is given in parenthesis next to the amino acid. These are specified in 3′-5′ direction. This in some cases leads to some unusual basepairings.

The major differences from the 'universal' code are:

- UGA codes for tryptophan rather than being a stop signal;
- initiating methionines are coded for by AUG, AUA, AUU and AUC (compared with the AUG for cytoplasmic protein synthesis);
- internal methionines are coded by both AUG and AUA (compared with AUG);
- there are four stop codons UAA, UAG, AGA and AGG (compared with UAA, UAG, UGA).

It might be helpful to mark these differences onto Table 9.6 so that you remember them.

Π In Table 9.6, we showed the anticodon sequence of the tRNAs and bracketed together the codons recognised by these tRNAs. We also mentioned in the legend to this table that these led to some unusual basepairings. See if you can identify any.

unusual base pairing during mitochondrial translation

There are, in fact, many such unusual pairings. Let us begin at the top left hand corner of the table. The codons 5′ UUU 3′ and 5′ UUC 3′ are recognised by a tRNA carrying the anticodon sequence 3′ AAG 5′, thus although this gives the usual basepairing with 5′ UUC 3′ it is unusual in pairing with 5′ UUU 3′. It is an inevitable consequence that unusual basepairing will arise if a single tRNA species can interprete more than a single codon triplet. In other words, if you had conducted carefully the intext activity described above, you should have identified a very large number of such unusual pairings.

Mitochondria can carry out protein synthesis using 22 tRNAs rather than the minimum 32 as in cytosolic synthesis. This is because groups of four codons that code for the same amino acid but differ in the third base of the codon (Table 9.6) are translated by a single type of tRNA. This tRNA has U in the 5′ position of the anticodon, which can pair with any of the four bases in the third codon position.

Plant mitochondria use CGG to encode tryptophan rather than arginine and UGA is used as a termination codon rather than as a code for tryptophan.

SAQ 9.6

Which of the following are true or false. Explain your answer.

1) Mitochondrial protein synthesis is inhibited by cycloheximide but not by chloramphenicol.

2) Mitochondrial protein synthesis is initiated by methionine.

3) Mitochondrial mRNA from humans has a poly(A) tail.

4) Protein biosynthesis in the mitochondrion takes place on 77S ribosomes in higher plants; 60S ribosomes in yeast and 50S ribosomes in mammals.

5) If an mRNA species is translated both in the cytsol and in the mitochondria, the same protein will be produced in each case.

9.7 The chloroplast genome

We now turn our attention to the other major organelle, the chloroplast. Higher plants contain circular double stranded molecules of DNA in their chloroplasts. The size of the DNA is typically in the range of 43-46 μm in circumference, molecular mass 9-10 x 10^7 Daltons and contains 140 000 bp. Each chloroplast has 50-100 copies of the genome. The alga *Chlamydomonas* has a larger genome (62 μm and molecular mass 12.6 x 10^7 Daltons). Figure 9.8 shows a map of the genome of spinach chloroplasts.

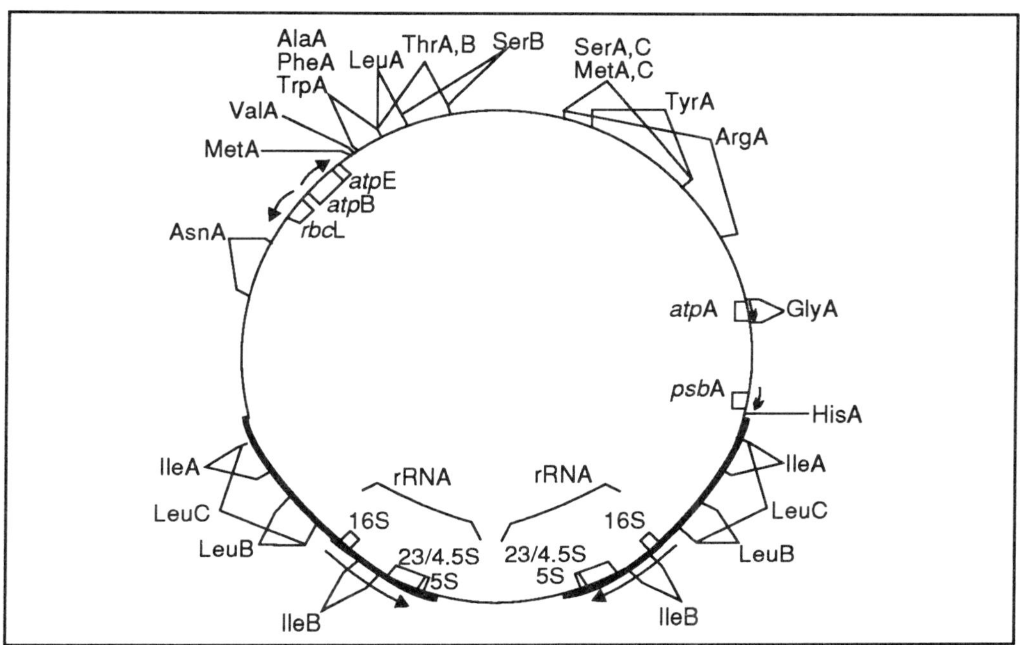

Figure 9.8 The genetic map of the spinach chloroplast genome. The genes for tRNAs, rRNAs and some proteins are shown. Arrows indicate the direction of transcription of genes and heavy lines show inverted repeats. atp, ABE are ATPase subunit genes, rbcL = gene for the light subunit of ribulose 1,5 bisphosphate carboxylase; psbA = gene for a thylakoid membrane protein of photosystem II. Note that not all genes have been shown on this map. Redrawn from 'Annual Review of Plant Physiology', 34, 1983, editor Briggs *et al*, p282 (see text for details).

The base composition of chloroplast DNA differs from that of nuclear DNA as demonstrated by their behaviour on density gradient centrifugation (Figure 9.9).

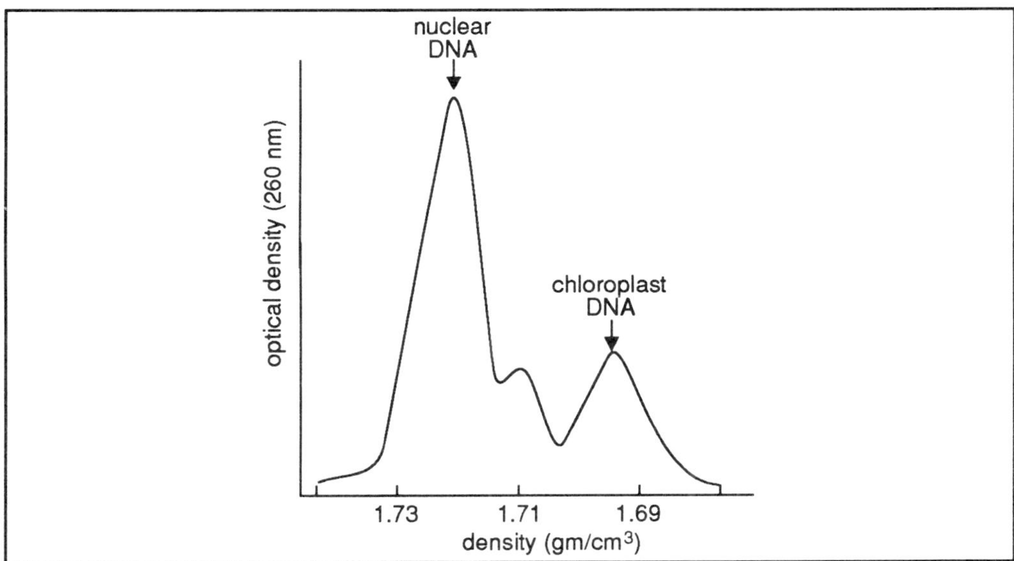

Figure 9.9 CsC1 density gradient separation of DNA from *Chlamydomonas reinhardii*. The chloroplast DNA has a density of 1.695 g cm^{-3} and separates from other DNA molecules. Redrawn from Chloroplasts, J.K. Hooker (1984), p149, Plenum Press, New York.

Two examples of the base compositions of chloroplast DNA are provided in Table 9.7. Note that the chloroplast DNA of many plants contain about 37% (G+C).

	Nucleotide composition (%)				
	T	G	A	C	(G + C)*
Spinach	31.4	18.0	32.1	18.5	36.5
Euglena	35.5	14.2	36.3	14.0	28.2

Table 9.7 The molar base composition of chloroplast DNA. * the chloroplasts DNAs of most higher plants have a (G+C) content of 37 ± 1%.

Π Use Figure 9.9 to determine whether plant nuclear DNA has a higher or lower (G+C) content than 37% given the fact that the chloroplast DNA of *Chlamydomonas* has a base composition which is typical of chloroplasts in general.

Since we are told that the base composition of the chloroplasts of *Chlamydomonas* is typical of chloroplasts in general, we must assume that it has a G+C content of about 37%. The nuclear DNA is denser than the chloroplast DNA indicating that it has greater G+C content.

9.8 The organisation of the chloroplast genome

inverted repeats

Plastids are immature chloroplasts and contain DNA able to code for 126 polypeptides of mean molecular mass 46 000 Daltons. About 30% of the base sequence of all chloroplast DNA is common to all the plant species so far studied. Chloroplast DNA from any species contains inverted repeats. These are sequences of 20-28 kb which are found twice in the genome. However, the second sequence runs in the opposite direction to the first.

sites of synthesis of plastid components

Interspersed within the repeats are the genes for proteins, rRNA and tRNA molecules (Figure 9.8). Table 9.8 lists the genes identified on plastid DNA. Many of the components of the plastid are coded by nuclear genes, synthesised in the cytosol and transported into the plastid. Table 9.9 identifies the site of synthesis of some plastid components.

Π Examine these two tables carefully. What general principle about the genes encoded by chloroplast DNA reflects the situation found in mitochondria?

We learnt that the mitochondria coded many of the components needed for translation within mitochondria (eg rRNA, tRNA) and for many of the peptides need to enable the mitochondria to fulfil their function of respiratory electron transport and oxidative phosphorylation. In chloroplasts, the situation is entirely analogous. Chloroplast DNA encodes many of the components needed to carry out translation (eg rRNA, tRNAs). But this organelle is primarily concerned with photosynthesis and carbon dioxide fixation. Thus we find genes coding for the components of photosystems I and II (PS-I and PS-II) and of carbon dioxide fixation (eg ribulose bisphosphate carboxylase - RubisCO). Note, however, that many of the essential components for both protein synthesis and photosynthesis are coded in the nuclear DNA.

Plastid DNA	Nuclear DNA
23S RNA	Small subunit of RubisCO*
16S RNA	Some PS-I components
5S RNA	Some PS-II components
4.5S RNA	Ferredoxin
tRNAs	Ferredoxin-NADP reductase
Large subunit of RubisCO*	Some ribosomal peptides
32 000 thylakoid component	Some aminoacyl tRNA synthetases
Some PS-I components	
Some PS-II components	
Cytochrome f and 553	
Coupling factor 1 subunits	
Elongation factors	

Table 9.8 The location of the DNA coding for some plastid components. * Ribulose bisphosphate carboxylase.

Plastid	Cytosol
Large subunit RubisCO*	Small subunit of RubisCO*
32 000 thylakoid component	PS-I components
PS-I components	PS-II components
PS-II components	Plastocyanin
Cytochrome f	Ferredoxin
Cytochrome b-559	RNA polymerase
Cytochrome b-563	Ribosomal polypeptides
Two elongation factors	Aminoacyl tRNA synthetase
Three coupling factor subunits	Many enzymes
Ribosomal polypeptides	

Table 9.9 The site of synthesis of some plastid polypeptides. * Ribulose bisphosphate carboxylase.

9.8.1 Chloroplast ribosomes

The large and small ribosomal subunits of chloroplasts contain 38 and 24 proteins respectively and have a similar structure to the animal mitochondrial ribosomes described in Section 9.6.2.

9.8.2 Ribosomal RNA

The cyanobacterium *Anacystis nidulans* has genes for 16S, 23S and 5S rRNA arranged in a similar manner to that found in the *E. coli* genome. Higher plant chloroplasts have an additional 4.5S molecule. The 23S, 5S and 4.5S RNAs are part of the 50S ribosomal subunit and 16S rRNA is a component of the 30S subunit. We remind you that the rRNAs are encoded within the inverted repeat regions of the genome. All of the rRNAs are transcribed as a single RNA copy which is then processed into individual molecules.

9.8.3 Ribonucleotides in chloroplast DNA

Chloroplast DNA can be degraded by alkali suggesting it contains ribonucleotides. Between 12 and 18 (+2) ribonucleotides have been found in chloroplast DNA from a range of species. This is similar to the number found in mitochondrial DNA. There is no evidence that ribonucleotides are associated with sites of replication in plant organelles.

9.8.4 Inverted repeats

spacer regions

The base sequences between inverted repeats are called spacer regions and the lengths of spacers and repeats are very similar in many species (Table 9.10). The genes for rRNA occupy only 20% of the sequence and the function of the other 80% is not known.

Species	Small spacer (bp)	Inverted repeat (bp)	Large space (bp)
Spinacia oleracea	18 500 +800	24 500 +500	86 000 +4500
Lactuca sativa	19 500 +800	24 500 +500	87 000 + 2500
Zea mays	12 600 +1000	22 500 +400	78 000 +3300

Table 9.10 The lengths of inverted repeats and spacer regions.

Now we have some idea of the numbers and functions of genes in mitochondria and chloroplasts, we can turn our attention to another important feature of organelle DNA - their replication. We will focus on the replication of DNA in mitochondria as there is much more known about this type of organelle than there is about chloroplasts.

9.9 Replication of the mitochondrial genome

origin consists of AT sequence flanked by GC sequences

A scheme for the replication of the mitochondrial genome is shown in Figure 9.10. Replication begins at a specific base sequence on the genome called the origin. The base sequence of the origin has been identified in a number of species and consists of an AT sequence flanked by GC sequences. Part of the base sequence of the origin of yeast is shown in Figure 9.11 and is similar in other species. (The origin of replication of mitochondrial DNAs has a similar structure to the initiation sequences for rRNA synthesis). The origin is the point of attachment of RNA polymerase which synthesises an RNA primer. A single RNA polymerase is found in mitochondria and makes all the RNA transcripts of DNA.

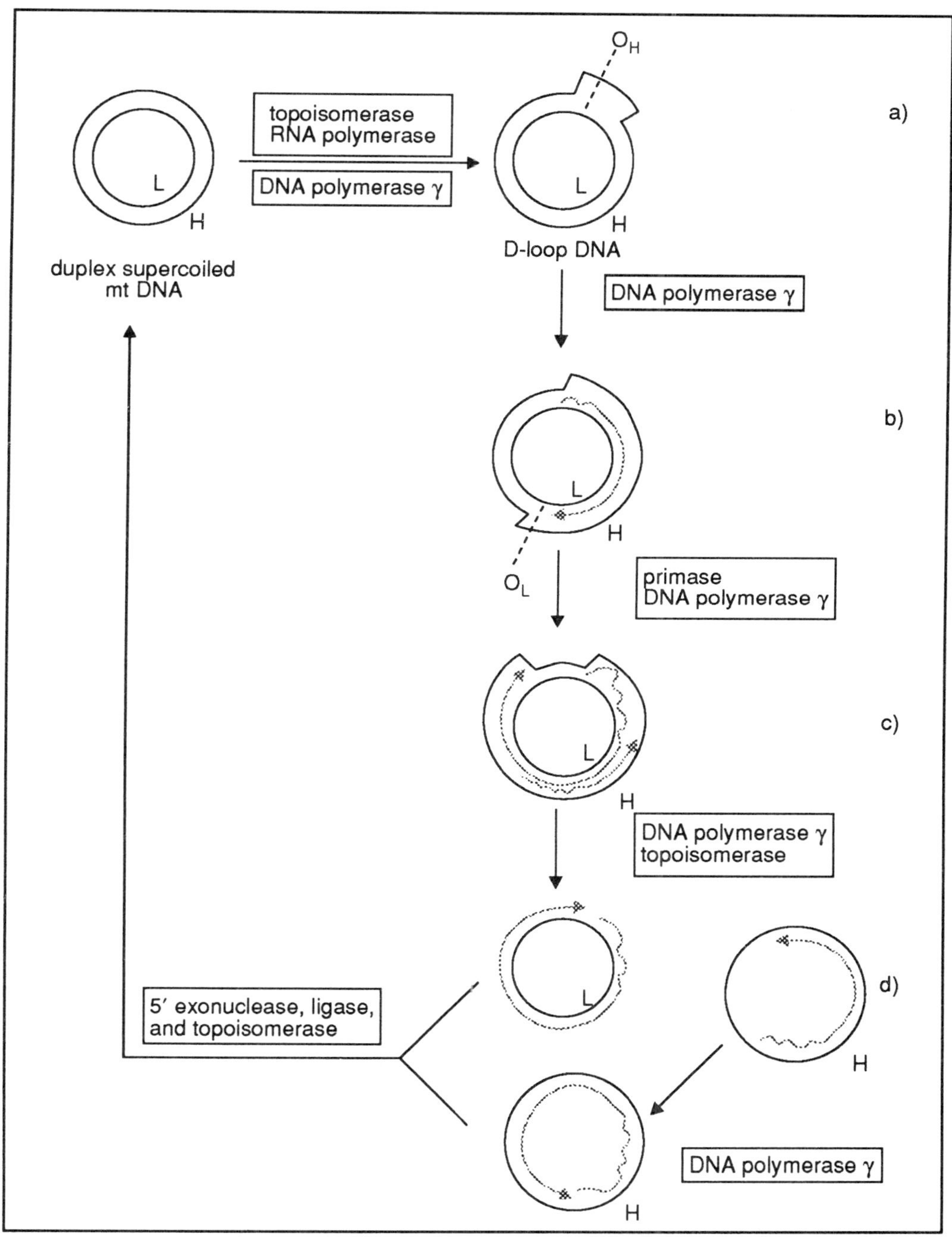

Figure 9.10 A possible scheme for mitochondrial DNA replication. a) RNA polymerase synthesises a primer of RNA. This provides a primer at the O_H for DNA polymerase to synthesise DNA. b) The DNA is extended and a D-loop generated c) When O_L is exposed an RNA primer is synthesised by RNA polymerase and extended by DNA polymerase. d) the molecules are separated while one strand is still incomplete. Redrawn from DNA Replication, R.L.P. Adams, (1991), p42, Oxford University Press.

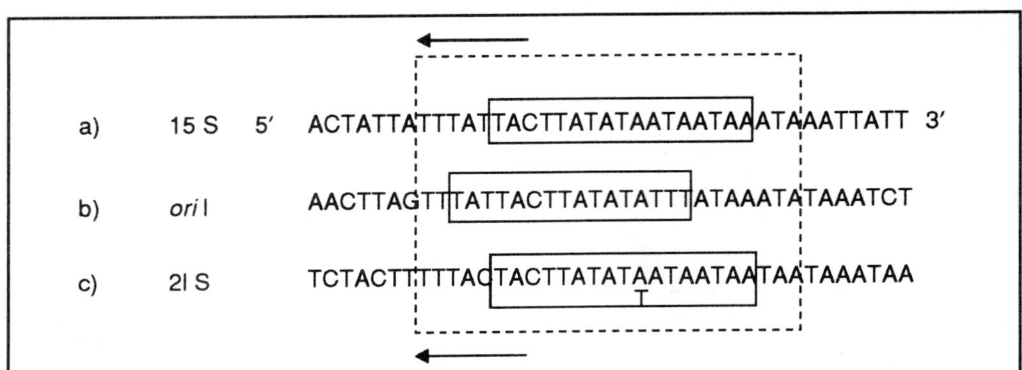

Figure 9.11 Comparison of the initiation sites of transcription of a) mitochondrial 15S ribosomal genes b) the origin of DNA replication in yeast and c) 21S ribosomal genes. The solid box is the transcription initiation sequence and the dotted line the homologous sequences. Redrawn from 'DNA makes RNA Makes Protein', edited T. Hunt *et al*, (1983), Elsevier.

The RNA primer is extended from its 3′ end by the catalysed addition of deoxyribonucleotides by DNA polymerase γ. This enzyme is synthesised in the cytosol, has a molecular mass of 150-300 000 Daltons and may also be found in chloroplasts.

topoisomerase

triple stranded

D-loop

The double stranded DNA is unwound by a topoisomerase and a daughter segment of 450-1000 bases synthesised. A new H-strand is synthesised initially producing a triple stranded molecule called a D-loop. The new H-strand is extended until the origin of the L-strand (O_L) is exposed and synthesis of a new L-strand commences. The L-strand is synthesised in the opposite direction to the H-strand. Because of the lag in the synthesis of the two strands the H-strand will be completed before the L-strand and about 70% of the H-strand will be synthesised before L-strand synthesis begins. The circles of parental and daughter DNA are separated by the action of a topoisomerase, which removes supercoiling, and a 5′-exonuclease which opens the circles and a ligase which reseals the circles after separation.

SAQ 9.7

Briefly describe the function of the following in mitochondrial DNA replication.

RNA polymerase, DNA polymerase, topoisomerase I, DNA ligase.

9.10 The import of proteins into the mitochondrion and chloroplast

It has already been emphasised that mitochondria and chlorplasts make only a small proportion of their total protein content and the remainder is coded for by the nuclear DNA, synthesised on ribosomes in the cytosol and imported into the organelles. In this section we will give you a few more details concerning the import of proteins into these organelles.

9.10.1 The import of proteins to the mitochondrion

transit signal

Proteins coded in the nucleus but destined for the mitochondrion are synthesised in the cytosol with an amino acid extension on the amino-terminus called a transit signal. This signal sequence is positively charged due to the presence of basic amino acids and binds to a receptor on the outer mitochondrial membrane. The proteins then move across the membrane. The signal is cleaved after entry into the mitochondrion by a protease synthesised in the cytosol. The cleavages of proteins destined for the inter-membrane space is more complex than proteins bound for the matrix or inner membrane but little is known of the detailed mechanisms involved. Metabolic energy is not required to insert a protein onto the outer mitochondrial membrane but an electrochemical gradient (a potential source of energy) is required to transport a protein to sites in the mitochondria other than the outer membrane. Figure 9.12 shows the involvement of cytoplasmic proteins in the generation of mitochondria.

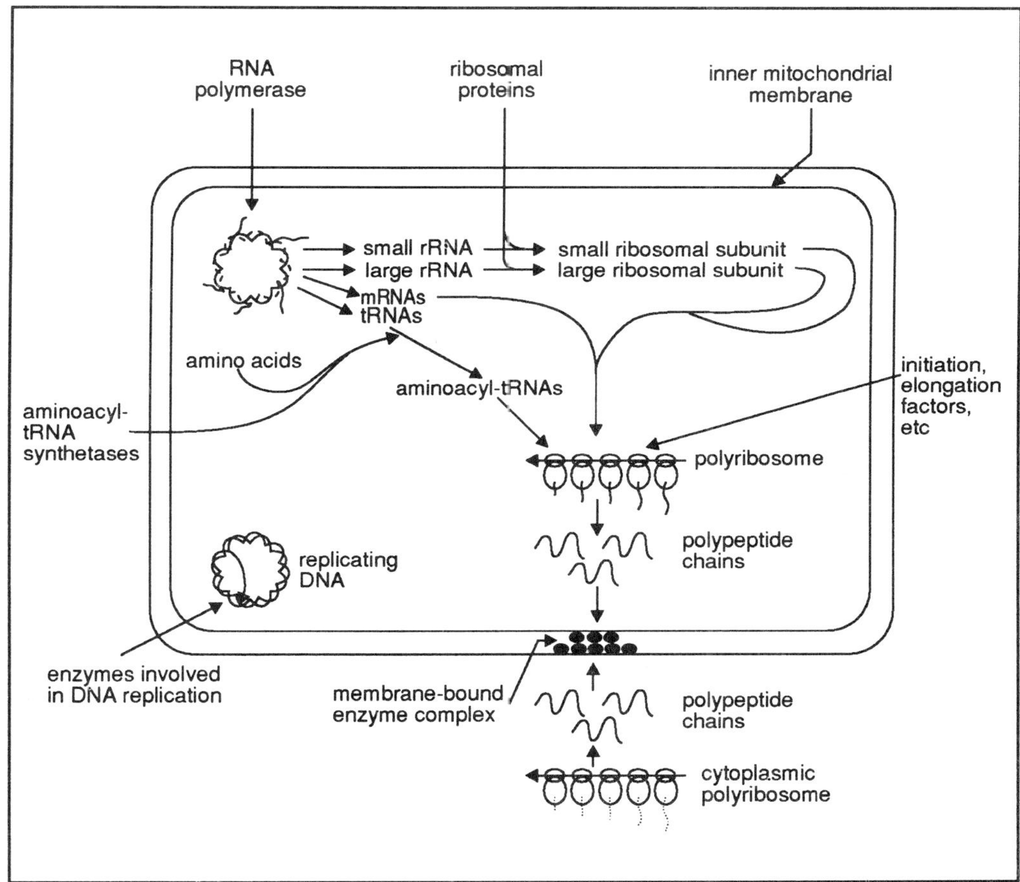

Figure 9.12 Many proteins required by the mitochondrion are synthesised in the cytosol. Redrawn from 'DNA makes RNA makes protein', edited T. Hunt *et al*, (1983), Elsevier.

9.10.2 The entry of proteins into the chloroplast

Proteins destined for import to the chloroplast are also synthesised with an amino acid signal at the amino-terminus. Entry is effected by binding to a receptor on the chloroplast envelope, the signal is cleaved by a protease. The entry of proteins into the chloroplast is stimulated by light.

SAQ 9.8

Account for the stimulation of entry of proteins into the chloroplast by light.

SAQ 9.9

1) Which of the mitochondrial or chloroplast proteins listed below (a-k) would you expect to have an amino-terminal extension?

2) Which are coded by plastid DNA?

a) Cytochrome oxidase II.

b) Cytochrome oxidase III.

c) ATP synthetase 6.

d) DNA polymerase γ.

e) RNA polymerase.

f) DNA ligase.

g) Initiation and elongation factors of protein synthesis.

h) NADH dehydrogenase subunits 1-6.

i) Ferredoxin.

j) Large subunit of RubisCO.

k) Cytochrome f.

9.11 The origin of chloroplasts and mitochondria

The genomes of mitochondria and chloroplasts resemble prokaryotic genomes in structure, organisation and sequence more than they resemble eukaryotic nuclear DNA. Other features of these organelles also indicate a marked resemblance to prokaryotes. Table 9.11 list some of these common molecular biological features. We could also add many other features of similarity such as the lipid compositions of their membranes.

Characrteristic	Mitochondria Chloroplast	Prokaryote	Eukaryote
DNA	circular	circular	linear
Ribosomes	70S	70S	80S
rRNAs	5S, 16S, 23S	5S, 16S, 23S	5S, 18S, 28S
DNA polymerase			
Inhibitors of protein synthesis	rifampicin puromycin chloramphenicol fusidic acid	rifampicin puromycin chloromphenicol fusidic acid	cycloheximide emetine
Initiation of protein synthesis	f-methionine	f-methionine	methionine

Table 9.11 Comparison of DNA and related functions in mitochondria, chloroplasts, prokaryotes and eukaryotes.

endosymbiotic origins of chloroplasts and mitochondria

It has been suggested that chloroplasts and mitochondria may have evolved following the invasion of a primitive anaerobic eukaryotic cell by aerobic prokaryotes. The first prokaryotes evolved about 3.5×10^9 years ago and would have been anaerobic. Shortly afterwards oxygen-releasing photosynthetic organisms such as cyanobacteria developed. Aerobic respiration would have been common about 2×10^9 years ago and some of these aerobic bacteria lost the ability to photosynthesise and became dependent on respiration. About 1.5×10^9 years ago such a bacterium may have invaded a primitive eukaryotic cell and established a symbiotic relationship enabling the primitive anaerobic eukaryotic cell to carry out aerobic respiration. Such an endosymbiotic relationship would then have evolved in such a way that genes were lost from the endosymbiont and their function taken over by genes in the nuclei of their hosts. Obviously the endosymbionts could lose their ability to make cell walls as they would be protected from damaging influences of the environment by being immersed in their host's cytoplasm. The similarities between mitochondria and prokaryotes provide convincing evidence of this evolutionary origins of mitochondria.

An analogous process is believed to have given rise to chloroplasts except in this case the endosymbiotic prokaryotes were capable of photosynthesis.

By studying the details of the similarities between the organelles and prokaryotes (for example the nucleotide sequences of their rRNA, nature of the photosynthesis pigments and respiratory chain electron carriers) it is possible to construct an evolutionary tree for the generation of mitochondria and chloroplasts. A simplified version of this is given in Figure 9.13.

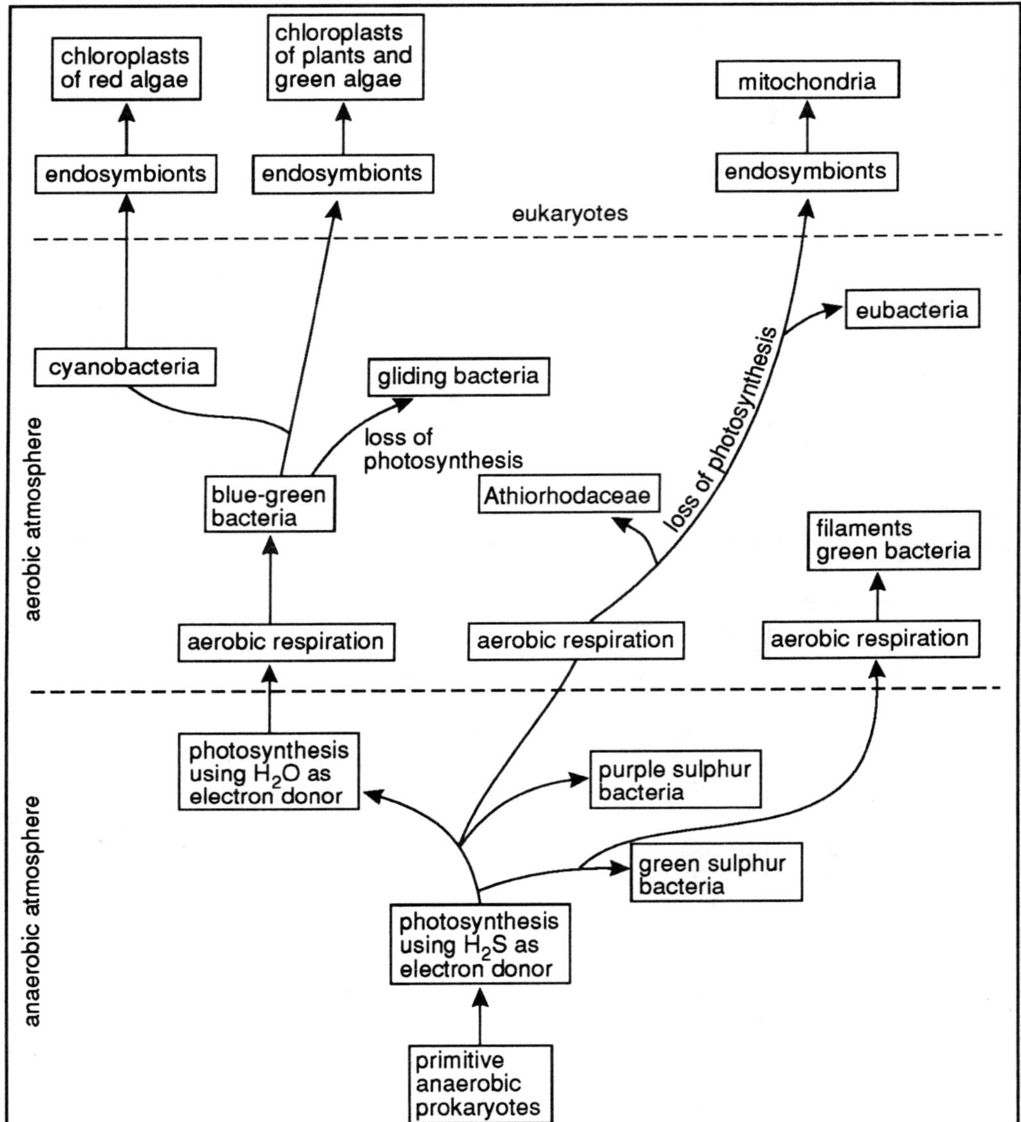

Figure 9.13 The probable evolution of chloroplasts and mitochondria from prokaryotic ancestors. It appears that aerobic respiration developed soon after the development of photosynthesis in which H_2O is used as the electron donor, evolving O_2 (about 2×10^9 years ago). Chloroplasts may have evolved separately from a variety of photosynthetic prokaryotes. It appears that mitochondria may have evolved from aerobic purple bacteria related to the Athiorhodaceae which had lost their ability to carry out photosynthesis.

Π What type of metabolism do you think primitive eukaryotes would have had before invasion by prokaryotes?

The earliest eukaryotes obtained their energy by anaerobic fermentation reactions, of which glycolysis is an example.

Mitochondria possibly descended from a purple photosynthetic bacterium which lost the ability to photosynthesise and was solely dependent on respiration as an energy source. The differences between protozoan, plant and animal organelle DNA suggests that invasion by aerobic bacteria may have occurred more than once during evolution. Chloroplasts possibly originated later than mitochondria by an oxygen-producing photosynthetic bacterium such as a cyanobacterium invading a cell. The evolutionary tree drawn in Figure 9.13 suggests that this event occurred several times giving rise to the chloroplasts of higher plants and green, red and brown algae.

9.12 The inheritance of mitochondria

Mitochondria (and mitochondrial DNA) are inherited only from the female gamete or oocyte, the male spermatozoa contributing only nuclear DNA. Mitochondria lack enzymes for repairing DNA and thus the DNA mutates at a rate about ten times higher than nuclear DNA. There is, therefore, a greater variation in base sequences in mitochondrial DNA and this variation can be estimated by the use of restriction fragment polymorphisms (RFLPs). We remind you that RFLPs are detected by digesting DNA with a restriction endonuclease and separating the fragments on gel electrophoresis. The size of the fragments will depend on the number of sites recognised by the enzyme. The fragments are separated by gel electrophoresis. The gel is read as a series of bands corresponding to DNA fragments. The more similar the banding patterns between two samples of mitochondrial DNA, the greater the similarity between the mitochondrial DNAs.

RFLPs

This procedure has been used to investigate the evolutionary origins of mankind. Mitochondrial DNA from individuals from Africa, Asia, Australia, New Guinea and Caucasians have been analysed using 12 restriction enzymes. The data suggested two strands of evolution, one group includes some Africans and the other four populations; the other Africans forms a separate group. The second group may be the older, as they have accumulated more mutations than the first. The results indicated that modern Man may have arisen in Africa 140 00-290 000 years ago.

Summary and objectives

Mitochondria and chloroplasts contain circular double stranded DNA molecules which are structurally distinct from the nuclear DNA. The mitochondrial DNA of metazoans is compact, lacks introns and has a small non-coding sequence. Plant mitochondrial DNA is larger, with some introns and larger non-coding sequences. Chloroplast DNA contains inverted repeats. The organelle DNA codes for only a small number of proteins, rRNA and tRNA molecules. The organelle DNA replicates independently of the nuclear DNA and by a mechanism resembling that of prokaryotes. The protein synthesis of mitochondria and chloroplasts is also similar to that of prokaryotes, in that smaller ribosomes are used, similar antibiotics inhibit the processes and protein factors are interchangeable between the two. There are deviations from the universal genetic code in both chloroplasts and mitochondria. The replication of organelle DNA is by means of a D-loop and involves sections of triple stranded DNA. It has been postulated that mitochondria and chloroplasts evolved by the invasion of primitive eukaryotic cells by prokaryotic organisms.

Now you have completed this chapter you should be able to:

- recognise that extranuclear DNA is found in the mitochondrion and chloroplast;
- understand the similarities between prokaryotic and organelle DNA;
- list the types of components coded on mitochondrion and chloroplast DNA and recognise that the majority of organelle proteins are coded on nuclear DNA;
- outline the mechanism of importing proteins into the organelle;
- describe the replication of organelle DNA and list the similarities to prokaryote DNA replication. You should also be able to identify the differences in replication of nuclear DNA;
- describe a possible scheme for the evolution of chloroplasts and mitochondrion from prokaryotic cells.

Responses to SAQS

Responses to Chapter 1 SAQs

1.1

1) False. The genetic information in the DNA of all the different cell types in an organism is the same (with a few exceptions). The distinction between cell types results from differential gene control.

2) False. It is defined as differentiation.

3) True.

4) False. Some prokaryotes are multicellular and include examples of morphological and physiological specialisation.

5) False. Some prokaryotes are multicellular and include examples of morphological and physiological specialisation.

1.2

1) False. Prokaryotic cells lack the organelles such as nuclei, mitrochondia, lysosomes and Golgi bodies that are characteristics of eukaryotes.

2) True. Yeasts are simple unicellular eukaryotes.

3) False. The nucleoid zone is formed by the compacted DNA in prokaryotes but this does not involve histones.

4) False. Introns were probably present in the first genes but have been lost during evolutions of the eubacteria to form a 'streamlined' genome.

5) True.

6) False. Operons are found in prokaryotes and are transcribed by RNA polymerase. Transcription factors are found in eukaryotes and are involved in the regulation of transcription by RNA polymerase.

7) True.

8) True.

Responses to Chapter 2 SAQs

2.1

1) The testcross parent will be homozygous recessive.

   Testcrossing always involves mating with an individual homozygous recessive at all relevant loci.

2) The testcross parent will be dwarf in height.

3) *T* and *t*. Remember that each gamete will contain only one of the alleles from the plant height locus.

4) *t* only, since the testcross parent is dwarf, of homozygous recessive genotype.

5) *Tt* and *tt*. These are combinations of each of the gamete types from the heterozygous tall pea plant with the *t* gamete from the dwarf testcross parent.

6) 1:1: tall : dwarf, since each of the gamete types from the heterozygous tall parent occur with equal frequency.

If you have incorrect answers to any of these questions, try going through the cross again, writing out the phenotypes, genotypes and gametes for both the parents and the F1 generation. It may also help to remind yourself what testcrossing is.

2.2

Using *T* for the dominant tall allele, and *t* for the recessive dwarf allele, *P* for the dominant purple flowered allele and *p* for the recessive white flower allele.

1) *Tt Pp* (by definition of the term dihybrid).

2) *tt pp* (must be a double homozygous recessive).

3) *TP, Tp, tP, tp*.

4) *tp*.

5) *TtPp, Ttpp, ttPp, ttpp*.

6) 1 tall, purple flowered : 1 tall, white flowered : 1 dwarf, purple flowered : 1 dwarf, white flowered pea plants.

2.3

1) *FBFB, FBFW, FBFW, FWFW*.

2) 1 black : 2 grey : 1 white.

Remember that the heterozygous genotype exhibits the grey phenotype.

2.4

1) $C\ c^{ch}\ c^{h}\ c$.

2) Wild type: *CC*, Cc^{ch}, Cc^{h}, *Cc*.

Chinchilla: $c^{ch}c^{ch}$, $c^{ch}c^{h}$, $c^{ch}c$.

Himalayan: $c^{h}c^{h}$, $c^{h}c$

Albino: cc.

2.5 1) 9:3:4 Recessive epistasis.

2) 15:1 Duplicate dominant epistasis.

3) 12:3:1 Dominant epistasis.

4) 13:3 Dominant and recessive interaction.

5) 9:7 Duplicate recessive epistasis.

6) 9:3:3:1 Independent assortment.

7) 9:6:1 Duplicate genes with cumulative effect.

In evaluating these sets of data, first of all decide how many phenotypic groups are involved, as this will permit you to eliminate several possibilities, leaving at most three possible explanations. Then try to match each of the remaining possible phenotypic ratios to the progeny data in the question. It may help you to express each of the groups as 16ths of the total. Consulting Table 2.2 may also assist you in forming conclusions.

2.6 1) Incompletely sex linked.

2) Holandric.

3) Completely sex linked.

4) Autosomal.

Consulting Figure 2.3 may assist your understanding of this question.

2.7 1) bb*, since there is a pattern bald daughter, who must be of genotype *b*b**, and had, therefore obtained a *b** allele from each parent.

2) *bb**, or *b*b**. Remember that *b** is dominant in males.

3) *b*b**, since no *bb* normal haired males were found amongst the 8 sons. Any *bb* sons would have had to have received the *b* allele from each parent.

2.8 1) Yes.

2) Only parental type products were formed. These parental types have the same allelic combinations as on the original homologues from the parents.

3) No.

4) Both parental and recombinant types were found in the F1 generation at equal frequency. The recombinant types have new allelic combinations on the same chromosome, not found in the parents. This means that the two loci are assorting independently, and cannot therefore be in the same linkage group.

2.9 1) a) *y w/y w* and *Y W/y w*.

b) 98.5%.

c) *y W/y w* and *Y w/y w*.

d) 1.5%.

e) 1.5 cM.

2)

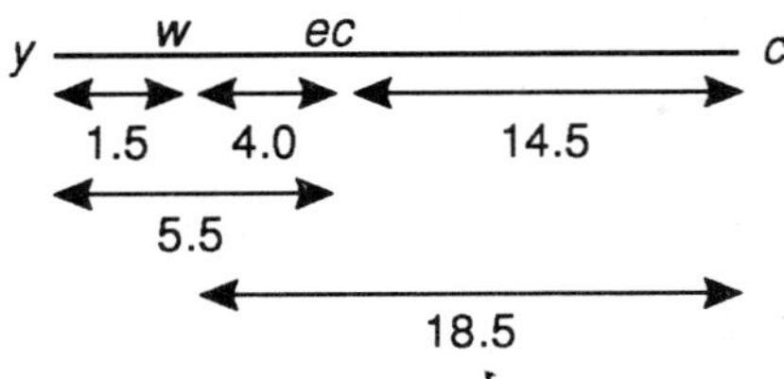

Responses to Chapter 3 SAQs

3.1

1) The base coding for Glu are changed to those that code for Val, GAA or GAG to GUA or GUG.

2) Since Hb^s has one less charge per molecule, it may be separated from normal Hb by electrophoresis.

3.2

1) The base sequences codes for the following amino acids:

CGA	AGT	GGC	GAT
Arg	Ser	Gly	Asp

2) a) If there was an A to G transition:

CGG	GGT	GGC	GGT
Arg	Gly	Gly	Gly

b) If there was a G to A transition:

CAA	AAT	AAG	AAT
Gln	Asn	Lys	Asn

3.3

1) It can be seen that the four codons for each amino acid have the first two bases the same but differ in the third base.

2) The most harmful mutations will be those involving a change from a polar amino acid to a nonpolar one or *vice versa*, or a negatively charged amino acid for a positively charged one or *vice versa*.

3.4

1) The mutagen X requires activation by liver enzymes.

2) The compound X is mutagenic.

3) The mutagen probable acts an intercalating agent.

4) The mutagen had caused deletion of four bases.

5) The bases deleted were purines.

6) The purines deleted were adenosines.

3.5

1) Ethylenimine produced 53.1 and triethyllenemelamine 24.0 revertants/µg respectively (Note that you may not have been able to determine these values so accurately from the figure, values of 50-54 mg-1 and 22-25 mg-1 would be acceptable.

2) Ethylenimine is the more mutagenic.

3.6 **Excision** repair involves base removal by enzymes called **endonucleases. Excision** repair also requires the enzymes **DNA *pol* I** and DNA **ligase.** Alkylated bases are removed by **glycolyase** enzymes and **thymine** dimers are excised by proteins coded for by the ***uvr*A, *uvr*B** and ***uvr*C** genes. The removal of alkylated bases may leave an **apyrimidinic** or **apurinic** site.

Post-replication repair occurs after substantial DNA damage and involves movement of sections of DNA from the undamaged to a damaged strand. The ***rec*A,** protein is essential for the transfer. Both **excision** and **post-replication** repair are **error**-free, and restore the correct sequence of bases in the DNA. If **mismatched bases** are introduced during repair then the process is **error-prone.** An example is **SOS-repair.**

Alkylated bases can be repaired by **alkyltransferases.** These enzymes catalyse the transfer of an **alkyl** group from the base to a **cysteine** residue on the enzyme. The enzyme may only be used **once** as it is then inactivated. Thymine dimers may also be repaired by an enzyme **photolyase.** The enzymes uses **visible light** to cleave the **cyclobutane** ring.

3.7 The *uvr* genes confer resistance to ultraviolet light and organisms deficient in these genes have a greater mutation rate than the wild type. Mutants deficient in *rec*A cannot perform recombination repair and are also sensitive to ultraviolet light. *pol*A mutants cannot carry out excision repair but are less sensitive to ultraviolet light. A mutant deficient in both *uvr*A and *rec*A has a very low tolerance to ultraviolet light.

3.8 AT patients cannot repair DNA and cells from such patients show increased sensitivity to ultraviolet light.

3.9 Percent increase = ((7 - 0.060) / (0.060) x 1000 = 11 567%.

3.10 1) See accompanying figure.

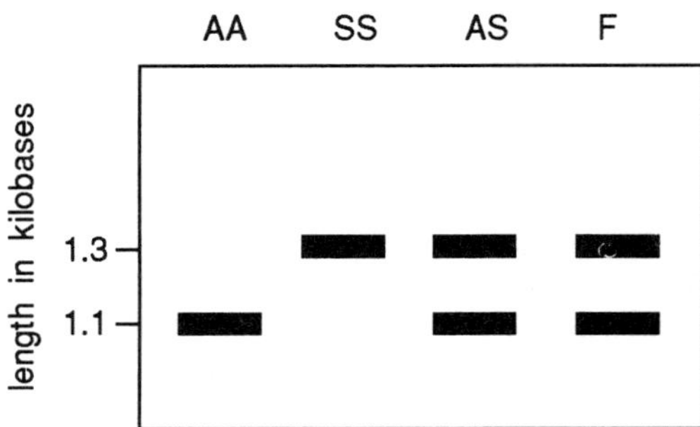

2) The foetus is a heterozygote.

3.11 Mutate the DNA for catabolase to give codons for cysteine residues instead of serine and alanine (UGU or UGC rather than UCU or GCU etc respectively). Since the cysteine are of similar size to serine and alanine residues, the effect on protein structure should not be great. The close proximity and orientation of the cysteine residues means, in oxidising conditions, they should form two disulphide bonds which hopefully will stabilise the enzyme at higher temperatures.

3.12 The codons for Pro are CCX where X is U, C, A or G. The codons for Ile are AUY where Y is U, C or A. If HSO_3- is used this will mutate C to U. The base sequences will then read UUX (Phe) and AUU or AUA (Ile). If the protein loses activity Pro is involved in catalysis.

3.13

1) True. Several mutations, perhaps as many as five, may be needed to transform a normal cell to a cancerous cell.

2) False. Oncogenes have a normal cellular counterpart called cellular or c-*onc* genes

3) False. Viruses and chromosomal translocation may be involved in the activation of oncogenes.

4) True. An example is the translocation accompanying chronic myelogenous leukaemia.

5) True. Some oncogenes code for growth factors or their receptors.

Responses to Chapter 4 SAQs

4.1 Length of DNA = 2×10^9 nm

length of loop = 60 000 x 0.34 nm

therefore number of loops = $2 \times 10^9/(60\,000 \times 0.34) = 98\,000$

4.2 $2\text{m} / 10^4 = 2 \times 10^{-4}\text{m} = 200\ \mu\text{m}$

4.3

1) The strands will not have fully separated if the temperature is below T_m. The absorbance will rise slightly as some bases will be exposed. As the complementary strands are still largely in a double helical conformation the absorbance will return to the initial level if the solution is slowly cooled.

2) The absorbance will rise (the hyperchromic effect) and will stay above the initial level as the strands will not be able to completely reassociate on rapid cooling.

4.4 Use the relationship given in the text. Substituting for % GC. 1) a) when GC is 20 the Tm is 77.5 °C b) when %GC is 40 the Tm is 85.7 °C.

2) % GC = 2.44 (Tm - 69.3)

% GC = 2.44(110 - 69.3)

% GC = 99.3

4.5

1) Three classes of DNA are present.

2) A (60%), B (20%) and C (20%). These value are obtained by horizontal extrapolation.

3) The complexity of the DNA is

   Complexity = $5 \times 10^5 . C_0t_{1/2(\text{pure})}$

   $C_0t_{1/2(\text{pure})} = C_0t_{1/2\,(\text{mixture})} . \text{GF}$

   The $C_0t_{1/2\text{mixture}}$ of A = ~ 2

   of B = 10^3,

   of C ~ 10^5 (approximately)

   The $C_0t_{1/2\text{pure}}$ of A is 2 x 0.6 = 1.2

   of B is $10^3 \times 0.2 = 2 \times 10^2$

   of C is $10^5 \times 0.2 = 2 \times 10^4$

4) Therefore the complexity is:

for A $5 \times 10^5 . 1.2 = 6 \times 10^5$ nucleotides

for B $5 \times 10^5 . 400 = 10^8$ nucleotides

for C $5 \times 10^5 . 2 \times 10^4 = 10^{10}$ nucleotides

4.6 The change in structure of DNA from a double to a **single stranded** form is called **denaturation**. This results in an increased **absorbance** at 260 nm, which is called the **hyperchromic** effect. The unwinding of the double helix is called **melting**. The $\mathbf{T_m}$ is the temperature at which **50%** of the DNA is single stranded. The richer the DNA in **GC** basepairs, the higher the $\mathbf{T_m}$.

If **denatured** DNA is slowly cooled below its $\mathbf{T_m}$ it **renatures**. The product of concentration x time when 50% of the DNA has renatured is called $\mathbf{C_0t_{1/2}}$. The $\mathbf{C_0t_{1/2}}$ of DNA is a measure of **unique sequences** in DNA and is directly proportional to its complexity. $C_0t_{1/2}$ analysis indicates there are **three** types of sequences in eukaryotic DNA. These are **highly repetitive** which are present in millions of copies, **moderately repetitive** which are present as thousands of copies and **non-repetitive** which are present as single copies.

4.7 From the data given the non-repetitive DNA expressed in the cell is $4.1/100 \times 9.4 \times 10^8 = 3.86 \times 10^7$ basepairs.

Since on average an mRNA molecule is 2000 nucleotides in length, the number of structural genes expressed is:

$3.86 \times 10^7/2000 = 19\,700$

Thus since each expressed gene will produce mRNA which will hybridise with the appropriate gene, we can assume that about 19 700 genes are being expressed in the cell.

4.8 All are false.

1) Is false, because restriction enzymes catalyse the hydrolysis of DNA not RNA.

2) Is false because these enzyme hydrolyse at unmethylated sites.

3) False, although this is the nucleotide sequence recognised by *Eco*RI, it hydrolyses the strands in a staggered way (see Table 4.1).

Responses to Chapter 5 SAQs

5.1 1) Transcription.

2) RNA polymerase.

3) Primer.

4) rRNA.

5) Initiation, termination.

6) Transcription units.

5.2 The key to successfully answering this question is to work out what RNA species will be present which have complementary sequences to the probe.

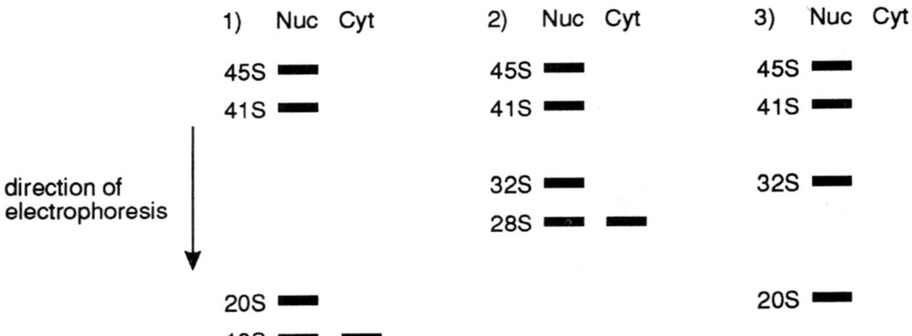

5.3

List A	List B
rRNA precursor	45S RNA
mRNA precursor	hnRNA
5′ end mRNA	cap
5′ end 5S rRNA	5′ triphosphate
5′ end 28S rRNA	5′ phosphate
3′ end mRNA	poly A tail
intervening sequence	intron
interrupted coding sequence	exon

5.4 U1 snRNP recognises the 5′ splice site

U2 snRNP recognises the branch site

U5 snRNP recognises the 3′ splice site

5.5 Different poly A sites are being used in the two alternative processing schemes. This leads to an additional intron being spliced out between part of exon 4 and exon 5.

There is also differential splicing out of exon 2 in the two processes.

5.6

1) Histone genes.

2) Interferon genes, many yeast genes.

3) Polyadenylation and splicing of introns.

4) Calcitonin.

5) Alternative splicing pathways. Troponin T.

Responses to Chapter 6 SAQs

6.1

1) histones.

2) 2, 2, 2A, 2B, 3, 4, nucleosome.

3) 1.

4) 200.

5) 140.

6.2

1) Analyse the DNase I sensitivity of genes which we know to be transcribed infrequently. Experiment c) provides the result of this analysis.

2) Analyse the DNase I sensitivity of genes which are known not to be active at the time of analysis in that particular cell, but may have been active at an earlier stage or could become active at a later stage. Experiment b) provides one result relevant to this.

3) Analyse the DNase I sensitivity of the regions of chromatin adjacent to an active gene (Experiment a)).

6.3

3) is the best summary of the results so far.

6.4

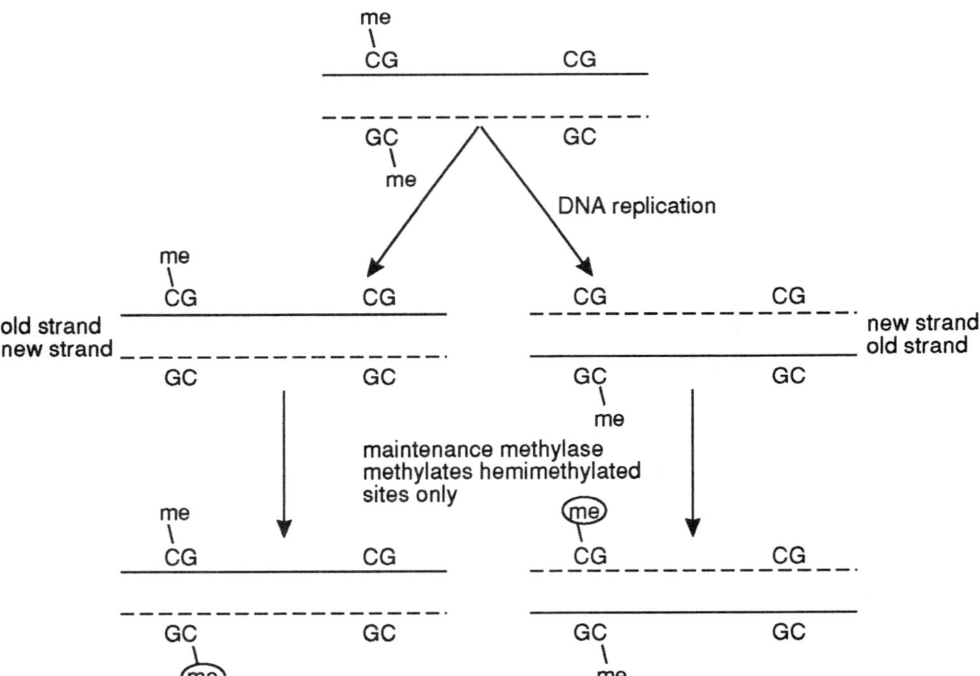

This figure shows how the methylation pattern may be maintained by comparing two sites, one unmethylated and one methylated on both strands. After DNA replication, each of the daughter molecules will be hemimethylated at the original methylated site. If the normal activity of the DNA methylase is to methylate only existing hemimethylated sites, the pattern of methylated will be inherited by both daughter cells.

6.5

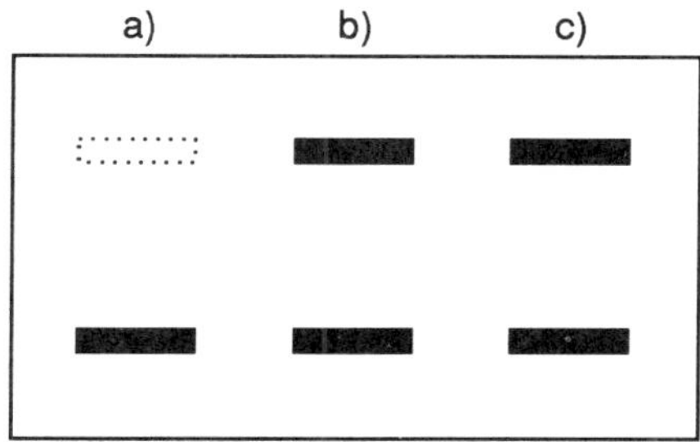

a) The unlabelled DNA competes for binding to the non-specific protein and because it is in large excess, very little protein retards the radioactive fragment.

b) The presence of the competitor DNA has no effect on the specific binding of the protein to the radioactive DNA, hence the retarded band is seen.

c) Control track to which no competitor DNA was added.

6.6 The bacterial RNA polymerase is essentially a coherent complex formed prior to association with the DNA and with an intrinsic DNA binding ability. The sigma factor is required for initiation and not elongation, but becomes part of the enzymic complex before DNA binding. In contrast, the specificity of RNA polymerase binding to the promoter in eukaryotes is determined by the independent activity of a number of separate transcription factors.

Responses to Chapter 7 SAQs

7.1

1) False. In the example we showed in Figure 7.1 transcription factors were illustrated binding to both the promoter and the enhancer regions. This is typical of most protein-coding genes in eukaryotes.

2) False. Although the promoter and enhancer regions bind their own set of transcription factors, some of the transcription factors bind to both of these regions. In the example shown in Figure 7.1, NF1-like and CACCC binding transcription factors interact with both the promoter and enhancer regions.

3) True.

4) True.

7.2

You should have selected 1, 2 and 3. Although single Zinc fingers either of the Cys_2/His_2 or of the Cys_4 types (options 1 and 2) cannot bind DNA, multiple copies of these are usually found arranged in tandem on a single molecule. In TFs with the homeobox motif, the DNA binding motif is provided by a single molecule folded into the correct conformation.

In TFs with basic domains (options 4 and 5) binding of these TFs to the DNA depends on the molecules forming dimers. This is achieved either through leucine zipper regions or by amphipathic helix-loop-helix arrangements.

7.3

1) It could be true. The introduction of this sequence means that the promoter will now bind glucocorticoid hormone receptors (see Table 7.2). Whether or not this would lead to transcription of the gene would depend on the exact location of the inserted sequence and the interaction of the glucocorticoid hormone receptors with other TFs and/or the initiation complex.

2) Thyroid hormone receptors (see Table 7.2).

3) No. The sequence given in the question is the consensus sequence which binds oestrogen receptors. But these receptors only bind with the consensus sequence in the presence of the hormone.

Thus:

a) in the absence of oestrogen

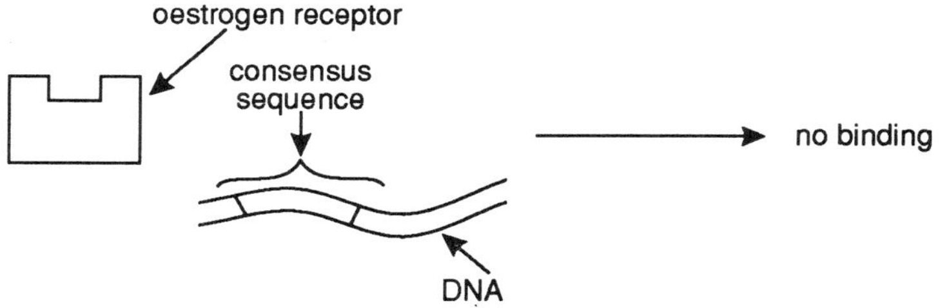

b) in the presence of oestrogen

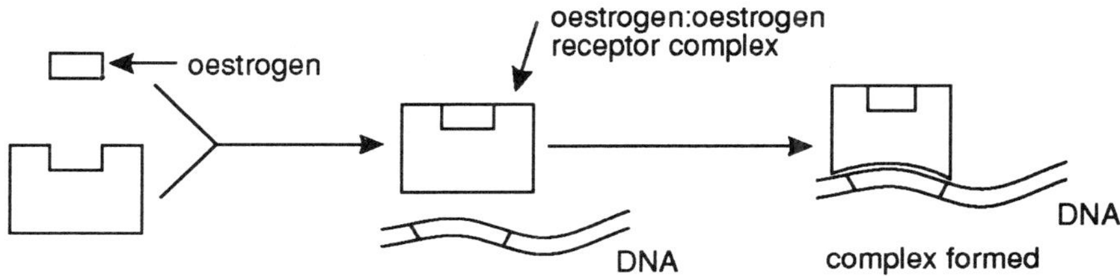

7.4 In all probability the yeast would still be unable to grow using lactose as its energy and carbon source at 25°C. If however it was heat shocked (for example its temperature raised to 30°C) then the heat shock factor may well bind with the promoter region of the inserted chimaeric gene. Thus this may lead to the transcription of the β-galactosidase gene. Thus under these circumstances we might anticipate that the yeast may be able to utilise lactose at these elevated temperatures.

7.5 1) a) Serum response element (SRE).

b) Serum response factor (SRF; p67 SRF).

c) TRE (TPA response element see Section 7.4.3).

2) *c-jun*. This contains the TRE consensus in its promoter. This is the response element to TPA.

7.6 Normally, we might expect erythoblasts to respond to thyroid hormone by differentiating into erythrocytes. This is because normally such cells produce the transcription factor c-Erb A which acts as a receptor for thyroid hormone. The c-Erb A thyroid hormone complex switch on those genes associated with erythrocyte differentiation. However, in the situation described in the question, we might anticipate that such cells would not produce c-Erb A unless they were heat shocked. Thus normally incubated (unshocked) cells would be non-responsive to thyroid hormone. On the other hand heat shocked cells might be anticipated to produce c-Erb A and thus behaviour normally in response to thyroid hormone. This is a somewhat simplification of what might happen because we are unable to predict all of the consequences of putting the heat shock consensus sequence into the c-Erb A promoter. It may well have a wide variety of secondary effects.

Responses to Chapter 8 SAQs

8.1 Polypeptides synthesis requires the participation of **ribosomes**, a source of metabolic **energy**, mRNA and a number of **proteins**. The **eukaryotic** ribosome has a **sedimentation coefficient** of 80S and contains about twice as much protein as the smaller, 60S **prokaryotic** ribosome.

The genetic information is carried from the **DNA** to the ribosome by **mRNA**. In prokaryotes, the mRNA codes for several **polypeptides**, while **eukaryotic mRNA** is a copy of a **single** gene. Initiation is more complex in eukaryotes than prokaryotes and requires the participation of a **7-methylguanine cap**. In prokaryotes, a **Shine-Dalgarno** sequence is needed. Many proteins are required for initiation, at least **10** in eukaryotes and three in prokaryotes. However, **termination** is an apparently simpler process in eukaryotes, requiring only a single protein factor compared to the **three** needed by prokaryotic cells.

8.2 Polypeptides synthesised using ^{14}C-labelled amino acids sediment with the polyribosomes. There is little radioactivity associated with free ribosomes indicating that the major portion of polypeptide synthesis is occurring on polyribosomes.

8.3 The radioactive amino acids will be found in the parts of the polypeptide chain synthesised last since at the time of addition of the amino acids part of the polypeptide will have been synthesised. ^{14}C-amino acids are added to this incomplete chain and can be seen to label the polypeptides towards the carboxyl terminus. Thus protein synthesis proceeds from the amino to the carboxyl terminus. We can represent this in the following way.

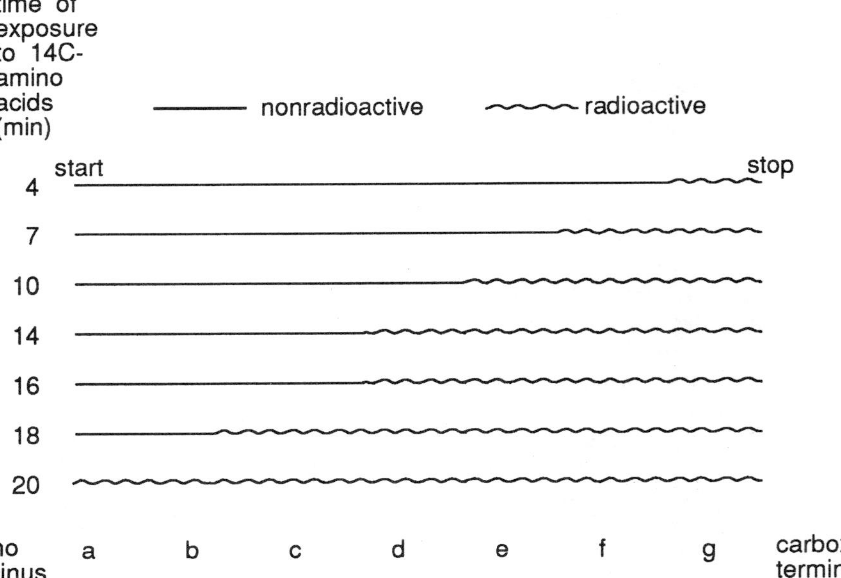

You can see that after only 4 minutes exposure to radioactive amino acids, the only completed globulin chains which will have incorporated radioactive amino acids will be those which had already linked a-f and that such molecules will contain radioactive g. If the exposure was extended to 7 minutes then chains which had previously contained a-e would be completed and these would contain radioactive f and g.

8.4

1) False. Protein initiation, termination and elongation factors are required.

2) False. GTP is the free energy source for translation. ATP is required for synthesis of the aminoacyl-tRNA molecule.

3) True.

4) True.

5) False. Eukaryotic ribosomes are aligned on the mRNA using a 7-methylguanine cap and cap binding proteins.

8.5

1) a) For mRNA the probability is that it will take $10^5/50 = 2000$ seconds to produce a defective molecule.

b) The time taken to produce a defective polypeptide is $10^4/15 = 666$ s.

2) a) The probability for mRNA is $1000 \div 10^5 = 1$ in 100. Note that not all of these will give rise to a change in the primary structure of the protein produced because of the degeneracy of the genetic code.

b) For proteins the figure is $330 \div 10^4 = 1$ in 30 (Remember that mRNA 1 kb long will give rise to a peptide that is 330 amino acids long. Thus if 1 in 10^4 amino acids are inserted in error, then 1 in 30 of the peptides produced would have an amino acid residue inserted in error.) Not all of these proteins will be biologically inactive since this will depend on the amino acid residue that is substituted.

8.6

1) True, although differences in uptake of the antibiotic may also be important.

2) True.

3) False, chloramphenicol blocks binding to the P site.

4) True.

5) False. Cycloheximide inhibits only eukaryotic protein synthesis.

8.7

Ribosomes translating the mRNA will become associated with the RER and the protein will be translocated into the lumen of the ER and will eventually be secreted.

8.8

Co-translational, it occurs as the rest of the mRNA is being translated.

8.9

1) a), b), c) and e) are all formed by post-translational modifications. d), glutamine is a standard amino acid coded for by CAA and CAG.

2) a), tyrosine; b), cysteine; c), glycine and e), alanine.

3) A reduction in the effectiveness of blood clotting (eg an extension to the time required for clotting). The lack of vitamin K would result in the reduction in the amount of post-translational changes to prothrombin. Thus the prothrombin would not function properly and blood clotting would be impaired.

8.10

Feature	Myristate	Palmitate
Chemical link	Amide	Thioester
Amino acid residue involved	Glycine	Cysteine
Half life	Equals protein	Less than protein
Post- or co-translational	Co-	Post-translational
Biological site of protein	Membrane bound	Membrane and cystosolic

8.11

Approximately 2-60%, but most are in the range of 10-20%.

8.12

Number of enzymes equals number of types of glycosidic bonds, therefore in Figure 8.18a i) two ii) three and iii) six enzymes are required respectively.

For Figure 8.18b, the number of enzymes required are i) ten, ii) eight and iii) eleven.

8.13

Removal of amino terminal Met	Co-translational
Removal of signal peptide	Co-translational
Cleavage between residues 32/33 and 60/61	Post-translational
Removal of residues 31, 32	Post-translational

Responses to Chapter 9 SAQs

9.1

1) Dividing the molecular weight by 642 Da allows the number of basepairs to be estimated.

2) Dividing the length of the molecule by 0.34×10^{-3} will also provide the number of basepairs. Both figures are included in the table and they do not always correspond because the average length and mass of a basepair is used.

Organism	length (μm)	mol wt (x 10^6 Da)	kbp	
Protozoa			a	b
*Tetrahymena**	15	35	54.5	44.1
*Paramecium**	14	30-35	50.6	41.2
Algae				
Chlamydomonas	4-5*	9.8	15.2	13.2
Fungi				
Saccharomyces	25	50	77.8	73.5
Neurospora	19-26*	40	62.3	66.2
Higher plants				
Pisum (Pea)	30	231	360	
Phaseolus (Bean)	20	37	58	
Zea (Maize)	193	365	570	
Platyhelminthes				
Hemenolepis	4.8	9.06	14.1	
Nematoda				
Ascaris	4.8	9.1	14.2	
Annelida				
Urechis	5.9	11.1	17.4	
Arthropoda				
Drosophila	6.2	12.4	19.3	18.2
Echinodermata				
Echinoidea	4.6-4.9*	8.97	14.0	
Chordata				
Fish	5.4	10.2	15.9	
Amphibia	4.9-5.8*	10.2	15.9	
Birds	5.1-5.4*	9.9	15.4	
Mammals	4.7-5.6*	9.7	15.1	
Human	4.7	9.8	15.2	13.8

* where a range of values is shown, the mean is used in the calculation

9.2 1) You should have calculated the following.

Species	kbp	% Nucleotide composition			
		A	G	C	T
Ascaris	14.28	49.8	20.4	7.6	22.2
Drosophilia	16.01	39.1	9.3	12.2	39.4
Stronglyocentrotus	15.65	30.2	19.4	22.7	28.7
Paracentrotus	15.7	29.5	17.2	22.5	30.8
Xenopus	17.55	30.0	13.5	23.5	33.0
Mus	16.295	28.7	12.4	24.4	34.5
Rattus	16.29	27.2	12.5	26.2	34.1
Bos	16.34	27.3	13.4	25.9	33.4
Homo sapiens	16.57	24.7	13.1	31.2	31.0

9.3

1) False. A reading frame designated URF6 (for unidentified open reading frame) is found on the light strand (see Figure 9.2).

2) True (see Table 9.4).

3) False. The 13 peptides are produced from 11 mRNA molecules. Two pairs of sequences (ND4 and ND4L sequence and the ATP6 and ATP8 sequences) are each carried on single mRNA molecules (see Section 9.2.1).

4) False. There are at least two clusters (Ala, Asn, Cys, Tyr and Leu, Ser, His) of tRNA sequences which are not separated by sequences coding for peptides or rRNA. Similar the tRNA sequence for Thr and Pro adjacent to the D-loop are not separated by such sequences.

5) True. They contain similar promoters and conserved sequences of nucleotides (see Figure 9.3).

9.4 The human mitochondrial genome contains **16 594** bp and is a **double-stranded** circular molecule. One strand (the **H-strand**) is heavier than the other (the **L-strand**) due to a difference in **thymidine** content. The DNA molecule contains 13 deoxyribonucleotide sequences called long open reading frames. These code for proteins such as **cytochrome b**, **ATP synthetase** and **NADH dehydrogenase**. The human mitochondrial genome also codes for two **rRNAs** and 22 **tRNAs**. Human mitochondrial DNA lacks **introns** but other species such as yeast have extensive **AT** rich sequences.

All mitochondrial DNA has **non-coding** regions which contain the D-loop, the point of **initiation** of DNA synthesis.

Plant mitochondrial DNA is substantially **larger** than animal DNA and is heterogenous with a number of smaller **circular** forms. It may also contain **promiscuous** sequences derived from other sources. **Kinetoplast** DNA is derived from trypanosomes and consists of maxicircle and **minicircle** DNA. The latter is **catenated**. The DNA of kinetoplasts has a high **AT** content.

9.5 The mRNA will hybridise to the DNA which contains complementary sequences. If the mRNA in the mitochondria binds only to mitochondrial DNA then it must have been synthesised using mitochondrial and not nuclear DNA as a template.

9.6

1) False. Mitochondrial protein biosynthesis is inhibited by chloramphenicol but not by cycloheximide. The reverse is true of eukaryotic protein synthesis.

2) False. Organelle and prokaryotic protein biosynthesis is initiated by N-formylmethionine.

3) True.

4) True except for yeast which has 72S ribosomes.

5) False. Because of the differences in the tRNAs in mitochondria and the cytosol, then the mRNA may be translated differently. For example an UGA codon in the mRNA would lead to trytophan being incorporated into the peptide within the mitochondron. In the cytosol, however, such a codon would be translated as a STOP signal and protein synthesis would be terminated at this point.

9.7 RNA polymerase catalyses the synthesis of primer RNA.

DNA polymerase catalyses the addition of deoxyribonucleotides to the primer forming DNA.

Topoisomerase I relaxes supercoiled DNA.

DNA ligase catalyses the linking of free ends of DNA to form an intact circular DNA molecule.

9.8 The stimulation of protein entry by light indicates that energy is required for the transport of protein into the chloroplast. In the presence of light electrochemical gradients (H^+ ions) and ATP are generated. These appear to be necessary for the transport of proteins into the chloroplast.

9.9

1) Proteins d), e), f), g), i) are coded for by nuclear DNA and synthesised in the cytosol and so would have amino-terminal transit signals.

2) Proteins j) and k) are coded for by plastid DNA.

Appendix 1 - Restriction endonucleases and their recognition sequences

This appendix includes a selection from the several hundred restriction enzymes that have been characterised. They are arranged in order of the sequences they recognise based on a system and nomenclature used by Boehringer Mannheim GmbH, Biochemica (PO Box 310-120, D-6800 Mannheim, Germany) - see also Kessler C, Höltke HJ (1986) Gene 47;1.

Recognition sequence	Restriction Enzyme	Recognition sequence	Restriction Enzyme
G↓$\overset{\circ}{A}\overset{+}{A}TT\overset{+}{C}$	*Eco* RI	A↓AGCTT	*Eco* VIII
↓G$\overset{\circ}{A}$TC	*Bce* 243	$\overset{+}{A}$↓AG$\overset{+}{C}$TT	*Hin* dIII
↓GATC	*Bsp* AI	A↓AGCTT	*Hsu* I
↓G$\overset{\circ}{A}$TC	*Cpf* I	G$\overset{\circ}{A}$G$\overset{+}{C}$T↓C	*Sac* I
GA↓TC	*Dpn* I	GAG$\overset{+}{C}$T↓C	*Sst*
↓G$\overset{+}{A}$TC	*Fnu* CI/*Hac* I	CAG↓$\overset{+}{C}$TC	*Pvu* II
↓G$\overset{+}{A}$T$\overset{\circ}{C}$	*Mbo* I	GG↓$\overset{+}{C}$C	*Bsu* RI
A↓G$\overset{\circ}{A}$T$\overset{+}{C}$T	*Bgl* II	GG↓CC	*Clt* I/*Fru* DI/*Sfa* I
A↓GATCT	*Nsp* MACI	GG↓$\overset{+}{C}\overset{\circ}{C}$	*Hae* III
G↓GATCC	*Ali* I	$\genfrac{}{}{0pt}{}{T}{C}$↓GGCC$\genfrac{}{}{0pt}{}{A}{G}$	*Cfr* 1/*Cfr* 141
G↓G$\overset{\circ}{A}$T$\overset{+}{C}\overset{\circ}{C}$	*Bam* HI	$\genfrac{}{}{0pt}{}{T}{C}$↓GG$\overset{+}{C}C\genfrac{}{}{0pt}{}{A}{G}$	*Eae* I
G↓GAT$\overset{+}{C}$C	*Bst* I	CG↓CG	*Fnu* DII
$\genfrac{}{}{0pt}{}{A}{G}$↓G$\overset{+}{A}TC\genfrac{}{}{0pt}{}{T}{C}$	*Mfl* I	$\overset{+}{C}$G↓$\overset{+}{C}$G	*Tha* I
$\genfrac{}{}{0pt}{}{A}{G}$↓G$\overset{\circ}{A}T\overset{+}{C}\genfrac{}{}{0pt}{}{T}{C}$	*Xho* II	G↓CGCGC	*Bss* HII
CGAT↓CG	*Nbl* I	CCGC↓GG	*Csc* I/*Gce* I/ *Gce* GLI/*Sst* II
CG$\overset{\circ}{A}$T↓$\overset{+}{C}$G	*Pvu* I	$\overset{+}{C}$CGC↓GG	*Sac* II
CGAT↓CG	*Rsh* I	C$\genfrac{}{}{0pt}{}{A}{C}$G↓C$\genfrac{}{}{0pt}{}{T}{G}$G	*Nsp* BII
CG$\overset{\circ}{A}$T↓$\overset{\circ}{C}$G	*Xor* II	TCG↓CG$\overset{+}{A}$	*Nru* I
T↓G$\overset{+}{A}$T$\overset{\circ}{C}$A	*Bcl* I	ATGCA↓T	*Eco* T22/*Nsi* I
CATG↓	*Nla* III	CTGCA↓G	*Ali* AJI/*Bsp* I/ *Cfl* I/*Sal* PI
GCATG↓C	*Pae* I	$\overset{+}{C}$TGC$\overset{+}{A}$↓G	*Pst* I
GCATG↓$\overset{\circ}{C}$	*Sph* I	CTGC$\overset{+}{A}$↓G	*Sfl* I
$\genfrac{}{}{0pt}{}{A}{G}$CATG↓$\genfrac{}{}{0pt}{}{T}{C}$	*Nsp* 75241/*Nsp* HI	A↓$\overset{+}{C}$GT	*Mae* II
$\overset{+}{C}$↓CATGG	*Nco* I	GACGT↓C	*Aat* II
AG↓$\overset{+}{C}$T	*Alu* I	TAC↓GTA	*Sna* BI
AG↓CT	*Mlt* I		

Recognition sequence	Restriction Enzyme	Recognition sequence	Restriction Enzyme
$G\overset{+}{C}G\downarrow C$	*Cfo* I	$G\downarrow GTAC\overset{+}{C}$	*Asp* 718
$GCG\downarrow C$	*Fnu* DIII	$C\downarrow TAG$	*Mae* I
$G\downarrow \overset{+}{C}GC$	*Hin* P1 I	$A\downarrow CTAGT$	*Spe* I
$AGC\downarrow GCT$	*Eco* 47111	$G\downarrow CTAGC$	*Nhe* I
$GGCGC\downarrow C$	*Bbe* I	$C\downarrow CTAGG$	*Avr* II
$GG\downarrow CGC\overset{\circ}{C}$	*Nar* I	$T\downarrow \overset{+}{C}TAG\overset{+}{A}$	*Xba* I
$GG\downarrow CGCC$	*Nda* I	$GTT\downarrow A\overset{+}{A}\overset{\circ}{C}$	*Hpa* I
$G\genfrac{}{}{0pt}{}{A}{G}\downarrow CG\genfrac{}{}{0pt}{}{T}{C}C$	*Acy* I/*Ast* WI/ *Asu* III/*Hgi* DI	$C\downarrow TTAAG$	*Afl* II
$G\genfrac{}{}{0pt}{}{A}{G}\downarrow \overset{+}{C}G\genfrac{}{}{0pt}{}{T}{C}\overset{+}{C}$	*Aha* II	$TTT\downarrow AAA$	*Aha* III/*Dra* I
$TGC\downarrow GCA$	*Aos* I/*Fdi* II/*Mst* I	$G\downarrow G\genfrac{}{}{0pt}{}{A}{T}CC$	*Afl* I
$C\downarrow \overset{+}{C}GG$	*Hap* II	$G\downarrow G\genfrac{}{}{0pt}{}{A}{T}C\overset{+}{C}$	*Ava* II
$C\downarrow CGG$	*Mno* I	$G\downarrow G\genfrac{}{}{0pt}{}{A}{T}CC$	*Bam* Nx/*Hgi* BI/ *Hgi* II/*Hgi* EI
$\overset{+}{C}\downarrow \overset{\circ}{C}GG$	*Msp* I	$GG\downarrow \genfrac{}{}{0pt}{}{A}{T}CC$	*Bam* 216
$G\overset{+}{C}C\downarrow GG\overset{+}{C}$	*Nae* I	$\genfrac{}{}{0pt}{}{A}{G}G\downarrow G\genfrac{}{}{0pt}{}{A}{T}CC\genfrac{}{}{0pt}{}{T}{C}$	*Ppu* MI
$\overset{+}{C}\downarrow \overset{+}{C}\overset{\circ}{C}GGG$	*Cfr* 91	$CG\downarrow G\genfrac{}{}{0pt}{}{A}{T}CCG$	*Rsr* II
$CC\overset{+}{C}\downarrow GGG$	*Sma* I	$C\overset{\circ}{C}\downarrow \genfrac{}{}{0pt}{}{A}{T}GG$	*Aor* I
$T\downarrow \overset{\circ}{C}G\overset{+}{A}$	*Taq* I/*Tth* HBBI	$\overset{\circ}{C}\overset{\circ}{C}\downarrow \genfrac{}{}{0pt}{}{A}{T}GG$	*Bst* NI
$G\downarrow TCGAC$	*Hgi* CIII/*Hgi* DII/*Nop* I	$\downarrow \overset{\circ}{C}\overset{+}{C}\genfrac{}{}{0pt}{}{A}{T}GG$	*Eco* RII
$G\downarrow T\overset{+}{C}G\overset{+}{A}C$	*Sal* I	$\overset{\circ}{C}\overset{\circ}{C}\genfrac{}{}{0pt}{}{G}{C}\downarrow GG$	*Bcn* I
$GT\downarrow \genfrac{}{}{0pt}{}{A}{C}\genfrac{}{}{0pt}{}{T}{G}\overset{+}{A}C$	*Acc* I	$CC\downarrow \genfrac{}{}{0pt}{}{G}{C}GG$	*Cau* II
$C\downarrow TCGAG$	*Blu* I/*Ccr* I	$G\downarrow ANTC$	*Fru* AI
$C\downarrow TCG\overset{+}{A}G$	*Pae* R7	$G\downarrow \overset{+}{A}NT\overset{+}{C}$	*Hin* fI
$C\downarrow T\overset{+}{C}G\overset{+}{A}G$	*Xho* I	$G\downarrow GNCC$	*Bsp* BII
$C\downarrow TCGAG$	*Xpa* I	$\downarrow C\overset{+}{C}NGG$	*Sso* 4711
$TT\downarrow CGAA$	*Asu* II/*Fsp* II/*Mla* I	$\downarrow GTNAC$	*Mae* III
$ATT\downarrow ATT$	*Ssp* I	$G\downarrow GTNACC$	*Asp* AI
$G\overset{+}{A}T\downarrow ATC$	*Eco* RV	$GC\downarrow TNAGC$	*Esp* I
$CA\downarrow TATG$	*Nde* I	$\overset{m}{C}$ restricts at all 5-methylcytosines	*Ala* JA11
$GT\downarrow AC$	*Rsa* I		
$AGT\downarrow ACT$	*Sca* I		

Symmetrical sequences are specified only for the 5′→3′ strand
Symbol N represents any nucleotide
$\overset{+}{A}$ and $\overset{+}{C}$ indicate that methylation of the bases inhibits the restriction enzymes
$\overset{\circ}{A}$ and $\overset{\circ}{C}$ indicate that restriction endonuclease activity is not affected by methylation of these bases

Index